INTRODUCTION TO CRYPTOGRAPHY WITH OPEN-SOURCE SOFTWARE

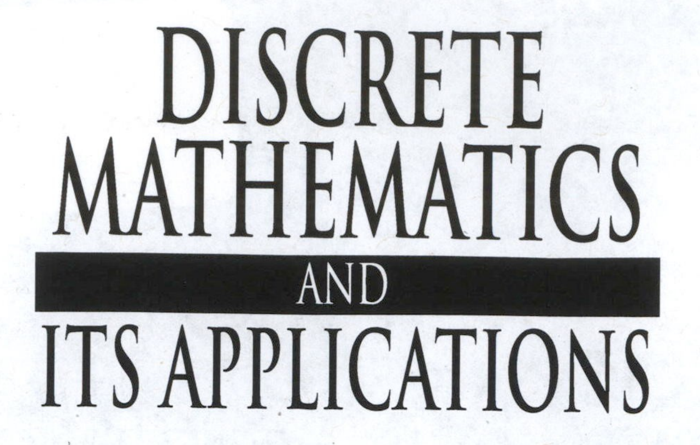

Series Editor
Kenneth H. Rosen, Ph.D.

R. B. J. T. Allenby and Alan Slomson, How to Count: An Introduction to Combinatorics, Third Edition

Juergen Bierbrauer, Introduction to Coding Theory

Donald Bindner and Martin Erickson, A Student's Guide to the Study, Practice, and Tools of Modern Mathematics

Francine Blanchet-Sadri, Algorithmic Combinatorics on Partial Words

Richard A. Brualdi and Dragoš Cvetković, A Combinatorial Approach to Matrix Theory and Its Applications

Kun-Mao Chao and Bang Ye Wu, Spanning Trees and Optimization Problems

Charalambos A. Charalambides, Enumerative Combinatorics

Gary Chartrand and Ping Zhang, Chromatic Graph Theory

Henri Cohen, Gerhard Frey, et al., Handbook of Elliptic and Hyperelliptic Curve Cryptography

Charles J. Colbourn and Jeffrey H. Dinitz, Handbook of Combinatorial Designs, Second Edition

Martin Erickson, Pearls of Discrete Mathematics

Martin Erickson and Anthony Vazzana, Introduction to Number Theory

Steven Furino, Ying Miao, and Jianxing Yin, Frames and Resolvable Designs: Uses, Constructions, and Existence

Mark S. Gockenbach, Finite-Dimensional Linear Algebra

Randy Goldberg and Lance Riek, A Practical Handbook of Speech Coders

Jacob E. Goodman and Joseph O'Rourke, Handbook of Discrete and Computational Geometry, Second Edition

Jonathan L. Gross, Combinatorial Methods with Computer Applications

Jonathan L. Gross and Jay Yellen, Graph Theory and Its Applications, Second Edition

Jonathan L. Gross and Jay Yellen, Handbook of Graph Theory

David S. Gunderson, Handbook of Mathematical Induction: Theory and Applications

Darrel R. Hankerson, Greg A. Harris, and Peter D. Johnson, Introduction to Information Theory and Data Compression, Second Edition

Darel W. Hardy, Fred Richman, and Carol L. Walker, Applied Algebra: Codes, Ciphers, and Discrete Algorithms, Second Edition

Titles (continued)

Daryl D. Harms, Miroslav Kraetzl, Charles J. Colbourn, and John S. Devitt, Network Reliability: Experiments with a Symbolic Algebra Environment

Silvia Heubach and Toufik Mansour, Combinatorics of Compositions and Words

Leslie Hogben, Handbook of Linear Algebra

Derek F. Holt with Bettina Eick and Eamonn A. O'Brien, Handbook of Computational Group Theory

David M. Jackson and Terry I. Visentin, An Atlas of Smaller Maps in Orientable and Nonorientable Surfaces

Richard E. Klima, Neil P. Sigmon, and Ernest L. Stitzinger, Applications of Abstract Algebra with Maple™ and MATLAB®, Second Edition

Patrick Knupp and Kambiz Salari, Verification of Computer Codes in Computational Science and Engineering

William Kocay and Donald L. Kreher, Graphs, Algorithms, and Optimization

Donald L. Kreher and Douglas R. Stinson, Combinatorial Algorithms: Generation Enumeration and Search

Hang T. Lau, A Java Library of Graph Algorithms and Optimization

C. C. Lindner and C. A. Rodger, Design Theory, Second Edition

Nicholas A. Loehr, Bijective Combinatorics

Alasdair McAndrew, Introduction to Cryptography with Open-Source Software

Elliott Mendelson, Introduction to Mathematical Logic, Fifth Edition

Alfred J. Menezes, Paul C. van Oorschot, and Scott A. Vanstone, Handbook of Applied Cryptography

Richard A. Mollin, Advanced Number Theory with Applications

Richard A. Mollin, Algebraic Number Theory, Second Edition

Richard A. Mollin, Codes: The Guide to Secrecy from Ancient to Modern Times

Richard A. Mollin, Fundamental Number Theory with Applications, Second Edition

Richard A. Mollin, An Introduction to Cryptography, Second Edition

Richard A. Mollin, Quadratics

Richard A. Mollin, RSA and Public-Key Cryptography

Carlos J. Moreno and Samuel S. Wagstaff, Jr., Sums of Squares of Integers

Dingyi Pei, Authentication Codes and Combinatorial Designs

Kenneth H. Rosen, Handbook of Discrete and Combinatorial Mathematics

Douglas R. Shier and K.T. Wallenius, Applied Mathematical Modeling: A Multidisciplinary Approach

Alexander Stanoyevitch, Introduction to Cryptography with Mathematical Foundations and Computer Implementations

Jörn Steuding, Diophantine Analysis

Douglas R. Stinson, Cryptography: Theory and Practice, Third Edition

Roberto Togneri and Christopher J. deSilva, Fundamentals of Information Theory and Coding Design

W. D. Wallis, Introduction to Combinatorial Designs, Second Edition

W. D. Wallis and J. C. George, Introduction to Combinatorics

Lawrence C. Washington, Elliptic Curves: Number Theory and Cryptography, Second Edition

DISCRETE MATHEMATICS AND ITS APPLICATIONS
Series Editor KENNETH H. ROSEN

INTRODUCTION TO CRYPTOGRAPHY WITH OPEN-SOURCE SOFTWARE

Alasdair McAndrew
Victoria University
Melbourne, Victoria, Australia

CRC Press is an imprint of the
Taylor & Francis Group, an **informa** business
A CHAPMAN & HALL BOOK

CRC Press
Taylor & Francis Group
6000 Broken Sound Parkway NW, Suite 300
Boca Raton, FL 33487-2742

Printed in the United States of America on acid-free paper
10 9 8 7 6 5 4 3 2 1

International Standard Book Number: 978-1-4398-2570-9 (Hardback)

Visit the Taylor & Francis Web site at
http://www.taylorandfrancis.com

and the CRC Press Web site at
http://www.crcpress.com

For my dear wife, Felicity

and my children:

Angus, Edward, Fenella,
William, and Finlay

Contents

Preface

The growth of cryptography as both an academic discipline and a teaching subject has exploded over the last few decades. Once the privilege of the very few—and those few working in secret—cryptography is now taught in some form at universities the world over, and new books are being published with amazing rapidity. Before 1990, only 58 books containing the word "cryptography" in their titles had ever been published, according to the catalogue of the Library of Congress. Since then, 329 cryptography books have been published, and in the decade 2001–2010 an average of about 21 books were published each year.[1]

Cryptography books tend to fall into two main classes: highly technical texts written for the postgraduate student or perhaps a professional practitioner; and texts written more for teaching, with an expository tone and a gentler pace. This text sits firmly in the second camp.

A note on terminology: the subject of this text is *cryptography*, or "secret writing." Allied to cryptography is *cryptanalysis*, which is the business of "breaking" cryptosystems. The combination of cryptography and cryptanalysis is called *cryptology*. These definitions aren't always rigorously followed; there are books titled Cryptology which seem to be mainly cryptography, and similarly for some books titled Cryptanalysis.

This book differs from other texts in several respects:

- The mathematics has been kept at a fairly manageable level. Although some mathematics is required to understand the algorithms involved, this book adopts the principle that too much mathematics can be as much a hindrance to understanding as a help. About a year's mathematics at a tertiary level should provide enough background for this text.

- The consistent use of a computer algebra system (CAS). Many texts involve some computer use: some a programming language such as Java or C; others a CAS such as Maple® or Mathematica®. The advantage of using a CAS over a programming language is that students have an environment in which exploration and experimentation is far easier than with the edit–compile–run cycle of programming.

  This text uses the CAS Sage [85] which has several advantages over the

[1] To put this figure into perspective, a total of 3187 calculus books have been published, 1389 of them since 1990.

systems mentioned above. First, it is open source, so can be downloaded and installed locally at no cost. Second, it can be used with a web-based interface, so, even without downloading, the student can create a (free) account on one of the public Sage servers, and use Sage in a browser. Third, Sage's support for cryptography and the associated mathematics is the best of any system, open-source, or commercial.

Sage is used throughout the text to illustrate algorithms and cryptosystems and to provide examples. With access to Sage either locally or on a public server, every example in this text can be produced by the reader.

For some general discussion on teaching cryptography with open-source software see [57].

- Copious exercises. This text is intended for *learning*. As well as many examples, there are many exercises. Most of them aren't very difficult, but the student is encouraged to attempt as many of them as possible. They have been carefully scaffolded so as to provide a highly constructive learning experience.

What this book is not

This book is not a compendium of modern cryptography. The field is now far too vast and diverse to encompass in a single volume (at least, a volume small enough to be picked up in one hand). It is thus highly selective in what topics are covered, and the depth to which they are covered. In general, however, the topics cover a reasonable range of modern cryptography.

It is not designed for beginning researchers. For such people, unless their existing knowledge is very small, there are more suitable and appropriate texts, which have a faster pace and are aimed at a higher level.

This book is not a cryptanalysis primer. The business of breaking cryptosystems is fascinating and deep, and is the topic of a huge number of papers and monographs. It is of course mentioned in places, and a few techniques briefly discussed, but in general the text concentrates on the systems themselves, rather than the means of breaking them.

What is in this book

The book consists of 14 chapters and two appendices. Chapter 1 "sets the scene," so to speak, by providing a context for modern cryptography: when and where its methods may be used, and how they can be used and misused. It is full of definitions and terminology, and also discusses some of the problems of cryptographic use, some communications problems which can be "solved" by the use of cryptographic methods, and some methods, or protocols, for applying cryptographic methods to communications. The chapter finishes with a sample of pencil-and-paper cryptosystems, which can be used independently of the rest of the text to illustrate some of the techniques and protocols discussed.

Chapter 2 begins the study of the theory of numbers: prime numbers and their properties, modular arithmetic, and various algorithms such as Euclid's algorithm for computing the greatest common divisor, modular exponentiation and inversion, and a brief discussion of quadratic residues. All modern public-key cryptography is based on these algorithms and methods, so this chapter may be considered as fundamental to the rest of the text.

Chapter 3 discusses what are now called "classical" cryptosystems; systems that predate the use of computers, so are designed to be used with hand calculations (although, of course, computer programs can make the investigation of them much easier). Although these systems have no cryptographic strength from a modern perspective, they are useful to study, as they provide insight into the methods of cryptosystem construction and of different types of attack. Because of their simplicity it is easy for the learner to gain valuable insight into basic cryptographic and cryptanalytic thinking.

Chapter 4 provides a brief introduction to information theory. This enables a foundation for the formal study of security, and how well a cryptosystem "hides" the plaintext. The chapter also includes discussion of entropy, or uncertainty, as well as a discussion of the redundancy of the English language, `wh--h c-n b- und--s---d ev-- wh-- m-st l--t-rs -r- m-s---g`.

Chapter 5 is the first of two chapters introducing some of the most important public-key cryptosystems. This chapter discusses those systems whose security is obtained from the difficulty of factorizing large numbers. In particular, the RSA system is introduced, along with some discussion of possible weaknesses, and ways in which its security can be undermined. The chapter also includes Rabin's cryptosystem, and some notes on factoring.

Chapter 6 continues the discussion of public-key systems, but concentrates on two classes: first, those whose security is obtained from the difficulty of solving the general discrete logarithm problem—in particular the El Gamal cryptosystem, and secondly, those based on knapsack problems. These last are of great historical importance, but all have been broken, so the chapter also includes some discussion of lattice methods for breaking knapsack cryptosystems.

Chapter 7 introduces digital signature algorithms, which are based on public key methods. In a public-key system, the public key is used for encryption and the private key for decryption. For a digital signature algorithm, the private key is used for signing, and the public key for verifying the signature. Most of the public-key systems of the previous two chapters can be modified to provide digital signature schemes. There is also some discussion of signature forgery.

In Chapter 8 the text leaves the purely number-theoretical aspect of cryptography, to move onto bit-oriented secret key, or symmetric, systems. These are some of the most used systems of all, being the fastest, and the most appropriate for encrypting large amounts of data, such as networked data, or for backups. Chapter 8 begins the investigation of such systems, with some general discussion of such block ciphers, and includes a full description of perhaps the most famous of all: the Data Encryption Standard. There is also some discussion of security issues, including strength against differential cryptanalysis, and a brief mention of some newer "lightweight" ciphers.

Chapter 9 provides an elementary and algorithmic introduction to the study of finite fields: mathematical objects which support addition, subtraction, multiplication and division (except by zero). Finite fields are fundamental to the understanding of the Advanced Encryption Standard, and for the development of cryptosystems using elliptic curves.

Having set the mathematical stage, the Advanced Encryption Standard, based on the system Rijndael, is explored in Chapter 10. As well as a full description, a simpler version—designed for hand calculations and to develop intuition into its workings—is given.

Hash functions, or more properly *cryptographic* hash functions, are the topic of Chapter 11. These provide a sort of digital "fingerprint" of the data, and are some of the most widely used algorithms of all. A hash function can be constructed either as a stand-alone algorithm, or from a block cipher. Hash functions are in general poorly represented in cryptographic texts, and at the time of writing there is no textbook devoted entirely to their construction, strengths, and attacks.

Chapter 12 introduces one of the hottest topics over the last few years: cryptosystems based on elliptic curves. These have the advantage of providing equivalent security to the public-key systems described in Chapters 5 and 6, but with smaller key sizes. As well as discussing the basic theory, and some cryptosystems and signature schemes, there is also a brief investigation of "pairing-based" methods. These are normally considered as an advanced topic, but treating a pairing as a black box, and by giving a simple example, some examples of pairing-based protocols are described.

Random number generation, which is at the heart of almost all cryptographic applications—such as the creation of keys—is investigated in Chapter 13. The chapter includes discussion of the basic different generators (true, pseudo, cryptographic), and ways of measuring their strengths. The chapter also shows how random number generators are linked to *stream ciphers*, where the data is encrypted one bit at a time. The chapter finishes with a public-key system which is uniquely designed to act like a stream cipher.

Finally, Chapter 14 provides a few examples of advanced uses of cryptographic systems. These include secure multi-party computation, where various parties provide inputs to a multi-valued function, without any party gaining any knowledge of any of the inputs; zero knowledge proofs, where a proof of an assertion is provided in a way that gives the verifier no information about the assertion itself; oblivious transfer, where the sending party is unable to determine exactly what the receiver obtains; digital cash; and voting protocols.

Finally two appendices provide an introduction to Sage: where to get it, installation, basic use and programming; and some advanced algorithmic number theory algorithms for factorization, computing discrete logarithms, and determining primality.

There is a very small and select bibliography. Even though this text is designed for teaching, space considerations mean that some topics get only cursory treatment. For deeper knowledge, the reader is strongly recommended to explore the other texts and articles listed. Many of the articles are available online.

Acknowledgments

First, to my many students over many years who have tested out early versions of the notes on which this text is based. They have taught me far more about good teaching and learning than I ever taught them.

I would especially like to acknowledge Minh Van Nguyen, who while a student read through the notes with a very close eye and made a number of extremely useful suggestions. He also first recommended that I investigate Sage, and has since been an energetic developer for, and promoter of Sage. He has greatly helped my understanding of it.

The many and various members of the `sage-support` newsgroup have been in every way helpful and supportive with my many questions. William Stein, the creator of Sage and its lead developer, has been vital in not only answering my questions, but those of others, and I have learned a great deal from reading his many contributions to the newsgroup. I can only deeply admire his tireless work on developing Sage and its application in many branches of mathematics.

I am delighted to acknowledge my editor at CRC Press, Li-Ming Leong, who first discussed the possibility of this book with me, and who was instrumental in helping me transform a vague idea into a solid publishing proposal. She has been unfailingly helpful at all stages of the text's preparation, and is an amazingly punctual answerer of emails.

Finally, heartfelt thanks to my long-suffering family, who for far too long have seen me only from the back view, hunched over a computer keyboard, and who have seen my face only in the glow from its screen.

Chapter 1

Introduction to cryptography

This chapter "sets the scene" for cryptography, and provides the basic security context into which much of cryptography is used. In particular, the following topics will be discussed:

- The need for information security.
- Basic definitions and terminology.
- Distinction between "secret key" (symmetric) and "public key" (asymmetric) cryptosystems.
- What it means to attack, and to break a cryptosystem, and some means for doing so.
- Some cryptographic problems and protocols.
- Some simple pencil-and-paper systems, designed to provide examples for the material of this chapter.

1.1 Hiding information: Confidentiality

Cryptography means "secret writing" and in its most fundamental form is just that: a means for somebody to hide some information so that only somebody else "in the know" can access it. Being "in the know" generally means being in possession of a secret key, which will "unlock" the data from its hiding place.

Why would anybody want to hide data? There are many examples where data is to be viewed only by a small group of people—perhaps only one person. Here are a few of them:

1. Bank statements and financial documents. If you send a letter to your bank requesting the withdrawal of some money, you don't want to make this public knowledge, do you? The letter will probably have some information which identifies you, and information which identifies your bank account (perhaps an account number).

2. Credit card numbers. It's quite common to buy online; perhaps you wish to provide your credit card details to an online merchant. How do you secure this so that only the merchant can access the details?
3. Medical records. By law, medical records are confidential documents, and should be accessed only by the patient, the medical staff, or legal persons—they should not be freely available.
4. Military orders. In order to achieve a suitable outcome, military orders are coded in some way.
5. Personal information. Do you want the world to know when you last visited the doctor, what medicines you last bought, what your last exam results were, when you last ate fast food, and how much...? Somehow all this information should be kept from prying eyes.
6. Personal letters. If you are sending a letter (either written on paper, or as an email), you generally want only the recipient to be able to see it. How do you stop others from seeing your letter?

Assume that the world is full of malicious people who will stop at nothing to get their hands on your precious information. If you are sending an email, it can be read by anybody with access to any of the nodes through which it passes. This text will follow the time-honored tradition in cryptographic writing in personalizing the protagonists as Alice and Bob, rather than as "sender" and "receiver." Along with Alice and Bob, there are various enemies, for example Eve, who will eavesdrop on the system in order to read their secret messages, and Mallory, a malicious attacker, who will attempt to modify messages and so subvert Alice and Bob's use of the system. This basic set-up is shown in Figure 1.1.

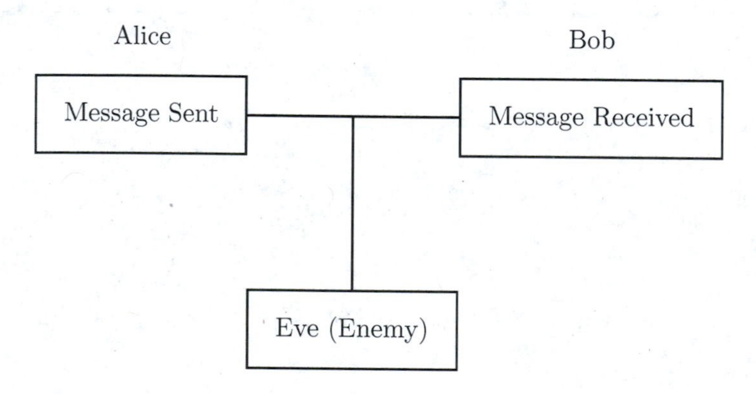

FIGURE 1.1: Basic set-up for the use of cryptography.

Here an "enemy" is any person or organization wishing to have unauthorized access to the communications. Knowing that their communications channel is insecure, Alice and Bob will want to *encrypt* their messages so that the only person who can decrypt it is the intended recipient, as shown in Figure 1.2.

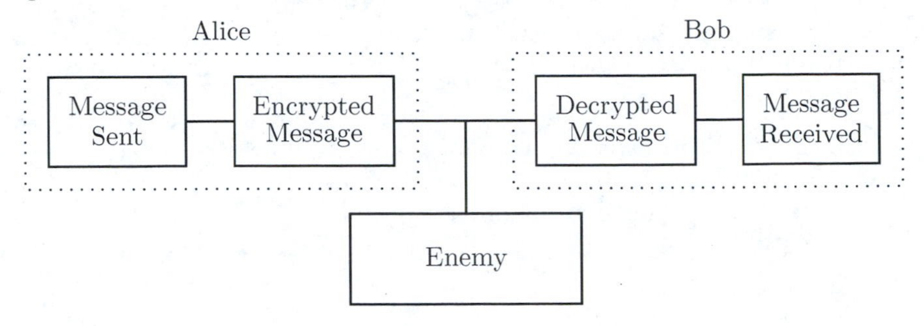

FIGURE 1.2: Encryption for security over an insecure channel.

Since the enemy doesn't know how to decrypt the message, all he or she gets is a load of apparent nonsense.

This aspect of cryptography, hiding information, is called *confidentiality*: ensuring that secret data does remain secret.

1.2 Some basic definitions

Assume here that all the data to be protected is to be sent from one computer to another, by email, or by direct file transfer.

Plaintext: This is the message to be sent.

Key: This is a piece of information or data that is used as part of the encryption process.

Encryption: This is "hiding" the message in some way by disguising its contents, so that only those "in the know" can read the message.

Ciphertext: This is the result when the encryption routine is applied to the plaintext.

Decryption: This is the process of undoing the encryption to turn the ciphertext back into the original plaintext.

Cryptosystem: This is the combined system of the encryption and decryption algorithms.

Cryptanalysis: This is the art and science (it seems to be a mixture of both) of "breaking" encryption schemes.

Attack: This is an attempt used to cryptanalyse a cryptosystem.

Generally an encryption or decryption algorithm will rely on a secret key, which may be a number with particular properties, or a sequence of bits; the algorithm itself may be well known, but to apply the decryption to a given ciphertext requires knowledge of the particular key used. There are two classes of cryptosystems:

Secret-key cryptosystems: (Also called *symmetric cryptosystems*) Here the same key is used for both encryption and decryption. So if someone gets hold of the key you are using for encryption, they can decrypt any message you send. Such a system is shown in Figure 1.3.

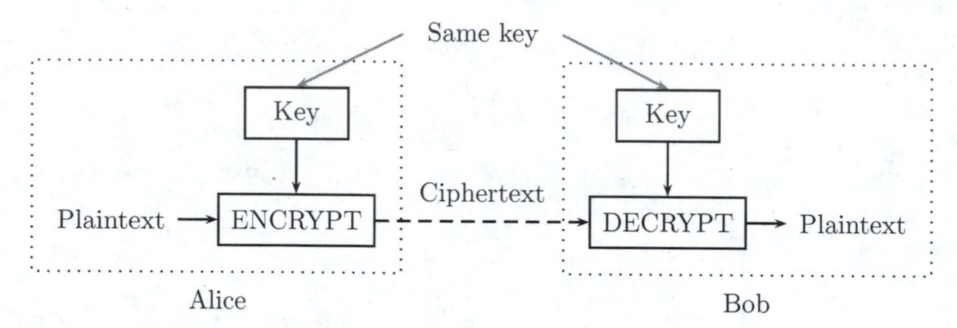

FIGURE 1.3: Using a secret-key system.

Public-key cryptosystems: Here the keys for encryption and decryption are different. So you can publicize a "public key" which anybody can use to encrypt messages to you, but only you know the appropriate corresponding private decryption key. For this method to work, there should be no way to compute (in a reasonable time) the decryption key from the encryption key. Such a system is shown in Figure 1.4.

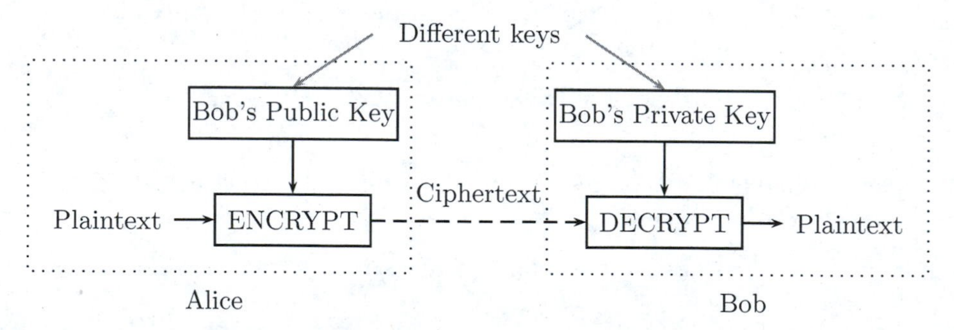

FIGURE 1.4: Using a public-key system.

Integrity, authentication and non-repudiation

So far only one aspect of the use of cryptography has been considered: confidentiality (or *privacy*), which ensures that the message or data does not fall into the wrong hands. However, there are other aspects, in some ways equally important. In particular:

integrity: ensuring that data is not changed;

authentication: meaning that the receiver (Bob) can verify that the message was indeed sent by the sender (Alice).

non-repudiation: meaning that the sender Alice can't deny sending the message.

Data integrity is vital to the workings of modern businesses; any breakdown in integrity (*data corruption*) can vastly diminish the value of the data. The loss or corruption of any financial, legal, or strategic data can vastly reduce not only income and profits, but public confidence in the company. This can be important at a personal level as well; if Alice is retrieving Bill's public key from a database, she would want to be sure that the key she receives is the correct key. And if she is sending Bill a message—for example, an offer to buy his car for $10,000—she would want to be confident that Bill receives her message as she sent it.

In pre-computer days, *authentication* could be established by the use of a signature, or even a wax seal. Barring possible forgeries, the presence of either guaranteed the receiver that the message was genuine. Fingerprints are another standard authentication procedure, being easy both to obtain and to verify. However, recently the reliability of fingerprinting has been questioned. Digital signatures, which are the topic of Chapter 7, provide not only greater strength, but greater flexibility.

Authenticity is also used for computer logins: when you have to convince the computer that it's really you logging into the system.

Non-repudiation is most commonly seen in the context of digital signatures. Having signed a document, Alice can't later deny signing it. But in general non-repudiation means that if Alice sends any message or transmission, she can't later deny or refute its validity.

1.3 Attacks on a cryptosystem

Any cryptosystem, no matter how secure looking, can be attacked. And in fact the only way to measure the strength or security of a cryptosystem is by how well it can withstand attacks. As attacks become more powerful and sophisticated, the creation of a secure cryptosystem becomes more difficult.

A vital principle for measuring the strength of a cryptosystem is provided by *Kerckhoff's principle*: that every aspect of the cryptosystem is known, *except the key*. In particular, the encryption and decryption routines are well known; the problem generally is to determine the key. In practice this is a very reasonable assumption. There are in fact only a small handful of cryptosystems strong enough to withstand modern attacks, and in almost every use of a system, it will be clear which cryptosystem is being used. The security of the system is then reliant on key management.

There are various classes of attacks against a cryptosystem:

Brute-force attack: The cryptanalyst simply tries every possible key until a key is found that works. Even though many cryptosystems have a large number of possible available keys, the speed of modern computing means that this attack can be quite feasible.

Ciphertext-only attack: The cryptanalyst has the ciphertext of various messages, all of which have been encrypted with the same algorithm. The job here is either to determine the key or keys being used or to recover as many plaintexts as possible.

Known-plaintext attack: Here the cryptanalyst not only has some ciphertexts, but also the corresponding plaintexts of at least some of them. The job is either to deduce the key(s) used, or to recover plaintexts from other ciphertexts produced by this system.

This attack is much stronger than the previous one, for the cryptanalyst can now exploit possible relationships between plaintext and ciphertext.

Chosen-plaintext attack: Here the cryptanalyst is in a position to choose the plaintext to be encrypted. This is even more powerful, as the cryptanalyst can choose plaintexts of which the encryptions will yield maximum information about the algorithm. Again, the object is to determine the key or the plaintext corresponding to a given ciphertext.

Adaptive-chosen-plaintext attack: This is a special case of the previous attack. Not only can the cryptanalyst choose the plaintexts to be encrypted, but he or she can modify the plaintexts based on the results of previous encryptions.

Chosen-ciphertext attack: The cryptanalyst has access to ciphertexts of his or her choice, and the corresponding plaintexts.

Differential cryptanalysis: This is a newer attack than the previous ones, and is a powerful variant of chosen-plaintext attack. Here the cryptanalyst chooses pairs of plaintexts that satisfy certain difference properties. By carefully investigating the way those differences appear in the corresponding ciphertexts, a very powerful attack can be mounted. This method has spelled the end for many cryptosystems previously held to be strong.

Linear cryptanalysis: A linear function f is one for which $f(x+y) = f(x)+f(y)$. Since such functions are easy to analyze and their properties easy to investigate, any reasonable cryptosystem will have much non-linearity built in. But it may be possible to approximate the results of the cryptosystem with a carefully designed linear system. In such a case another powerful attack can be mounted. Again, this is a very new method.

Since new methods of cryptanalysis are being developed all the time, it pays to be very careful when choosing a cryptosystem. It also makes the creation of new cryptosystems very hard indeed, as any cryptosystem must be shown to be invulnerable to all the attacks listed above.

1.4 Some cryptographic problems

There are many problems inherent in using cryptosystems. The very fact that such a system is being used may alert enemies to the fact that some sensitive information is being sent, and that they should concentrate all efforts on cryptanalysis of the messages, or somehow intercepting them. As well, there are difficulties in just using a given cryptosystem. Here are some examples:

Key exchange: Alice and Bob wish to communicate using a secret key cryptosystem. They must agree on a key known only to them both before they start sending messages to each other. (If the key is known to anybody else, that person or persons could send messages from Alice to Bob, pretending to be Alice, or the other way round.) How do Alice and Bob agree on a key, without making that key known to any other eavesdroppers on the system?

Man in the middle attack: Suppose Alice and Bob decide to get around the problem of key exchange by using public-key cryptography. Suppose also that on the system there is a malicious person, Mallory, intent on destroying communication between Alice and Bob. Mallory can encrypt messages to Bob, pretending to be Alice, and also encrypt messages to Alice, pretending to be Bob. He can also intercept messages, so that the only ones which reach either Alice or Bob are those sent by Mallory. How can Alice and Bob guard against this attack?

Digital signatures: Bob receives a message from Alice. How does he know that the message is actually from Alice, and not from some other person pretending to be Alice? If Alice can digitally "sign" her message so that it is definitely from her, then Bob has no trouble knowing that the message is indeed from Alice. But how can Alice sign her message in a way that can't be "forged" by anyone else?

Timestamps: A related problem to that of signatures is that of digital timestamps, where a message is to be "stamped" with the time of its sending. Such a stamp may be vital, for example, to a foreign exchange transaction, where the exchange rates may change quickly.

Authentication: When logging onto a computer system, how can you indicate to the system that it really is you who is logging in, and not some malicious intruder pretending to be you?

Coin flipping: Alice and Bob want to decide to which restaurant they should go at the end of the week, when they can finally be together. Alice wants to go Chinese, and Bob wants Italian. So they agree to flip a coin: heads they go Chinese; tails it's Italian. But how can they achieve this securely (that is, with both of them being sure that the other isn't cheating) over the computer system?

These problems can in part be solved by the use of *cryptographic protocols*, which are rules detailing how a cryptosystem is to be used to achieve a certain result. A great many fundamental protocols are given by Schneier [77]; a few of the more elementary will be introduced in the next section.

1.5 Cryptographic protocols

Key exchange

Generally for each communication, Alice and Bob will use a different key. (If they keep on using the same key, time after time, an eavesdropper has more material to use for decryption.) These different keys are called *session keys*. There are in fact several different protocols for this problem, but a very simple and practical one uses public-key cryptography.

Since the use of public-key cryptography allows for Alice and Bob to place their public keys on some publicly accessible database (which again will be called the KDC). Here's how this protocol works:

1. Alice gets Bob's public key from the KDC.
2. Alice generates a suitable random session key, encrypts it using Bob's public key, and sends the result to Bob.
3. Bob decrypts this using his private key.
4. They can now both use this session key securely.

Since secret-key cryptosystems are generally much faster (by the order of about 1000) than public-key systems, this is a very practical and appropriate use of a public-key cryptosystem.

Man in the middle attack

Now suppose that on the network is Mallory, a malicious intruder, who will stop at nothing to foil communication between Alice and Bob. Not only can he intercept messages between them, he can also replace messages from one to the other with messages from him. Here's how he might attack a public-key based communication between them:

1. Alice sends Bob her public key. Mallory intercepts this message and sends Bob his own public key instead.

2. Mallory does the same for the public key which Bob sends to Alice.

3. Alice sends a message to Bob, encrypted in what she assumes to be Bob's public key. However, since she is using Mallory's public key, Mallory can intercept this message and decrypt it, using his own private key. He can then change the message in whatever way will cause him the most satisfaction, and send that message on to Bob, using Bob's own public key.

4. Mallory can of course do the same thing for messages being sent by Bob to Alice.

Even if the keys are stored on a database, Mallory can still attack. He could, for example, intercept their requests to the database, and substitute the replies with his public key. Even more simply, he may just replace their keys on the database with copies of his own key.

One of the best solutions to use this is to invoke a *public key infrastructure* (PKI) which consists of all the software, hardware, policies, protocols and people necessary to manage public keys and their authentication. A properly set-up PKI will ensure that a key is associated with only a particular entity, and a level of trust is provided so that the users can be sure that all keys are genuine.

Keys are usually bundled into *digital certificates*, which are password-protected files providing, among other things, the person's (or entity's) name, their public key, the certificate authority and its digital signature, the period for which the key is valid, and a unique serial number.

The working of a PKI is based on a trusted *certificate authority* (CA). The job of the CA is to manage digital certificates, by issuing and revoking them. "Revoking" a certificate means taking that certificate, and the corresponding public key, out of service—this may be necessary if the private key has been compromised, or if there has been a name change. A CA may be a public authority which acts as a trusted third party for Internet users: Thawte and Verisign are examples; these are called *public certificate authorities*, and they are used by many web browsers.

Some large companies manage their own internal CAs. A problem with CAs is that there is no single, universally accepted CA.

Digital signatures

What is required here is a method by which a message or document can be signed by Alice in such a way that the signature is

- *Authentic*—it convinces the receiver that the message is indeed from Alice;
- *Unforgeable*—nobody else but Alice could have signed the message;
- *Not reusable*—even if the message is sent on to a third party, the signature can't travel with it.

Other requirements are

- The signed message is *unalterable*.
- The signature can't be *repudiated*—Alice cannot later deny that she signed the document.

Using a public-key cryptosystem is one way to provide a digital signature.

1. Alice encrypts her message with her private key. This encryption is her signature.
2. Alice sends this to Bob.
3. Bob decrypts the message with Alice's public key, and thus verifies that the message is indeed from Alice.

For this method to work, the cryptosystem must be one in which either of the two keys (public or private) can be used for encryption, and the corresponding other key for decryption. And in fact there are public-key cryptosystems which do indeed satisfy this requirement. Because public-key systems tend to be slow, it is preferable not to sign the message itself, but a *cryptographic hash* of the message. This is a sort of "digital fingerprint" of the message, much smaller in size, and will be explored in Chapter 11.

In fact there are several very powerful digital signature algorithms, which while based on public-key systems, are different in their construction. Some of them will be studied in detail in Chapter 7.

Authentication

The first step is the notion of a *one-way function*, that is, a function $f(x)$ that is easy to compute, but for which the inverse is much harder to compute. That is, given x, the value $y = f(x)$ can be easily determined, but given y alone, it is not easy to determine the original value of x. It is very difficult to show that a particular function is one-way; a simple example is given by the function $y = x^2$. Although this is easy to calculate, the inverse, $\sqrt{x}$, is in general much harder. The creation of suitable one-way functions is crucial to much cryptography.

Using one-way functions

When Alice first logs onto a system, she creates a password for herself. The system applies a suitable one-way function to her password, and stores the result in a password file. Now for every successive login

1. Alice sends the system her password;

2. the system applies the one-way function to her password and compares the result with the result it calculated originally;

3. if the two agree, Alice is allowed into the system.

The possibility is open for a dictionary attack; to get onto the system, a hacker (say, Mallory) need only compile a list of lots of possible passwords, and try each of them. To prevent this, most computer systems add *salt*; this is a random string appended to the password prior to its first operation by the one-way function. Then the system must store both the salt and the value of the function in the password file. Since Mallory can't test every possible password with every possible salt string, this method of attack is effectively eliminated. Adding salt also provides some protection to those people unwise enough to use the same password on two or more computer systems.

However, there is no way for the computer to know if it is in fact Alice using the system, or somebody else pretending to be Alice. Moreover, if somebody else has access to the transmission path between Alice's keyboard and the system, that person can read the password before the one-way function operates on it. (It may be, for example, that Alice is in one country, and logging onto a system in a different country.) There are various solutions to this.

Using public-key cryptography

Here are the steps the system takes to allow Alice access.

1. The system sends Alice a random string.

2. Alice encrypts the string with her private key, and sends the result to the system along with her name.

3. The system looks up Alice's public key in its database, and decrypts the above message using that key.

4. If the decrypted string matches that which the system sent to Alice in step 1, then all is well and Alice is allowed in.

Since only Alice knows her private key, she can't be impersonated by anybody else. And since her private key is never transmitted to the system, it is impossible for anyone to tap the transmission path to get it. This assumes that the terminal at which Alice logs on is intelligent enough to perform the encryption, but the transmission path may be totally insecure and open.

Coin flipping

One-way functions can be used here. Alice and Bob must agree on a suitable one-way function f to use. Then:

1. Alice chooses a random number x and computes $y = f(x)$. She then sends y to Bob.
2. Bob guesses whether x is even or odd, and sends his guess to Alice.
3. If Bob's guess is correct, then the result of the coin toss is designated to be heads; if his guess is incorrect the result of the coin toss is tails.
4. Alice announces the result of the coin toss and sends x to Bob.
5. Bob confirms that $y = f(x)$.

The one-way function must be strong enough so that Alice can't find two numbers x_1 and x_2 with one even and the other odd, and with both $f(x_1) = y$ and $f(x_2) = y$.

1.6 Some simple ciphers

This section introduces some simple "pencil and paper" ciphers to illustrate the points which have been discussed so far.

Kid Krypto

"Kid Krypto" [26] is a family of cryptosystems developed by Michael Fellows and Neal Koblitz for the teaching of cryptography without using advanced mathematics. Here is "Kid RSA," which is a simplified version of the RSA cryptosystem which will be investigated in Chapter 5.

Kid RSA is a public-key system, so it uses two different but related keys for encryption and decryption.

To set up Kid RSA, Alice chooses four random integers a, b, A and B. She then computes:

$$\begin{aligned} M &= ab - 1 \\ e &= AM + a \\ d &= BM + b \\ n &= \frac{ed - 1}{M}. \end{aligned}$$

She makes the pair (n, e) available as her public key and keeps d as her secret key. All the other numbers can be discarded; at any rate they should never be revealed.

Messages in this system are integers $x < n$. Suppose Bob wishes to send a message to Alice. He encrypts it by first multiplying x by e, and dividing the product xe by n. The *remainder* of this division is the ciphertext y.

Alice decrypts y by multiplying by d to obtain yd, then dividing this product by n. The remainder of this division is the plaintext.

To see this in action, suppose that Alice chooses

$$a = 3, \quad b = 4, \quad A = 5, \quad B = 6.$$

Then it is easy to determine

$$M = 11, \quad e = 58, \quad d = 70, \quad n = 369.$$

Her public key is $(369, 58)$, and her private key is 70.

Suppose Bob wants to encrypt $x = 200$. He multiplies by $e = 58$ to obtain $xe = 11600$. Dividing by $n = 369$ leaves a remainder of 161. This is the ciphertext y he sends to Alice.

Alice multiplies this ciphertext $y = 161$ by $d = 70$ to obtain $yd = 11270$. She then divides by n to obtain a remainder of 200, which is the required plaintext.

For this system to be secure, there should be no way that Bob (or anyone else) can easily determine the value of d from the public values n and e. He knows that $ed - 1 = Mn$, but he doesn't know M.

To quickly demonstrate this in Sage:

```
sage: a,b,A,B = 3,4,5,6
sage: M = a*b-1; e = A*M+a; d = B*M+b; n = (e*d-1)/M
```

Now take a plaintext value and check that it is less than n:

```
sage: x = 200
sage: x < n
  True
```

Now the encryption and decryption, using the percentage symbol for remainder after division:

```
sage: y = (x*e)%n; y
  161
sage: (y*d)%n
  200
```

The Bifid cipher

The Bifid cipher was created in about 1901 by a French cryptographer, Félix Delastalle. Although it has never been used for military or any other "serious" purpose, it has a very elegant design, is easy to implement, and quite hard to break given its simplicity.

The key for this cipher is any permutation of the alphabet (except for the letter J). One way to remember a key is to choose a word with no repeating letters such as "ENCRYPT" to start the permutation, and finish with the remaining letters. This permutation is placed in a 5×5 array called the *tableau.* This produces:

	0	1	2	3	4
0	E	N	C	R	Y
1	P	T	A	B	D
2	F	G	H	I	K
3	L	M	O	Q	S
4	U	V	W	X	Z

Alternatively, a 6×6 tableau can be constructed using all 26 letters and the ten digits. Each letter thus corresponds to a row and column index. For this tableau, `A` has indices 1 and 2; while `X` has indices 4 and 3.

Such a tableau, by which a letter of the alphabet can be encoded as two symbols from a smaller set, is an example of a *Polybius square*, named after the Ancient Greek historian and scholar. Polybius envisaged that this method of encoding the alphabet using a smaller set of symbols would be useful for communications signalling. Expressing a large symbol set (such as an alphabet) into groups of symbols from a smaller set (here the numbers 0–4) is called *fractionation.*

To use the Bifid cipher, encode the message using the indices from the tableau. So that, for example, the message "`MEET ME ON FRIDAY`" would be encoded as

M	E	E	T	M	E	O	N	F	R	I	D	A	Y
3	0	0	1	3	0	3	0	2	0	2	1	1	0
1	0	0	1	1	0	2	1	0	3	3	4	2	4

The indices are then read off row by row:

```
3 0 0 1 3 0 3 0 2 0 2 1 1 0 1 0 0 1 1 0 2 1 0 3 3 4 2 4
```

These indices are then grouped back into pairs and turned into letters by using the original tableau:

30	01	30	30	20	21	10	10	01	10	21	03	34	24
L	N	L	L	F	G	P	P	N	P	G	R	S	K

The ciphertext is thus "`LNLLFGPPNPGRSK`."

The cipher is easily implemented in Sage. Rather than using a 5×5 array, use a string consisting of the permutation below.

```
sage: kw = 'ENCRYPTABDFGHIKLMOQSUVWXZ'
```

To find the indices, first find the position of the character in the key. For example, `A`:

```
sage: i = kw.index('A'); i
  7
```

Now the two indices can be obtained from `i` very easily:

```
sage: i//5, i%5
  (1, 2)
```

The first value simply finds the row by performing the integer division of the index by 5; the column value is the remainder after division by 5 (as given by the percentage operator).

The rest of the program is straightforward:

```
def bifid(pl,key):
    n = len(pl)
    pairs = []
    for x in pl:
        kx = key.index(x)
        pairs += [[kx//5,kx%5]]
    print pairs
    tmp = flatten([x[0] for x in pairs]+[x[1] for x in pairs])
    ct = ''
    for i in range(n):
        pair=[tmp[2*i],tmp[2*i+1]]
        ct += key[5*pair[0]+pair[1]]
    return ct
```

For example:

```
sage: bifid('MEETMEONFRIDAY',kw)
  'LNLLFGPPNPGRSK'
```

Decryption would apply the above steps in reverse.

The Bifid cipher obtains its strength from using two components: fractionation and transposition.

The Playfair cipher

The Playfair cipher is a secret key, symmetric cryptosystem which is a nice example of the use of a key and of block encryption, where a block of plaintext (in this case two letters) is used as input to the encryption procedure. Like all ciphers pre-dating the use of computers, it has since been broken.

Historical note: The Playfair cipher was developed by Charles Wheatstone in 1854, and popularized and promoted by his friend Lord Playfair. Although Playfair was scrupulous in crediting Wheatstone with the cipher, it is his name now which is associated with it. It was used by the British in both the Boer War and the First World War. It was also used during the Second World War in 1943, by the future US President John F. Kennedy.

The key for this cipher is any permutation of the alphabet without J, put into a 5×5 array, as in the Bifid cipher. Also as in the Bifid cipher, a word with no repeated letters is chosen to start the array, with the remaining letters of the alphabet (except for `J`) placed in order into the array to produce an *enciphering array.* For example:

```
E N C R Y
P T A B D
F G H I K
L M O Q S
U V W X Z
```

Plaintexts are broken into two letter blocks in such a way that no block consists of repeated letters. Then each block is encrypted separately:

1. If the the two letters of a block are in different rows and columns of the 5×5 enciphering array, they are replaced by the letters at the other corners of their rectangle. The letter in the same row as the first letter in the block comes first. For example `FR` is encrypted to `IE`, and `EX` to `RU`.

2. If the two letters are in the same row, they are replaced with the corresponding next letters in that row, if necessary cycling around to the first letter. So that `GI` is encrypted to `HK`, and `TD` to `AP`.

3. If the two letters are in the same column, they are each replaced with the next letter that follows it in that column, if necessary cycling around to the first letter. So that `PL` is encrypted to `FU`, and `HW` to `OC`.

So for example, the plaintext "`MEET ME ON FRIDAY`" is encrypted as

ME	ET	ME	ON	FR	ID	AY
LN	NP	LN	MC	IE	KB	DC

Decryption uses the enciphering array, and just works backwards through the blocks of the ciphertext.

The Playfair cipher has several characteristics that help its cryptanalysis. First, no letter is ever encrypted to itself. Second, for the first case of encryption, if for example `FR` is encrypted as `IE`, then `IE` will be encrypted as `FR`.

The existence of such plaintext/ciphertext pairs can greatly simplify a known plaintext attack. Thirdly, any letter will only be encrypted either to a different letter in its row, or the letter immediately below it. That is, any letter will be encrypted only to five possible ciphertext letters. For a fascinating discussion of the cipher and its cryptanalysis, see Gaines [29].

The ADFGVX Cipher

This secret-key cipher shows the strength that can be obtained by combining two different operations. The Playfair cipher works entirely by *substitution*: each two-letter block of plaintext is substituted by a two-letter block of ciphertext. The ADFGVX cipher works both by substitution, and also by *permutation*, where the values are rearranged. The result is a cipher that is one of the strongest "pencil and paper" ciphers, and was used extensively by the Germans during the First World War.

It has two keys: an arrangement of the letters A to Z and the digits 0 to 9 in a 6×6 array, and a word or phrase in which no letters are repeated. For example, the first array may be:

	A	D	F	G	V	X
A	F	L	1	A	0	2
D	J	D	W	3	G	U
F	C	I	Y	B	4	P
G	R	5	Q	8	V	E
V	6	K	7	Z	M	X
X	S	N	H	Ø	T	9

and the word `ENCRYPT` can be used as the second key. Notice that the first array is an example of a Polybius square, where the indices are the letters ADFGVX instead of the integers 0–4.

A message is encrypted in two stages. First, each letter or digit is replaced by its row and column index from the first array:

S	E	E	M	E	A	T	1	0	T	O	M	O	R	R	O	W
X	G	G	V	G	A	X	A	X	X	A	V	A	G	G	A	D
A	X	X	V	X	G	V	F	G	V	V	V	V	A	A	V	F

These letters are now put into an array whose columns are indexed by the second keyword:

E	N	C	R	Y	P	T
X	A	G	X	G	X	V
V	G	X	A	G	X	V
A	F	X	G	X	V	A
V	V	V	A	V	G	A
G	A	A	V	D	F	X

and the array is filled with Xs. This new array is now reordered so that the indices of the columns are in alphabetical order:

C	E	N	P	R	T	Y
G	X	A	X	X	V	G
X	V	G	X	A	V	G
X	A	F	V	G	A	X
V	V	V	G	A	A	V
A	G	A	F	V	X	D

The resulting array is read column by column, and split into groups of some prearranged fixed amount, say 6, which also obscures the length of the second keyword:

```
GXXVAX VAVGAG FVAXXV GFXAGA VVVAAX GGXVDX
```

with as many extra Xs as needed on the end to balance these blocks.

Historical note: In spite of the strength of this cipher, it was broken by a French cryptanalyst, Georges Painvin, who in 48 hours of unremitting labor in 1918, solved a German military telegram that used this code. A method of cracking general ADFGVX ciphers was finally published in 1933.

The letters A, D, F, G, V, and X were chosen so that the ciphertext could be sent in Morse code, in which those six letters are very different.

More information about these (and many other ciphers), and their historical use, can be found in Kahn's *The Codebreakers*[43], a magisterial history of cryptography before computers.

1.7 Cryptography and computer security

It is important to realize that cryptography is only one of many aspects that make up computer security. In fact, choosing a secure cryptosystem is only part of the work; it is just as important to make sure the system is used in a manner that does not compromise its security. And this can be the really hard part. There are plenty of cryptosystems available that are very strong and hence supposedly secure. But using them appropriately can be a very difficult matter.

Then there are the added problems of password choice and storage, physical security of the system, and so on. Computer security can be described as analogous to house security: a house is secure only if it is not possible to gain access at any point. A cryptosystem then is the front door lock to the house.

No matter how strong the lock, the house is insecure if a window is left open, or the door surrounds are so weak that the door can be pushed in, or if the key is left somewhere easy to find.

A cryptosystem is often the strongest part of any computer security system, but the use of a strong cryptosystem does not guarantee system security.

1.8 Glossary

Since there are many definitions in this chapter, this glossary will contain only a few of the most commonly used terms.

Attack. An attempt at compromising the security of a cryptosystem, by obtaining either the key or the plaintext.

Integrity and confidentiality. Aspects of message security.

Key exchange problem. The problem of exchanging a secret key over an insecure channel.

Man-in-the-middle attack. An attack by which a malicious middleman can subvert a public-key system by replacing Alice and Bob's keys with his own.

One-way function. A function that is easy to compute but (computationally) hard to invert.

Plaintext, ciphertext, key. Inputs and outputs to a cryptosystem (full definitions on page 3).

Public-key cryptosystem. A system with two different, but related, keys: one for encryption and one for decryption.

Secret-key cryptosystem. A system that uses the same key for encryption and decryption.

Symmetric and asymmetric cryptosystems. Alternative names for secret-key and public-key cryptosystems, respectively.

Exercises

Review Questions

1. What is the distinction between a private and a public-key cryptosystem?

2. What does it mean to *attack* a cryptosystem?

3. In the context of cryptography, what is a protocol?

4. What is the man-in-the-middle attack?

5. What are some of the situations in which cryptography may play an important role?

6. What is a situation where it is more important that a message be authenticated (that is, digitally signed), than it be secret?

7. What are some of your day-to-day activities that will involve the use of a cryptosystem?

Beginning Exercises

8. Suppose Alice is logging on to her computer from a distant terminal, using the public-key authentication protocol. In what ways can Mallory, who may tamper with any messages between the host computer and Alice, foil Alice's attempts to log on?

9. How can the protocol be improved to foil Mallory?

10. How can the digital signature protocol be improved so that the document can be signed by two people?

11. Using Kid-RSA and the values $a = 5$, $b = 6$, $A = 7$ and $B = 8$, determine the values of M, e, d and n. Send the plaintexts (a) 10, (b) 750, (c) 1000, and decrypt the resulting ciphertexts.

12. Repeat the above question, but using values $a = 20$, $b = 31$, $A = 17$ and $B = 27$, and plaintexts (a) 10, (b) 1000, (c) 200,000

13. Use the Bifid cipher with the tableau as given on page 14 to

    (a) encrypt `BRING ALL YOUR MONEY`

    (b) decrypt `PDRRNGBENOPNIAGGF`

14. Use the Bifid cipher with a permutation beginning with the word `CRYPTO` to

(a) encrypt `THIS IS A SECRET MESSAGE`

(b) decrypt `MDQHBOEOIZZEPZGDFE`

15. Suppose the Bifid cipher was applied to two characters of the plaintext at a time, rather than to the entire plaintext. How does this differ from the Playfair cipher?

16. Use the Playfair cipher with keyword `ENCRYPT` to

    (a) encrypt `SELL MY CAR`

    (b) decrypt `PX CD QU ML AM YH`

17. Use the Playfair cipher with keyword `SURFING` to

    (a) encrypt `MOVE NORTH TONIGHT`

    (b) decrypt `UD DM OP UH OP GD ZY`

18. Use the ADFGVX cipher with the array given on page 17 and the keyword `ENCRYPT` to

    (a) encrypt `GO SOUTH`

    (b) decrypt `AVAV VXXD VAVD VXDX VDVX GVFV GXXF`

19. Use the ADFGVX cipher with the array given on page 17 and the keyword `CODE` to

    (a) encrypt `BRING A FRIEND`

    (b) decrypt `XAAGGX GAAAXX XDVXGG ADAADG`

Sage Exercises

20. Using the Bifid code given, write some code to perform decryption. Test it on the example in the notes, and on the examples in questions 13 and 14.

21. Edit your Bifid code to allow for the version where the plaintext is divided into equal-sized blocks, and each block encrypted separately.

22. Delastalle also developed a "Trifid" cipher, where 27 characters (all letters and a space, say) are put into a $3 \times 3 \times 3$ grid. The indices of each plaintext letter are written column by column, and the ciphertext is obtained by reading groups of three indices at a time row by row.

    Write a program to implement this.

23. Develop a simple Playfair-like cipher with the following encryption: for a pair of letters `XY` in a Polybius square, if the indices of `X` and **Y** are (a, b) and (c, d) respectively, then the ciphertext is the pair of letters whose indices are (a, d) and (b, c).

Chapter 2

Basic number theory

This chapter provides the mathematical background for much of the rest of the book. In particular, it investigates:

- Prime numbers, their definition and uses.
- Factorization.
- Modular arithmetic, including powers and inverses.
- Fermat's theorem, Euler's totient function and Euler's generalization of Fermat's theorem.
- The Chinese remainder theorem.
- The Euclidean algorithm, both standard and extended forms.
- Quadratic residues and the Legendre symbol.
- Some methods of primality testing.

2.1 Introduction

Much modern cryptography is based around the theory of numbers, in particular prime numbers and their properties. This chapter will include investigating some of the basic properties and developing enough theory to enable future investigation of various public-key cryptosystems. Recall from Chapter 1 that the availability of a public-key cryptosystem allows neat protocols for digital signatures.

2.2 Some basic definitions

In this chapter, and for the remainder of this text, all numbers will be integers. One of the most important properties for this text is that of *divisibility*.

Given two integers n and k, then n *is divisible by* k, written

$$k|n$$

if there is an integer m for which

$$mk = n.$$

Another way of saying the same thing is that k *is a divisor (or factor) of* n, or that k *divides* n. The following properties of divisibility are standard:

1. If $k|n$ and $n|m$ then $k|m$.

2. If $k|n$ and $k|m$ then $k|(am + bn)$ for any integers a and b.

More properties will be discussed later.

A *prime number* is an integer greater than 1 whose only divisors are itself and 1. For example: 3, 5, 7, 11, 257, 65537 are all prime numbers; but 391 is not a prime number (because $391 = 17 \times 23$). It is a fact that there are infinitely many prime numbers, so that there are prime numbers as big as you like. One way of showing this is attributed to Euclid: suppose contrary to fact that there is only a finite number of primes:

$$p_1, p_2, \ldots, p_n.$$

Then consider their product, plus 1:

$$m = p_1 p_2 \cdots p_n + 1.$$

This number m cannot be divisible by any of the primes p_1 to p_n, so either it is prime itself, or it is divisible by a new prime.

Prime numbers satisfy the following divisibility properties:

1. If p and q are distinct primes, and if p and q are both factors of n, then $(pq)|n$.

2. If $p|(mn)$ then either $p|n$ or $p|m$, or both.

Prime numbers receive importance in the *fundamental theorem of arithmetic*, which states that the factoring of any integer $n > 1$ into prime factors is unique apart from the order of the factors. For example,

$$49,000 = 2^3 \times 5^3 \times 7^2$$

and there is no other combination of prime numbers whose product is $49,000$.

Factorization of an integer n can be performed by *trial division*. Starting with the lowest prime, 2, divide n by 2 as many times as possible. Next divide by the next prime, 3, as many times as possible, and then by 5, 7, 11, and so on. Note that if an integer n is *composite* (that is, not prime), it must have a factor not greater than $\sqrt{n}$. For if all of its factors were greater than $\sqrt{n}$, then their product would be greater than n, which is a contradiction.

A slightly more sophisticated method of factoring, which is particularly powerful if the number to be factored is a product of two primes of similar size, starts with noting that if

$$n = pq$$

with p and q both large (and hence odd) primes, then the sum $p + q$ and difference $p - q$ will both be even. Then:

$$\begin{aligned} p + q &= 2r \\ p - q &= 2k \end{aligned}$$

with k a small value. From these two equations it follows that

$$\begin{aligned} p &= r + k \\ q &= r - k \end{aligned}$$

and so

$$\begin{aligned} n &= pq \\ &= (r + k)(r - k) \\ &= r^2 - k^2 \end{aligned}$$

and so

$$n + k^2 = r^2.$$

To factor by this method simply requires adding square numbers k^2 to n until another square r^2 is reached. Then the factors will be $r \pm k$. For example, suppose $n = 5609$. Then:

k	$n + k^2$	Square root
1	5610	Not a square
2	5613	Not a square
3	5618	Not a square
4	5625	75

This means the two factors of n are

$$75 \pm 4 = 71, 79.$$

There are several variants of this basic method, but they all rely on expressing n as a difference of squares. Such a method is called *Fermat's method*, after Pierre de Fermat, who developed the method in the early 17th century.

Historical note: Pierre de Fermat (early 1600s–1665) was a French jurist and amateur mathematician, whose mathematical discoveries rival those of Descartes. He developed much of the early theory of calculus, as well as number theory and the solution of equations. Although he did not provide formal proofs for many of his results, his status as one of the foremost mathematicians of his time was never doubted.

Given two integers m and n, their *greatest common divisor*, denoted $\gcd(m,n)$, is the largest integer that is a divisor of both m and n. For example, $\gcd(60,195) = 15$, as 15 is the largest number that is a divisor of both 60 and 195.

Two integers m and n are *relatively prime* if $\gcd(m,n) = 1$. For example, 60 and 77 are relatively prime (check it).

If the remainder when a is divided by n is r (that is, $a = qn + r$ for some r), then a is said to be *equal to* r *modulo* n, and that r *is a residue of* a *modulo* n. This can be written as

$$a = b \pmod{n}.$$

For example, if 105 is divided by 17 there is a remainder of 3. Thus

$$105 = 3 \pmod{17}.$$

Given the equation

$$a = qn + r$$

with $0 \leq r \leq n-1$ then r is the *remainder* when a is divided by n, and q is the *quotient*.

Modular arithmetic satisfies the following properties:

1. $(a \pm b) \pmod{n} = ((a \bmod n) \pm (b \bmod n)) \pmod{n}$,
2. $(a \times b) \pmod{n} = ((a \bmod n) \times (b \bmod n)) \pmod{n}$.

These follow immediately from the definition. In other words, the modulus of a sum or difference is equal to the sum or difference of the individual moduli (reduced mod n), and the same goes for products.

If $a = b \pmod{n}$ then a and b are said to be *congruent mod* n, and the value $b \pmod{n}$ is referred to as a *congruence*. The values 0 to $n-1$ are called the *residues* modulo n. The set of these residues is denoted

$$\mathbb{Z}_n.$$

As with the set of integers, elements of this set can be added, subtracted, and multiplied, but not necessarily divided. Such a set, with addition, subtraction, multiplication, is called a *ring*[1].

---

[1]There is much more to rings than this: more formally addition and multiplication must be commutative; there must be an additive identity and an additive inverse for every element; and multiplication must be distributive over addition.

For example, $105 = 3 \pmod{17}$ from above; it is also easy to compute $211 = 7 \pmod{17}$. To find the value of $(105 \times 211) \bmod 17$, use the second property above, and just multiply the moduli, and reduce mod 17: $3 \times 7 = 21 = 4 \pmod{17}$. Notice that there was no need to compute 105×211 in this calculation. Modular arithmetic is extensively used in much modern cryptography; many nasty looking expressions can be readily simplified by the use of these properties.

Modular arithmetic in Sage can be performed in two ways; by using the percentage operator, or the `mod` command. However, the outputs of these have different properties:

```
sage: a = 105%17; a
  3
sage: a^20
  3486784401
sage: b = mod(105,17); b
  3
sage: b^20
  13
sage: a+100
  103
sage: b+100
  1
```

The reason for this is that Sage attaches a "type" to each object. For the two objects `a` and `b`, the first is an ordinary integer, but the second is an integer modulo 17, and so all subsequent operations on that variable will return results modulo 17. The `parent` command is a handy way of demonstrating these distinctions:

```
sage: parent(a)
  Integer Ring
sage: parent(b)
  Ring of integers modulo 17
```

2.3 Some number theoretic calculations

Testing numbers for primality

First notice that if n is not prime, it must have a factor less than or equal to $\sqrt{n}$. So just test for divisibility by all prime numbers less than or equal to $\sqrt{n}$. For example, if $n = 1487$, then $38 < \sqrt{1487} < 39$, so divisibility need be tested only by all prime numbers up to 38; they are 2, 3, 5, 7, 11, 13, 17, 19, 23, 29, 31, 37. Since none of these are a divisor of 1487, it follows that 1487 is indeed prime.

This method will work for any number n, no matter how large. However, it becomes very clumsy and inefficient in n is very large. Since most useful prime numbers (from a cryptographic perspective) are in the region of several hundred digits long, better and faster methods are required. The creation of algorithms for efficient primality testing of large integers is thus a major area of mathematical research. Two methods will be discussed in Section 2.4.

Calculating the greatest common divisor

One very popular and fast algorithm for calculating $\gcd(m, n)$ is *Euclid's algorithm* (although in fact it wasn't discovered by Euclid, just popularized by him). It is over 2000 years old, but is still holding up well. Here is how it works:

1. Find the remainder when m is divided by n. (That is, compute $m \bmod n$.) If $m < n$, the result will be just m.

2. Now set $m = n$ and $n = m \bmod n$.

3. Repeat the above steps until $m \bmod n = 0$. Then the gcd is n.

Rather than give a formal proof, note that the algorithm can be described as a sequence of steps:

$$\begin{aligned} m &= q_1 n + r_1 \\ n &= q_2 r_1 + r_2 \\ r_1 &= q_3 r_2 + r_3 \\ &\vdots \\ r_{k-2} &= q_k r_{k-1} + r_k \\ r_{k-1} &= q_{k+1} r_k \end{aligned}$$

where the final remainder, r_{k+1}, is zero. By definition r_k is a divisor of r_{k-1}, and hence by the previous line it is a divisor of r_{k-2}. Moving back up through the list of equations it is easy to see that r_k must be a divisor of both m and n. This means that r_k must be a divisor of $\gcd(m, n)$. But, suppose that some other value g also divides both m and n. By moving down through the equations, starting with $m = xg$ and $n = yg$ it can be seen that g must divide r_k. Thus r_k is equal to $\gcd(m, n)$.

This algorithm can be presented as an iteration. Start off with $m_0 = m$ and $n_0 = n$ and for $i \geq 1$ define

$$m_i = n_{i-1} \quad \text{and} \quad n_i = m_{i-1} \bmod n_{i-1}.$$

When $n_k = 0$ for some k, then the gcd is m_k. An example: $m = 195$, $n = 60$. The iteration can be given as a table:

i	m_i	n_i	
0	195	60	
1	60	15	(= 195 mod 60)
2	15	0	(= 60 mod 15)

At this stage the algorithm stops with the result $\gcd(195, 60) = 15$.

Here's another example, finding $\gcd(2697, 2553)$:

i	m_i	n_i	
0	2697	2553	
1	2553	144	(= 2697 mod 2553)
2	144	105	(= 2553 mod 144)
3	105	39	(= 144 mod 105)
4	39	27	(= 105 mod 39)
4	27	12	(= 39 mod 27)
5	12	3	(= 27 mod 12)
6	3	0	(= 12 mod 3)

and so $\gcd(2697, 2553) = 3$.

Euclid's algorithm can be written as a recursive function for programming:

$$gcd(m, n) = \begin{cases} m & \text{if } m = n, \\ \gcd(n, m \bmod n) & \text{otherwise.} \end{cases}$$

The extended Euclidean algorithm

It can be shown that for any integers m and n, there are two other integers s and t for which

$$sm + tn = \gcd(m, n).$$

For example, if $m = 45$ and $n = 39$ their gcd is 3, and if $s = -6$ and $t = 7$ then:

$$(-6)45 + (7)39 = 3.$$

To find these values s and t, Euclid's algorithm can be extended by using a tabular method. Start by setting up a table with five columns:

i	q	r	u	v
-1		m	1	0
0		n	0	1

The table can be continued as follows; for each row $i \geq 1$ define:

$$q_i = \left\lfloor \frac{r_{i-2}}{r_{i-1}} \right\rfloor.$$

That is, divide the two previous values of r and throw away the remainder: the symbol $\lfloor x \rfloor$ means the integer part of x. Then define:

$$r_i = r_{i-2} - r_{i-1}q_i$$
$$u_i = u_{i-2} - u_{i-1}q_i$$
$$v_i = v_{i-2} - v_{i-1}q_i$$

Stop when $r_k = 0$ for some k; at which point:

$$s = u_{k-1}$$
$$t = v_{k-1}.$$

This algorithm can be shown to work by the same process used for the Euclidean algorithm above. First note that

$$r_i = mu_i + nv_i \tag{2.1}$$

for every i; this is easily shown by induction. Since also at each step

$$r_{i-2} = q_i r_{i-1} + r_i,$$

it follows that the final value r_{k-1} (for which $r_k = 0$) is equal to $\gcd(m, n)$. Then by equation 2.1 with the index $i - 1$ it follows that

$$s = u_{k-1}$$
$$t = v_{k-1}.$$

For example, suppose $m = 45$ and $n = 39$. The initial state is

i	q	r	u	v
-1		45	1	0
0		39	0	1

For the next row

$$q_1 = \left\lfloor \frac{45}{39} \right\rfloor = 1$$

and then

$$r_1 = r_{-1} - r_0 q_1 = 45 - (39)(1) = 6$$
$$u_1 = u_{-1} - u_0 q_1 = 1 - (0)(1) = 1$$
$$v_1 = v_{-1} - v_0 q_1 = 0 - (1)(1) = -1.$$

At this stage the table is

i	q	r	u	v
-1		45	1	0
0		39	0	1
1	1	6	1	-1

For the next row

$$q_2 = \left\lfloor \frac{r_0}{r_1} \right\rfloor = \left\lfloor \frac{39}{6} \right\rfloor = 6$$

and then

$$\begin{aligned} r_2 &= r_0 - r_1 q_2 &&= 39 - (6)(6) &&= 3 \\ u_2 &= u_0 - u_1 q_2 &&= 0 - (1)(6) &&= -6 \\ v_2 &= v_0 - v_1 q_2 &&= 1 - (-1)(6) &&= 7. \end{aligned}$$

This produces

i	q	r	u	v
-1		45	1	0
0		39	0	1
1	1	6	1	-1
2	6	3	-6	7

Continuing on,

$$q_3 = \left\lfloor \frac{r_0}{r_1} \right\rfloor = \left\lfloor \frac{6}{3} \right\rfloor = 2.$$

This row is completed with

$$r_3 = r_1 - r_2 q_3 = 6 - (3)(2) = 0.$$

Since $r = 0$, the algorithm stops here with the table

i	q	r	u	v
-1		45	1	0
0		39	0	1
1	1	6	1	-1
2	6	3	-6	7
3	2	0		

from which

$$s = u_2 = -6, \qquad t = v_2 = 7$$

can be read.

The Euclidean algorithm and its extended version are available in Sage as `gcd` and `xgcd` respectively:

```
sage: gcd(49,36)
  1
sage: xgcd(49,36)
  (1, -11, 15)
```

The Chinese remainder theorem

Consider the congruence

$$x = 24 \pmod{35}.$$

This can be broken into smaller congruences by factoring $35 = 5 \times 7$. Since

$$x = 35k + 24$$

for some k, then

$$x \bmod 5 = 24 \bmod 5 = 4$$

and

$$x \bmod 7 = 24 \bmod 7 = 3.$$

Thus the above congruence can be simplified using smaller numbers as

$$\begin{aligned} x &= 4 \pmod{5} \\ x &= 3 \pmod{7}. \end{aligned}$$

The *Chinese remainder theorem* provides a method of going the other way: to solve a system of simultaneous congruences. Most cryptographic algorithms require the version with just two congruences, although the result is generalizable to an arbitrary number of simultaneous congruences—see exercise 35.

> Suppose that m and n are relatively prime. Then there is a single solution $\pmod{mn}$ to the congruences
>
> $$\begin{aligned} x &= a \pmod{m} \\ x &= b \pmod{n}. \end{aligned}$$

Since m and n are relatively prime, $\gcd(m, n) = 1$. Then by the extended Euclidean algorithm, s and t can be computed so that

$$sm + tn = 1$$

and so

$$\begin{aligned} sm &= 1 \pmod{n} \\ tn &= 1 \pmod{m}. \end{aligned}$$

Let

$$x = bsm + atn.$$

Then

$$x \bmod m = atn \bmod m = a,$$
$$x \bmod n = bsm \bmod m = b.$$

For example, consider the two congruences above: $x = 4 \pmod 5$ and $x = 3 \pmod 7$. Since $n = 5$ and $m = 7$ are relatively prime, the extended Euclidean algorithm can be used to obtain the table:

i	q	r	u	v
-1		7	1	0
0		5	0	1
1	1	2	1	-1
2	2	1	-2	3
3	2	0	5	-7

Thus $s = -2$ and $t = 3$, and it can be readily checked that $(-2)7 + (3)5 = 1$. With $a = 3$ and $b = 4$, then

$$\begin{aligned} x &= bsm + atn \\ &= (4)(-2)(7) + (3)(3)(5) \\ &= -11 \\ &= 24 \pmod{35}. \end{aligned}$$

In Sage the congruences

$$x = a \pmod m$$
$$x = b \pmod n$$

can be computed using the command `crt(a,b,m,n)`. For example, with the values above

```
sage: crt(4,3,5,7)
  -11
```

Note that the result is not necessarily positive, but

```
sage: _%35
  24
```

Quadratic residues

For many cryptosystems, it is necessary to be able to solve the congruence

$$x^2 = a \pmod p$$

or to determine whether such a congruence is solvable. Consider the list of all non-zero values of $x^2 \pmod{17}$:

x	1	2	3	4	5	6	7	8	9	10	11	12	13	14	15	16
x^2	1	4	9	16	8	2	15	13	13	17	2	8	16	9	4	1

It can be seen from this table that the equation

$$x^2 = a \pmod{19}$$

is solvable for the values $a = 1, 2, 4, 8, 9, 13, 15, 16$ and not solvable for $a = 3, 5, 6, 7, 10, 11, 12, 14$. The "solvable" values are called *quadratic residues modulo p*, and the "unsolvable" values are called *quadratic non-residues modulo p*. For the study of quadratic residues a useful notation is provided by the *Legendre symbol*

$$\left(\frac{a}{p}\right)$$

and defined by

$$\left(\frac{a}{p}\right) = \begin{cases} -1 & \text{if } a \text{ is a quadratic non-residue modulo } p \\ 0 & \text{if } a \text{ is a multiple of } p \\ 1 & \text{if } a \text{ is a quadratic residue modulo } p. \end{cases}$$

From the table above, for example:

$$\left(\frac{2}{17}\right) = \left(\frac{13}{17}\right) = 1$$

and

$$\left(\frac{5}{17}\right) = \left(\frac{11}{17}\right) = -1.$$

One not very efficient way of calculating the Legendre symbol is to use *Euler's criterion* which states that

$$\left(\frac{a}{p}\right) = a^{(p-1)/2} \pmod{p}.$$

It will be shown later that for any prime p and $a < p$ that $a^{p-1} = 1 \pmod{p}$, and so $a^{(p-1)/2} = \pm 1 \pmod{p}$. If a is a quadratic residue, then $a = x^2 \pmod{p}$ for some x and so

$$a^{(p-1)/2} = x^{p-1} = 1.$$

For a full proof, see Yan [97].

The Legendre symbol satisfies many useful properties, many of which can be proved using Euler's criterion. For example:

1. If $a = b \pmod p$ then
$$\left(\frac{a}{p}\right) = \left(\frac{b}{p}\right).$$
2. If a and b are both relatively prime to p, then
$$\left(\frac{ab}{p}\right) = \left(\frac{a}{p}\right)\left(\frac{b}{p}\right).$$
3. For any $a \neq 0 \pmod p$:
$$\left(\frac{a^2}{p}\right) = 1.$$

Solving quadratic congruences

One very important and practical application of the Chinese remainder theorem is solving quadratic congruences, for example

$$x^2 = 11 \pmod{35}.$$

Note that in general a quadratic congruence

$$x^2 = a \pmod n$$

with n being a product of two distinct primes pq is solvable only if

$$\left(\frac{a}{p}\right) = \left(\frac{a}{q}\right) = 1.$$

This is readily checked:

$$\left(\frac{11}{5}\right) = \left(\frac{1}{5}\right) = 1.$$

and

$$\left(\frac{11}{7}\right) = \left(\frac{4}{7}\right) = \left(\frac{2}{7}\right)^2 = 1.$$

So given

$$\begin{aligned} 11 &= 1 \pmod 5 \\ 11 &= 4 \pmod 7 \end{aligned}$$

the above congruence can be reduced to the two simpler congruences

$$\begin{aligned} x^2 &= 1 \pmod 5, \\ x^2 &= 4 \pmod 7. \end{aligned}$$

By inspection, the solutions to these are 1, −1, and 2, −2. The Chinese remainder theorem can now be used to solve the following systems of congruences:

$$\begin{array}{llll} x = 1 \pmod 5 & \text{and} & x = 2 \pmod 7 \\ x = -1 \pmod 5 & \text{and} & x = 2 \pmod 7 \\ x = 1 \pmod 5 & \text{and} & x = -2 \pmod 7 \\ x = -1 \pmod 5 & \text{and} & x = -2 \pmod 7. \end{array}$$

Letting $m = 5$ and $n = 7$ the Euclidean algorithm produces $s = 3$ and $t = -2$. The solutions to the original equation are thus the four values

$$bsm + atn \pmod{35}$$

for each $a = \pm 1$ and $b = \pm 2$. These values are

$$\begin{array}{rl} (2)(3)(5) + (1)(2)(7) = 9 & \pmod{35} \\ (2)(3)(5) + (-1)(2)(7) = 16 & \pmod{35} \\ (-2)(3)(5) + (1)(2)(7) = 19 & \pmod{35} \\ (-2)(3)(5) + (-1)(2)(7) = 26 & \pmod{35} \end{array}$$

and it can be checked that each of these four values is a solution to the original quadratic congruence.

Modular exponentiation

The problem here is to calculate $a^k \bmod n$, without actually having to determine the value of a^k first. This can be achieved by using property 2 of moduli, that the modulus of a product is equal to the product of the moduli.

First notice that property 2 implies that $a^2 \bmod n = (a \bmod n)^2$. As an example, consider the problem of computing $39^8 \pmod{53}$. First notice that a power of 8 can be obtained by repeated squaring:

$$39^8 = ((39^2)^2)^2.$$

However, the value of 39^8 is not required, but only the remainder of 39^8 after division by 53. This remainder can be obtained by performing three repeated squarings, and *taking the modular value each time.* Here's how:

$$\begin{array}{lll} 39^2 & = 1521 & = 37 \pmod{53}, \\ 37^2 & = 1569 & = 44 \pmod{53}, \\ 44^2 & = 1936 & = 28 \pmod{53} \end{array}$$

and so finally:

$$39^8 \pmod{53} = 28.$$

Notice that this result was computed using only fairly small numbers; the biggest number in the calculation above was 1936. If the computation is started by calculating 39^8 first, this number is 5352009260481, which is far bigger!

This approach can be generalized to provide a very neat algorithm for modular exponentiation, that is, to find $a^k \bmod n$. First express k as a binary integer. Now, start from the left end of this binary number and work towards the right. Begin with a current modular value of 1. Whenever there's a 0, square the current modulus and reduce; whenever there's a 1, square the current modulus, multiply by $a \bmod n$, and reduce.

Here's an example; to calculate $39^{25} \bmod 53$. First express 25 as a binary number; it is 11001. Now starting with a modular value of 1; the left hand digit is 1, so first compute $((1^2) \times 39) \bmod 53$; this result is 39. Now the next digit in the binary number is 1, so compute $(39^2 \times 39) \bmod 53 = 12$. The next digit is 0, so compute $12^2 \bmod 53 = 38$. The next digit is also 0, so compute $38^2 \bmod 53 = 13$. The final digit is the rightmost 1, and so the final result is $(13^2 \times 39) \bmod 53 = 19$.

This can be more neatly displayed in a tabular arrangement; the left-most column is the binary representation of 25, going top to bottom.

$$\begin{array}{llrcrcl}
1 & 1^2 \times 39 & = & 39 & = & 39 & \pmod{53}, \\
1 & 39^2 \times 39 & = & 59319 & = & 12 & \pmod{53}, \\
0 & 12^2 & = & 144 & = & 38 & \pmod{53}, \\
0 & 38^2 & = & 1444 & = & 13 & \pmod{53}, \\
1 & 13^2 \times 39 & = & 6591 & = & 19 & \pmod{53}.
\end{array}$$

Note again that the largest number in the entire computation was 59319. For comparison,

$$39^{25} = 5978815829156164049805202768774674720999.$$

One more example: $83^{37} \bmod 191$. First determine that the binary representation of 37, which is 100101. Then:

$$\begin{array}{llrcrcl}
1 & 1^2 \times 83 & = & 83 & = & 83 & \pmod{191}, \\
0 & 83^2 & = & 6889 & = & 13 & \pmod{191}, \\
0 & 13^2 & = & 169 & = & 169 & \pmod{191}, \\
1 & 169^2 \times 83 & = & 2370563 & = & 62 & \pmod{191}, \\
0 & 62^2 & = & 3844 & = & 24 & \pmod{191}, \\
1 & 24^2 \times 83 & = & 47808 & = & 58 & \pmod{191}.
\end{array}$$

One way of obtaining the result without first computing the binary representation requires a table of three columns, containing values k_i, a_i and m_i, starting with the values of $a_1 = a$, $k_1 = k$, and $m_0 = 1$. The first column is obtained by dividing the previous value by two and keeping the integer part only, so that

$$k_i = k_{i-1}//2$$

stopping when a value of zero is reached. The second column is obtained by modular squaring, so that

$$a_i = a_{i-1}^2 \pmod{n}.$$

For the final column:

$$\text{if } k_i \text{ is } \begin{cases} \text{even, then } m_i = m_{i-1} \\ \text{odd, then } m_i = m_{i-1}a_i. \end{cases}$$

For the previous example, this produces

i	k_i	a_i	m_i	
0			1	
1	37	83	83	k_1 is odd, so $m_1 = m_0a_1 \pmod{n}$
2	18	13	83	k_2 is even, so $m_2 = m_1 \pmod{n}$
3	9	169	84	k_3 is odd, so $m_3 = m_2a_3 \pmod{n}$
4	4	102	84	k_4 is even, so $m_4 = m_3 \pmod{n}$
5	2	90	84	k_5 is even, so $m_5 = m_4 \pmod{n}$
6	1	78	59	k_6 is odd, so $m_7 = m_6a_6 \pmod{n}$
7	0			k_7 is zero, so stop.

The final m_i value is the result. This particular format is particularly suitable for programming.

Modular powers can be calculated efficiently using this method in Sage, with several different commands. The simplest is `power_mod`:

```
sage: power_mod(83,37,191)
  58
```

Or if an object is known to be of type "modulo 191," then an exponential can be done directly:

```
sage: mod(83,191)^37
  58
```

Note that `power_mod` returns an integer, but the second command returns an element of $\mathbb{Z}_{191}$. A command such as

```
sage: (83^37)%191
  58
```

should *never* be used: what this command does is first compute the integer power 83^{37}, and then reduce that value modulo 191. For small values this will work; for large values the integer power may be so big as to cause an error:

```
sage: a,k,n = 2^100, 3^100, 5^20
sage: mod(a,n)^k
```

```
  48679334736626
sage: power_mod(a,k,n)
  48679334736626
sage: (a^k)%n
---------------------------------------------------------
  RuntimeError
```

Modular inverses

The problem here, given a and a modulus n, is to find a value k for which $ak = 1 \pmod{n}$. Such a value k will exist only if $\gcd(a, n) = 1$, that is, if a and n are relatively prime. The value k is called the *modular inverse* of a, and is written $k = a^{-1} \pmod{n}$. For example, using trial and error, $9^{-1} = 11 \pmod{14}$ since $9 \times 11 = 1 \pmod{14}$.

Modular inverses can be computed in a number of different ways. The easiest is by trial and error. For example, to find $9^{-1} \pmod{14}$ requires testing all values $9k \pmod{14}$ for $k = 1, 2, 3, \ldots, 13$ to find a value equal to $1 \pmod{14}$:

$$\begin{aligned}
9 \times 1 &= 9 \pmod{14} \\
9 \times 2 &= 4 \pmod{14} \\
9 \times 3 &= 13 \pmod{14} \\
9 \times 4 &= 8 \pmod{14} \\
9 \times 5 &= 3 \pmod{14} \\
9 \times 6 &= 12 \pmod{14} \\
9 \times 7 &= 7 \pmod{14} \\
9 \times 8 &= 2 \pmod{14} \\
9 \times 9 &= 11 \pmod{14} \\
9 \times 10 &= 6 \pmod{14} \\
9 \times 11 &= 1 \pmod{14}
\end{aligned}$$

Thus $9^{-1} = 11 \pmod{14}$. Clearly this method will not be of much use for very large integers.

A better way is to use the extended Euclidean algorithm. To find $m^{-1} \pmod{n}$, with $\gcd(m, n) = 1$, first compute the values s and t for which

$$sm + tn = 1.$$

This can be written as either

$$\begin{aligned}
sm &= (-t)n + 1 \\
tn &= (-s)m + 1
\end{aligned}$$

and so by definition

$$s = m^{-1} \pmod{n},$$
$$t = n^{-1} \pmod{m}.$$

For example, to find 46^{-1} mod 99:

i	q	r	u	v
−1		99	1	0
0		46	0	1
1	2	7	1	−2
2	6	4	−6	13
3	1	3	7	−15
4	1	1	−13	28
5	3	0		

and so 46^{-1} mod $99 = 28$. Also, 99^{-1} mod $46 = -13 = 33 \pmod{46}$.

In Sage there are, as for modular powers, several methods for calculating inverses, each with a different type of output:

```
sage: inverse_mod(46,99)
  28
sage: parent(_)
  Integer Ring
sage: 1/mod(46,99)
  28
sage: parent(_)
  Ring of integers modulo 99
```

These both have alternative forms:

```
sage: 1/46%99
  28
sage: mod(1/46,99)
  28
```

Another method will be introduced later.

Discrete logarithms

Given any integers a, k and n, modular exponentiation $a^k \pmod{n}$ can be computed very efficiently. An example above showed that $83^{37} = 58 \pmod{191}$ in only a few steps, and using only small numbers. But difficulties arise going in the other direction; that is, given 83 and 58, find the exponent. In general, how can this equation

$$a^x = b \pmod{n}$$

be solved for x, given the values a, b and n? This is called the *discrete logarithm problem* and is in general very difficult to solve. The fact that one direction (exponentiation) is very easy, but the other (discrete logarithms) is very hard, makes this the foundation of some very powerful cryptosystems.

Fermat's and Euler's theorems

Fermat's theorem claims that if p is a prime number, then for any integer a

$$a^p = a \pmod p.$$

An alternative formulation is obtained by dividing each side of this equation by a (assuming that a is not a multiple of p); then the theorem reads

$$a^{p-1} = 1 \pmod p.$$

To show this, consider all the values $a, 2a, 3a, \ldots (p-1)a$. Since a is not a multiple of p, all these values must have different residues when reduced mod p. For if any two were equal, then

$$ma = na \pmod p$$

for some values m and n between 1 and $p-1$. That is,

$$(m-n)a = 0 \pmod p.$$

But this means that either a or $m-n$ is divisible by p. Since a is not divisible by p, the only way for $m-n$ to be divisible by p is to have $m=n$. But this is impossible since all the coefficients of a are different. This means that the residues of $a, 2a, 3a, \ldots (p-1)a \pmod p$ are just the values $1, 2, 3, \ldots, p-1$, in some order. Thus:

$$(a)(2a)(3a)...((p-1)a)) = 1.2.3...(p-1) \pmod p.$$

Since $1.2.3...(p-1)$ is not divisible by p, each side of this equation can be divided by this value to obtain

$$a^{p-1} = 1 \pmod p.$$

which is the required result.

Fermat's theorem can be used to simplify modular powers. For example, consider the power

$$10^{27} \pmod{13}.$$

Since by Fermat's theorem

$$10^{12} = 1 \pmod{13},$$

the original power can be expressed using multiples of 12 as

$$\begin{aligned}10^{27} &= 10^{2(12)+3}\\ &= (10^{12})^2\,10^3.\end{aligned}$$

Hence

$$\begin{aligned}10^{27} \pmod{13} &= (10^{12})^2\,10^3 \pmod{13}\\ &= 10^3 \pmod{13}\end{aligned}$$

and this last power can be evaluated with much less trouble than the original. Note that the final power 3 is just the original power 27 modulo 12. This means that in general, if p is prime and $a < p$ then

$$a^n = a^{n \bmod (p-1)} \pmod{p}.$$

There is an important generalization of this theorem due to Euler, which requires first a useful number theoretical function.

Historical note: Leonhard Euler (1707–1783) was a Swiss mathematician, and one of the greatest and most prolific mathematicians of all time. He made deep and important discoveries in all branches of mathematics, pure and applied, and vastly enriched the subject. His collected works, which has been published since 1907, consists so far of eighty volumes, and is not yet complete!

Definition: The *Euler totient function of* n, denoted $\phi(n)$, is defined to be the number of integers less than n which are relatively prime to n.

For example, if $n = 15$, the integers less than 15 which are relatively prime to 15 are all those not divisible by either 3 or 5 (which are the prime factors of 15). They are: 1, 2, 4, 7, 8, 11, 13, 14. There are 8 numbers in this list, so $\phi(15) = 8$. Here are some properties of this function:

1. If p is prime, then $\phi(p) = p - 1$.
2. If m and n are relatively prime, then $\phi(mn) = \phi(m)\phi(n)$.

   In particular, if p and q are primes, then $\phi(pq) = \phi(p)\phi(q) = (p-1)(q-1)$.
3. If p is prime, then $\phi(p^e) = p^{e-1}(p-1)$.
4. For any integer $n = p_1^{e_1} p_2^{e_2} \cdots p_k^{e_k}$,

$$\phi(n) = p_1^{e_1-1}(p_1 - 1)p_2^{e_2-1}(p_2 - 1) \cdots p_k^{e_k-1}(p_k - 1).$$

Some examples:

- $\phi(37)$. Since 37 is prime, $\phi(37) = 36$.
- $\phi(77)$. Since $77 = 7 \times 11$, then $\phi(77) = \phi(7)\phi(11) = 6 \times 10 = 60$.
- $\phi(243)$. Since $243 = 3^5$ then $\phi(243) = 3^4(3-1) = 81 \times 2 = 162$.
- $\phi(49,000)$. Since $49,000 = 2^3 \times 5^3 \times 7^2$ then

$$\phi(49,000) = 2^2(2-1) \times 5^2(5-1) \times 7^1(7-1) = 16,800.$$

Euler's generalization of Fermat's theorem states that for any integers a and n with $\gcd(a, n) = 1$, then

$$a^{\phi(n)} = 1 \pmod{n}.$$

For example, take $n = 15$. Then $\phi(15) = 8$ and since $\gcd(4, 15) = 1$, the theorem claims that

$$4^8 = 1 \pmod{15}$$

and this is in fact true, as can easily be checked. The proof follows the same line as for that of Fermat's theorem. Let $r_1, r_2, \ldots, r_{\phi(n)}$ be the numbers less then n and relatively prime to n. Then the residues $r_1a, r_2a, \ldots, r_{\phi(n)}a \pmod{n}$ are all distinct. For, if any two were equal, then

$$r_ia = r_ja \pmod{n}$$

for some numbers r_i and r_j less then n and relatively prime to it. Thus,

$$r_ia - r_ja = (r_i - r_j)a = 0 \pmod{n}.$$

This means that either a or $r_i - r_j$ is divisible by n. Since $\gcd(a, n) = 1$, the first is impossible, and the only way for the second to be possible is to have $r_i = r_j$, which is also impossible since all the r_l values are different. This means that the values $r_1a, r_2a, \ldots, r_{\phi(n)}a \pmod{n}$ are only the values $r_1, r_2, \ldots, r_{\phi(n)}$ in some order. Thus:

$$(r_1a)(r_2a)\cdots(r_{\phi(n)}a) = r_1r_2\cdots r_{\phi(n)} \pmod{n}.$$

But since $r_1r_2\cdots r_{\phi(n)}$ is not divisible by n, both sides of this equation can be divided by this value to obtain

$$a^{\phi(n)} = 1 \pmod{n}$$

which is the result.

As with Fermat's theorem, Euler's theorem can also be used to simplify powers. The power can be reduced modulo $\phi(n)$ to obtain

$$a^m = a^{m \bmod \phi(n)} \pmod{n}.$$

For example, to compute

$$27^{100} \pmod{64},$$

first calculate $\phi(64) = 32$, and then

$$100 \bmod 32 = 4.$$

Thus

$$27^{100} = 27^4 \pmod{64}.$$

A further word on inverses

Euler's theorem can be used for finding modular inverses. Dividing the result by a produces

$$a^{-1} = a^{\phi(n)-1} \pmod{n}.$$

For example, to find $9^{-1} \pmod{14}$ start by using the properties of $\phi(n)$, to obtain $\phi(14) = 6$. Then

$$\begin{aligned} 9^{-1} &= 9^5 \pmod{14} \\ &= 11 \pmod{14}. \end{aligned}$$

2.4 Primality testing

Trial division discussed earlier is one simple way of testing a number for primality. However, the more recently acquired powerful number-theoretical tools such as Fermat's theorem, Euler's generalization of it, and efficient modular exponentiation provide the means to develop some more sophisticated methods. The topic of this section is to investigate ways of using these tools for more efficient primality testing algorithms.

The Miller–Rabin test

Recall from Fermat's theorem that if p is prime, then for any integer a, the power $a^{p-1} = 1 \pmod{p}$. However, the converse is not true; if

$$a^{n-1} = 1 \pmod{n}$$

then it does not necessarily follow that n is prime. For example

$$2^{340} = 1 \pmod{341}$$

but $341 = 11 \cdot 31$ is not prime. In general, a composite number n for which

$$2^{n-1} = 1 \pmod{n}$$

is called a *pseudoprime to base 2*, and there are an infinite number of them. There are also pseudoprimes to other bases; in general if n is composite and

$$a^{n-1} = 1 \pmod{n}$$

then n is called a *pseudoprime to base a*. To make matters worse there are numbers that are pseudoprimes to *every* base; the smallest is 561:

$$a^{560} = 1 \pmod{561}$$

for all a. These numbers are called *Carmichael numbers*, after the mathematician who first discussed them. Only quite recently has it been shown that the number of Carmichael numbers is infinite. It can be shown that if $n = p_1 p_2 \dots p_k$ is a product of distinct primes, and if $(p_i - 1)|(n-1)$ for all i, then n is a Carmichael number.

In order to use Fermat's theorem for primality testing, some method is needed for avoiding the test being "tricked" by Carmichael numbers or other pseudoprimes. First note that the contrapositive of Fermat's theorem is that if

$$a^{n-1} \neq 1 \pmod{n}$$

then n is composite. The standard test for primality testing is the Miller–Rabin test, which is based on the result that if p is prime, then $x^2 = 1 \pmod{p}$ if and only if $x = \pm 1$. This follows from writing the modular equation as

$$(x-1)(x+1) = 0 \pmod{p}$$

from which it follows that p must divide either $x - 1$ or $x + 1$. This means that $x - 1 = 0 \pmod{p}$ or $x + 1 = 0 \pmod{p}$, which gives the result.

From this result it follows immediately that for a given n, if there exists a number $r \neq \pm 1 \pmod{n}$ for which $r^2 = 1 \pmod{n}$, then n is composite. Such an r is called a *non-trivial square root of* 1 (the trivial square roots are 1 and −1.)

Suppose that n is the odd number to be tested for primality, and is written as

$$n = 1 + 2^s d$$

where d is odd. Choose a value b, and compute the "b sequence":

$$[b_0, b_1, b_2, \dots, b_s] = [b^d, b^{2d}, b^{2^2 d}, \dots, b^{2^s d}] \pmod{n}.$$

Note that $b_k = b_{k-1}^2$ for all terms in the sequence. If n is prime, then by

Fermat's theorem the last value must be 1, since $2^s d = n - 1$. Taking square roots backwards through the sequence will produce either all ones,

$$[1, 1, 1, \ldots, 1]$$

or a sequence with a -1 in it,

$$[*, *, \ldots, *, -1, 1, \ldots, 1],$$

where "$*$" indicates a value different from 1 or -1. These sequences can be explored by writing a small program in Sage, using the `trailing_zero_bits` method to find the value of s, and using a helper function called `bmod` which produces "balanced" moduli: between $\lfloor (n-1)/2 \rfloor$ and $\lfloor n/2 \rfloor$ instead of between 0 and $n - 1$:

```
sage: def bmod(x,n):
....:     return (x+(n-1)//2)%n-(n-1)//2
sage: def bseq(b,n):
....:     s = (n-1).trailing_zero_bits()
....:     d = (n-1)>>s
....:     return [bmod(power_mod(b,2^j*d,n),n)\
....:       for j in range(s+1)]
```

This can now be tested, with the prime $n = 3889$:

```
sage: bseq(2,3889)
  [592, 454, -1, 1, 1]
sage: bseq(3,3889)
  [1, 1, 1, 1, 1]
```

There are other possibilities of the b-sequence; ending in -1,

$$[*, *, \ldots, *, -1],$$

ending in 1's but starting with a non-trivial square root of 1,

$$[*, *, \ldots, *, 1, 1, \ldots, 1]$$

or containing all values neither 1 or -1,

$$[*, *, \ldots, *, *].$$

Each of these forms proves n to be composite.

The use of this b-sequence provides a much stronger test for primality than using Fermat's theorem alone. However, it can still be tricked; there are composite numbers n whose b-sequences are of prime form. For example:

```
sage: bseq(2,4033)
  [3521, -1, 1, 1, 1, 1, 1]
```

but $4033 = 37 \cdot 109$. Such a number is called a *strong pseudoprime to base* b. Strong pseudoprimes exist for every base, but are much rarer than pseudoprimes. Also, for any composite number n, there exists a base b for which n is not a strong pseudoprime. This means that there are no numbers that are strong pseudoprimes to all bases; in other words there is no strong pseudoprime equivalent to the Carmichael numbers.

The Miller–Rabin primality test for an odd number n works like this:

1. Choose a random (generally small) base b.
2. Determine the b-sequence. If the sequence is of composite form, output COMPOSITE and stop.
3. If the sequence is of prime form, consider n to be "possibly prime."
4. Choose another base b and go back to step 2.

The idea is that the more bases that produce the result "possibly prime," then the more likely it is that n is prime. Note that this is not a deterministic test—it doesn't prove without doubt that n is prime, but it does provide a very high probability of its primality. The more bases tried (and passed), the more likely it is that the number is prime. This test is both fast and practical—for most purposes passing the test for about 20 bases is sufficient.

Consider the composite number 4033 above, which passed the test for the base $b = 2$. For $b = 3$:

```
sage: bseq(3,4033)
  [-482, -1590, -591, -1590, -591, -1590, -591]
```

and so it is shown to be composite. Some numbers require several tests: 341550071728321 passes the test for the first eight primes 2, 3, 5, 7, 11, 13, 17, 19, but fails for 23 (and hence is shown to be composite). It is conjectured that the first composite number that passes all tests for the first 20 primes is greater than 10^{36}.

A further and newer method, which has the advantage of being completely deterministic as well as running in polynomial time, will be described in appendix B.

2.5 Glossary

Congruence. Equality in terms of remainder after division by a fixed integer, the modulus.

Discrete logarithm. The value x in the equation $a^x = b \pmod{n}$.

Discrete logarithm problem. The problem (in general with no easy solution) of solving a discrete logarithm equation.

Euclidean algorithm. An efficient algorithm for finding the greatest common divisor of two numbers.

Factorization. The problem (in general with no easy solution) of decomposing a number into its prime factors.

Non-deterministic primality test. A primality test, such as the Miller–Rabin test, that does not prove a number to be prime, but produces a very high probability of its primality.

Prime number. An integer with no factors other than itself and 1.

Primality testing. Determining whether or not a given number is prime.

Relatively prime. Two numbers are relatively prime if their greatest common divisor is 1.

Trial division. A means of testing for primality by checking divisibility by all numbers up to the square root.

Exercises

Review Questions

1. Define the following terms: prime, composite, divisible, congruence, remainder, residue.

2. What is Fermat's theorem?

3. How does Euler's theorem generalize Fermat's theorem?

4. For what problem does the Chinese Remainder Theorem provide a solution?

5. What is a quadratic residue?

6. What is a Legendre symbol?

7. Define what it means for x and y to be inverses modulo n.

8. What is a pseudoprime?

9. How does the existence of pseudoprimes prevent an easy primality test by using the converse of Fermat's theorem?

10. What is a Carmichael number?

11. What is a strong pseudoprime?

Beginning Exercises

12. Prove the following basic properties about divisibility and prime numbers:

    (a) If $k|n$ and $n|m$ then $k|m$.

    (b) If $k|n$ and $k|m$ then $k|(am + bn)$ for any integers a and b.

    (c) If p and q are distinct primes, and if p and q are both factors of n, then $(pq)|n$.

    (d) If $p|(mn)$ then either $p|n$ or $p|m$, or both.

All the following exercises are to be done using pencil and paper, or a hand-held calculator only.

13. Express each of the following numbers as a product of prime powers:

    (a) 100, (b) 10000, (c) 1728, (d) 3025, (e) 10829.

14. By trial division, test each of the following numbers for primality:

    (a) 289, (b) 541, (c) 2813, (d) 1583, (e) 14803, (f) 7919.

15. Apply the "Sieve of Eratosthenes" as follows. First write down all the numbers from 2 to 100. Circle the 2 and cross out all its multiples. Circle the next uncrossed number (this will be 3), and cross out all its multiples. Continue in this way circling the next uncrossed number and crossing out its multiples until all numbers are either circled or crossed out.

    (a) What do you notice about the circled numbers?

    (b) How many are there?

16. Use Fermat's method to factorize

    (a) 74513, (b) 379447, (c) 1700407, (d) 2951422913.

17. Evaluate the following congruences:

    (a) 254 (mod 29), (b) 2967 (mod 97), (c) 24596 (mod 141), (d) −654 (mod 23), (e) −221 (mod 47).

18. How can you can you use your calculator to evaluate

    $$987654321987654321 \pmod{12345678}$$

    when the first number is too big to be entered?

19. Evaluate the following congruences, by first evaluating the individual terms:

    (a) $83+47 \pmod{17}$, (b) $83-47 \pmod{17}$, (c) $83\times 47 \pmod{17}$, (d) $217 \times 219 \pmod{43}$.

20. Convert the following numbers into binary:

    (a) 37, (b) 526, (c) 12345, (d) 71625341.

21. Evaluate the following modular powers:

    (a) $47^{23} \pmod{53}$, (b) $13^{47} \pmod{61}$, (c) $2^{80} \pmod{83}$, (d) $217^{219} \pmod{344}$.

22. Determine the last 3 digits in 23^{59}.

23. Use Fermat's theorem to reduce the power in the following:

    (a) $47^{1000} \pmod{53}$, (b) $17^{12781} \pmod{61}$, (c) $23^{5811} \pmod{71}$, (d) $51^{9219} \pmod{97}$.

24. Use Fermat's theorem *without a calculator* to reduce the powers and hence calculate the following modular powers:

    (a) $3^{10003} \pmod{101}$, (b) $7^{902} \pmod{91}$, (c) $2^{1603} \pmod{401}$.

25. Using Euclid's algorithm, find the greatest common divisors of the following pairs:

    (a) 83, 47 (b) 315, 165 (c) 4171, 3193 (d) 4773, 4929.

26. Prove that Euclid's method works by proving that $\gcd(a, b) = \gcd(b, a \bmod b)$. This can be done by showing that

    (a) If $b|a$ then $\gcd(a, b) = b$,

    (b) If $a = qb + r$ with $r < b$ then $\gcd(a, b) = \gcd(b, r)$.

    (c) Use the extended Euclidean algorithm to find the values s and t for which $sm + tn = g$ for each of the pairs m, n in question 25, where g is the greatest common divisor.

27. Using Euclid's algorithm, find the following modular inverses:

    (a) $29^{-1} \pmod{47}$, (b) $89^{-1} \pmod{315}$, (c) $105^{-1} \pmod{143}$, (d) $64^{-1} \pmod{81}$.

28. Show that the extended Euclidean algorithm works by showing

    (a) $r_i = r_{i-2} \pmod{r_{i-1}}$,

    (b) For each i, $mu_i + nv_i = r_i$.

29. Show that the Kid-RSA system described in section 1.6 is not at all secure. What is $ed \bmod n$? Given (n, e), how do you calculate d? Check it with the examples given in that section.

30. Calculate $\phi(n)$ for each of the following values of n:

    (a) 64, (b) 27, (c) 60, (d) 1000, (e) 343, (f) 385.

31. Suppose that n is a product of two primes. Show that given n and $\phi = \phi(n)$ the factors can be obtained. (Hint: show that if p and q are the factors, then $p+q$ can be expressed in terms of n and ϕ. Then solve a quadratic equation to obtain p and q.)

32. Use your result from the previous question to factor n given

    (a) $n = 2173$, $\phi = 2080$

    (b) $n = 8051$, $\phi = 7872$

    (c) $n = 68422027$, $\phi = 68405400$

33. Using Euler's formula, evaluate the following modular inverses:

    (a) $8^{-1} \pmod{27}$, (b) $25^{-1} \pmod{27}$, (c) $27^{-1} \pmod{64}$,
    (d) $35^{-1} \pmod{64}$, (e) $35^{-1} \pmod{81}$, (f) $64^{-1} \pmod{81}$.

34. By hand, use the Chinese remainder theorem to solve the following simultaneous congruences:

    (a) $x = 2 \pmod 5$, $x = 3 \pmod 7$,

    (b) $x = 3 \pmod{11}$, $x = 5 \pmod{13}$,

    (c) $x = 2 \pmod 7$, $x = 10 \pmod{17}$.

    Check your answers with Sage's `crt` function.

35. The Chinese remainder theorem can be extended to an arbitrary number of simultaneous congruences

    $$x = a_i \pmod{n_i}, \quad i = 1, 2, \ldots, k$$

    where the n_i values are pairwise relatively prime. Let $N = n_1 n_2 \cdots n_k$ and $N_i = N/n_i$. Since n_i and N/n_i are relatively prime, Euclid's algorithm can be used to find r_i and s_i so that $rn_i + sN/n_i = 1$. Then $x = \sum a_i sN/n_i \pmod N$ is a solution.

    (a) Prove that this is correct.

    (b) Solve the congruences

    $$\begin{aligned} x &= 2 \pmod 5 \\ x &= 3 \pmod 7 \\ x &= 5 \pmod{11}. \end{aligned}$$

36. Show that a solution to the general congruences

$$x = u \pmod{p}, \quad x = v \pmod{q}$$

can be given by the formula

$$x = u + (p^{-1}(v - u) \bmod q)p.$$

This is sometimes known as *Garner's formula.*

37. Use Garner's formula to verify the results in question 34.

38. List all the values of $n^2 \pmod{23}$ for all $1 \leq n \leq 22$. Use your results to list all quadratic residues and quadratic non-residues modulo 23.

39. Use Euler's criterion to calculate the following Legendre symbols:

(a) $\left(\frac{2}{23}\right)$, (b) $\left(\frac{5}{23}\right)$, (c) $\left(\frac{10}{31}\right)$, (d) $\left(\frac{11}{31}\right)$.

40. Show that each of the following numbers is a Carmichael number:

(a) 1729, (b) 825265, (c) 5394826801.

41. Show that 25326001 passes the Miller–Rabin test for the primes 2, 3 and 5, but fails it for 7.

42. Solve each of the following congruences $x^2 = s \bmod n$ by the following steps:

    (i) Factorize $n = pq$.

    (ii) Solve each of the congruences $x^2 = (s \bmod p) \bmod p$ and $y^2 = (s \bmod q) \bmod q$ by inspection.

    (iii) Put the results together using the Chinese remainder theorem.

(a) $x^2 = 92 \pmod{143}$, (b) $x^2 = 123 \pmod{323}$,

(c) $x^2 = 534 \pmod{1517}$.

43. A cryptosystem developed by Pohlig and Hellman works like this: a large prime p is chosen, and the secret shared key consists of values e, d so that $ed = 1 \bmod (p - 1)$. Given a plaintext m, the ciphertext c is obtained by

$$c = m^e \pmod{p}$$

and decryption is performed by

$$m = c^d \pmod{p}.$$

    (i) Prove that decryption recovers the plaintext.

(ii) Check it with the values $p = 71$, $e = 11$, $d = 51$, $m = 40$.

(iii) What problem provides the security for this system? Suppose an attacker, using a known plaintext attack, has p, m and c. How can the attacker find e (and d)?

44. *Lucas's primality test* (named for Edouard Lucas, 1842–1891, a French mathematician) works as follows: for a number n to be tested, first find all prime factors of $n - 1$. If a number $a < n$ can be found for which

$$a^{n-1} = 1 \pmod{n}$$

and

$$a^{(n-1)/q} \neq 1 \pmod{n}$$

for all prime factors q of $n - 1$, then n is prime.

(a) Show that 79 is prime by using $a = 3$.

(b) Show that 1009 is prime by using $a = 11$.

(c) What happens if you use any of the primes 2, 3, 5, 7 for a in the test for 1009?

(d) If $a^{n-1} \neq 1 \pmod{n}$, what does Fermat's theorem tell you about whether n is prime or composite?

Sage Exercises

45. Create random 10 digit primes:

```
sage: p = next_prime(randint(10^9,10^10-1))
sage: q = next_prime(randint(10^9,10^10-1))
```

Now multiply them and factor the product:

```
sage: n = p*q
sage: %time factor(n)
```

How long did it take?

46. Try the same thing with 20 digit primes and 30 digit primes. How does the time change with the size of the numbers being factored?

47. Determine the gcd of $3^{100} + 4$ and $7^{60} + 14$.

48. Solve the following simultaneous congruences (all moduli are primes):

$$\begin{aligned} x &= 1000 \pmod{2^{13} - 1} \\ x &= 2000 \pmod{3^8 + 2} \\ x &= 5000 \pmod{5^6 + 4}. \end{aligned}$$

49. Using Sage's `crt` function, find an integer which has residues 1, 2, 3, 4, 5 modulo 3, 5, 7, 11 and 13 respectively.

50. Let $p = 3^{10} + 2$ and $q = 5^7 - 4$, both of which are prime. Solve for x the equation

$$x^2 = 3550677580 \pmod{pq}$$

by using equations modulo p and q separately and the Chinese remainder theorem.

51. Use the Miller–Rabin algorithm and the Sage function `bseq` to test the primality of

(a) $5^7 + 12$, (b) $3^{20} + 8$, (c) $7^{11} + 10$.

Chapter 3

Classical cryptosystems

This chapter will consider

- The use of simple pencil-and-paper systems to give an insight into the working, construction and uses of cryptosystems.
- Some simple means of cryptanalysis.
- A tiny bit of modular linear algebra.

3.1 Introduction

This chapter will be concerned with some simple cryptosystems. None of the systems described in this chapter are ever used in practice, except in puzzles and games. However, they are worth studying because they provide very good insight into the structure and behavior of cryptosystems. Partly this is because these systems work with letters of the alphabet, rather than with large numbers of strings of bits, so it is easier to see how these systems work.

For the cryptosystems to be investigated here, the letters of the alphabet will be given numeric values, so that arithmetic operations can be performed on these values. For the sake of simplicity, only upper case letters will be used. The values are

A	B	C	D	E	F	G	H	I	J	K	L	M	N	O	P	Q	R	S	T	U	V	W	X	Y	Z
0	1	2	3	4	5	6	7	8	9	10	11	12	13	14	15	16	17	18	19	20	21	22	23	24	25

Since all arithmetic will be performed modulo 26, it is reasonable to use values 0 to 25, rather than 1 to 26.

Sage contains commands to perform many of these classical ciphers, but for the sake of investigating them, they will be developed "from scratch."

3.2 The Caesar cipher

In this cipher, each letter of the alphabet is replaced by another letter, according to this scheme:

```
A B C D E F G H I J K L M N O P Q R S T U V W X Y Z
D E F G H I J K L M N O P Q R S T U V W X Y Z A B C
```

Historical note: According to the Roman writer and biographer Suetonius, Julius Caesar (100BC–44BC) was the first to use this cipher to protect messages of military significance. Although there is no known evidence to suggest how effective the cipher was at protecting his messages, it is believed to have been very secure for the time.

Note that each letter is replaced by the letter three along in the alphabet, and the alphabet "wraps around," in the sense that the third letter after `Y`, say, is `B`. (`Y`, `Z`, `A`, `B`). This can be described in terms of the numeric values of the letters as

$$c = p + 3 \pmod{26}$$

where p is a plaintext letter, and c is the corresponding ciphertext letter. Each letter is identified with its numeric value, and treating the letter and its numeric value as being the same. Using this cipher, the message

```
WITHDRAW ONE HUNDRED DOLLARS
```

becomes

```
ZLWKGUDZ RQH KXQGUHG GROODUV.
```

Decryption is straightforward; it can be described as

$$p = c - 3 \pmod{26}.$$

To implement this in Sage, first the letters must be translated into numbers; this is easily done with the `ord` command, which returns the ASCII value of a character. The ASCII values of "A" and "Z" are 65 and 90 respectively. So to obtain values 0 to 25, 65 has to be subtracted from each ASCII value. And this operation needs to be applied to every character in a string:

```
sage: pl = "WITHDRAWONEHUNDREDDOLLARS"
sage: pln = [ord(x)-65 for x in pl]
sage: print pln
[22, 8, 19, 7, 3, 17, 0, 22, 14, 13, 4, 7, 20, 13, 3, 17,
 4, 3, 3, 14, 11, 11, 0, 17, 18]
```

It will be convenient to have a little command that takes a string and turns it into a numeric list:

```
def str2lst(s):
   return [ord(x)-65 for x in s]
```

and similarly a command that takes a list and turns it back into a string. For this there is the `chr` command, which produces the character corresponding to a given ASCII value, and the `join` command which joins all the elements of a list of characters into a single string:

```
def lst2str(lst):
   return join([chr(int(x)+65) for x in lst],'')
```

The `reduce` command takes a single operation (in this case concatenation of characters) and applies it across the entire list of characters, producing a single string.

This can all be put together as a little program, where the actual encryption is done with adding 3 modulo 26, by `(x+3)%26`:

```
def caesar(pl):
   pln = str2lst(pl)
   pln2 = [(x+3)%26 for x in pln]
   return lst2str(pln2)
```

3.3 Translation ciphers

These are a simple generalization of the Caesar cipher, in which encryption involves shifting letters along the alphabet by a prearranged value n. Thus the encryption and decryption algorithms are:

$$\begin{aligned} c &= p+n \pmod{26} \\ p &= c-n \pmod{26}. \end{aligned}$$

The value n can be considered as being the *key* for this cryptosystem. The Caesar cipher is thus a translation cipher with $n = 3$. The translation cipher with $n = 13$ is found on some computer systems as **ROT13**. It is used in some Usenet postings, either to hide material that may be considered offensive to some, or to hide the solution of a puzzle or problem.

It is a trivial matter to turn the Caesar cipher program above into translation cipher; all that's necessary is to replace the 3 with a value `n` to be entered by the user:

```
def trans(pl,n):
   pln = str2lst(pl)
   pln2 = [(x+n)%26 for x in pln]
   return lst2str(pln2)
```

This cipher can be quickly broken with a brute-force attack. Since there is only a small number of keys (26), all that's needed is to try them all until the message is obtained.

Example

Suppose a ciphertext:

```
SECUJEJXUFQHJOJEDYWXJ
```

is known to have been produced with a translation cipher. To break this cipher, simply make a table listing all possible translations of this ciphertext, until one is obtained that makes sense:

```
sage: pl = 'SECUJEJXUFQHJOJEDYWXJ'
sage: for i in range(25): print i,trans(pl,i)

0 SECUJEJXUFQHJOJEDYWXJ
1 TFDVKFKYVGRIKPKFEZXYK
2 UGEWLGLZWHSJLQLGFAYZL
3 VHFXMHMAXITKMRMHGBZAM
4 WIGYNINBYJULNSNIHCABN
5 XJHZOJOCZKVMOTOJIDBCO
6 YKIAPKPDALWNPUPKJECDP
7 ZLJBQLQEBMXOQVQLKFDEQ
8 AMKCRMRFCNYPRWRMLGEFR
9 BNLDSNSGDOZQSXSNMHFGS
10 COMETOTHEPARTYTONIGHT
```

Since a meaningful text has been produced, the plaintext is recovered.

3.4 Transposition ciphers

These are a further generalization. In all the above ciphers, the alphabetical order has been maintained (with wrap around). But this ordering can be changed. For example, messages could be encrypted according to this scheme:

```
A B C D E F G H I J K L M N O P Q R S T U V W X Y Z
Z A Y B X C W D V E U F T G S H R I Q J P K O L N M
```

so that

```
WITHDRAW ONE HUNDRED DOLLARS
```

becomes

```
OVJDBIZO SGX DPGBIXB BSFFZIQ
```

For this cipher, the permutation is its key. Since there are

$$26! = 403291461126605635584000000$$

different permutations, a brute-force attack is no longer possible.

This cipher can be easily implemented in Sage. As previously, `pl` will be the plaintext, and now `perm` will be a permutation of the numbers 1 through 26:

```
sage: p = 'ZAYBXCWDVEUFTGSHRIQJPKOLNM'
sage: perm = [ord(x)-64 for x in p]
```

Applying this permutation to the plaintext is easy:

```
sage: pln = str2lst(pl)
sage: plr = [perm[pln[i]]-1 for i in range(len(pl))]
sage: lst2str(plr)
  'OVJDBIZOSGXDPGBIXBBSFFZIQ'
```

The extra "`-1`" in the second command is required because in Sage permutations start at 1, but list indices start at 0.

Sage has the functionality to produce inverse permutation, thus enabling the decryption of the ciphertext:

```
sage: invperm = Permutation(perm).inverse()
sage: pls = [invperm[plr[i]]-1 for i in range(len(pl))]
sage: lst2str(pls)
  'WITHDRAWONEHUNDREDDOLLARS'
```

In each of these ciphers, Caesar, translation, and transposition, the plaintext alphabet is replaced, letter by letter, by a ciphertext alphabet. Since only one ciphertext alphabet is used, these ciphers are called *mono-alphabetic ciphers*.

Cryptanalysis

With all these ciphers, cryptanalysis is very easy, and can be done speedily using a ciphertext-only attack. Since in English text letters have known frequencies[1]: the letter E is the most common, occurring somewhere between 11% and 12.7% of the time, followed by either E or T, with somewhere between 8% and 9%, all that's needed is to make a list of all the frequencies of the letters in the ciphertext, and compare that list with the standard frequencies. Decryption is then immediate.

Historical note: Frequency analysis is believed to have been first considered in the 9th century by the Arabic writer Al-Kindi, in one of the first books on cryptography, which predates similar works in the West by over 800 years.

The following tables of letter frequencies, for a sample of 1000 letters given both alphabetically and by frequency, is taken from Sinkov [82].

A	71	E	130
B	9	T	93
C	30	N	78
D	44	R	77
E	130	I	74
F	28	O	74
G	16	A	73
H	35	S	63
I	74	D	44
J	2	H	35
K	3	L	35
L	35	C	30
M	25	F	28
N	78	P	27
O	74	U	27
P	27	M	25
Q	3	Y	19
R	77	G	16
S	63	W	16
T	93	V	13
U	27	B	9
V	13	X	5
W	16	K	3
X	5	Q	3
Y	19	J	2
Z	1	Z	1

[1]The exact values of the frequencies are not standardized; depending on which source you consult, you'll get slightly different values.

With a known plaintext attack, the problem is even easier, since there is a simple correspondence between ciphertext and plaintext.

Since these ciphers are so trivial to break, they are now never used, except for puzzles and games.

3.5 The Vigenère cipher

Historical note: This cipher is attributed to Blaise Vigenère (1532–1596) who was a diplomat to the Vatican, although it was in fact developed not by him but by Giovan Battista Bellaso in 1553. By a curious stroke of historical fate, this cipher is now called the "Vigenère Cipher," whereas in fact the cipher that Vigenère *did* create, known as an "autokey cipher" is much stronger. It was considered a great improvement over mono-alphabetic ciphers, and was thought at the time to be unbreakable. It was finally broken in 1832.

In this cipher, encryption depends not on a single value giving the shift, but on a keyword.

The encryption and decryption algorithms can be described as

$$c_i = p_i + k_i \pmod{26}$$
$$p_i = c_i - k_i \pmod{26}$$

where p_i, k_i and c_i are the corresponding plaintext, keyword and ciphertext letters. For example, using the same plaintext as before, and the keyword `CODE`, the encryption is:

```
 plaintext: W I T H D R A W O N E H U N D R E D D O L L A R S
   keyword: C O D E C O D E C O D E C O D E C O D E C O D E C
ciphertext: Y W W L F F D A Q B H L W B G V G R G S N Z D V U
```

To see what's going on, look at the first two letters:

$$\begin{aligned} \mathtt{W} + \mathtt{C} &= 22 + 2 \\ &= 24 \\ &= \mathtt{Y} \\ \mathtt{I} + \mathtt{O} &= 8 + 14 \\ &= 22 \\ &= \mathtt{W} \end{aligned}$$

and additions are performed modulo 26, so for the sixth letter:

$$\begin{aligned} \mathtt{R}+\mathtt{O} &= 17+14 \\ &= 31 \\ &= 5 \pmod{26} \\ &= \mathtt{F} \end{aligned}$$

The Vigenère cipher may be considered as a number of translation ciphers operating simultaneously. In the example given above, the first, fifth, ninth, thirteenth (and so on) letters are all encrypted by being shifted forward by 3 (since `C` has value 3). Similarly, the second, sixth, tenth and so on letters are all encrypted by being shifted forward by 14 (since `O` has value 14). The Vigenère cipher is thus an example of a *polyalphabetic cipher*.

To implement this cipher in Sage, the correct letter of the key needs to be added to the current letter of the plaintext:

Index of plaintext i:	0	1	2	3	4	5	6	7
Plaintext:	W	I	T	H	D	R	A	W
Key	C	O	D	E	C	O	D	E
Index of key j:	0	1	2	3	0	1	2	3

The indices of the key repeat 0, 1, 2, 3, which are just the residues, modulo 4, of the key indices. This means that in the Vigenère cipher, we add the i-th character of the plaintext, and the j-th character of the key, where

$$j = i \pmod{n}$$

with n being the length of the key. This can all be put together as:

```
def vige(s,key):
   pn = len(s)
   kn = len(key)
   pl = str2lst(s)
   kl = str2lst(key)
   cl = [mod(pl[i]+kl[mod(i,kn)],26) for i in range(pn)]
   return lst2str(cl)
```

A similar program called `vigd` can be written for decryption, the main difference being in the next-to-last line, where the key value will be *subtracted* from the ciphertext value:

```
   pl = [mod(cl[i]-kl[mod(i,kn)],26) for i in range(pn)]
```

Note that in this cipher repeated occurrences of letters in the plaintext are not necessarily encrypted to the same letters in the ciphertext. Cryptanalysis by counting frequencies is thus not possible. Also, since the number of keywords is effectively infinite, a brute-force attack is not feasible. For these reasons, the cipher was considered at the time, and for many years after, to be "undecipherable."

There are several variants of the Vigenère cipher that have been developed (and all broken), but one elegant variant is that of Sir Francis Beaufort[2], and thus called the *Beaufort cipher*. Its encryption and decryption equations are

$$c_i = k_i - p_i \pmod{26}$$
$$p_i = k_i - c_i \pmod{26}.$$

Note that encryption and decryption use exactly the same equations: in each case subtraction from the key. This is an example of a *reciprocal cipher*.

Cryptanalysis

Cryptanalysis of the Vigenère cipher by a ciphertext-only attack consists of two steps:

1. finding the length of the keyword,
2. breaking up the ciphertext according to the length of the keyword, and performing a frequency analysis on each group to determine the shift.

To find the length of the keyword, first examine the ciphertext for repeating groups of letters. If there are groups of length m, whose distance apart is some multiple of m, then it is highly likely that the length of the keyword is m. In a long ciphertext, such repeated groups to make this method work are almost certain to be found. This method of cryptanalysis of the Vigenère cipher is called *Kasiski's method* after its discoverer. Another method, called the *index of coincidences*, uses statistical analysis of frequencies of letter groups.

Historical note: Friedrich Wilhelm Kasiski (1805-1881) was an officer in the Prussian army, and the author of a seminal work of cryptanalysis. His method of breaking polyalphabetic ciphers was a huge advance on currently available techniques.

To see how this works, take a ciphertext created with the Vigenère cipher, and read it into Sage:

```
sage: f = open("vigenereciphertext.txt","r")
sage: ct = f.read()
```

The next step is to note that a cyclic shift by `i` can be obtained on a string `st` with `st[i:]+st[:i]`. Now compare the ciphertext with different shifts of itself, to see which shift produces the greatest number of "coincidences": the same letters in the ciphertext and its shift.

[2] Dates 1774–1857; he is better known for the *Beaufort scale* for measuring wind intensity.

```
sage: for i in range(1,11):
....: ctx = ct[i:]+ct[:i]
....: print i, sum(x==y for x,y in zip(ct,ctx))
1 52
2 44
3 42
4 38
5 93
6 43
7 49
8 58
9 43
10 68
```

(The `zip` function produces all ordered pairs of objects from its two operands.) There's a large "spike" in values for shifts of 5 and 10, so it is reasonable to assume that the keyword has length 5. Now the ciphertext can be divided into five groups, and the most common letter in each determined:

```
sage: ct5s = ['','','','','']
sage: for i in range(clen):
   ct5s[i%5] = ct5s[i%5]+ct[i]
sage: alph = 'ABCDEFGHIJKLMNOPQRSTUVWXYZ'
sage: print [ct5s[0].count(i) for i in alph]
[9, 0, 1, 9, 6, 13, 18, 4, 0, 16, 10, 22, 3, 3, 6, 0, 2,
0, 30, 5, 6, 11, 31, 7, 3, 15]
```

The most commonly occurring letter here seems to be in the fourth from last place, in other words `W`, which must correspond with the most commonly occurring English letter `E`. Applying this procedure to the other four groups produces the following list of most common letters:

Group	Letter
0	W
1	A
2	M
3	J
4	X

Since each of these letters correspond to `E` in the plaintext, which has a value of 4, this value can be simply subtracted from each of the ASCII values of these letters to reveal the keyword `SWIFT`. And decrypting the ciphertext with this keyword produces:

```
'FOURSCOREANDSEVENYEARSAGOOURFATHERSBROUGHTFORTHONTHIS
CONTINENTANEWNATIONCONCEIVEDINLIBERTYANDDEDICATEDTOTHE
PROPOSITIONTHATALLMENARECREATEDEQUAL
```

```
...
THISNATIONUNDERGODSHALLHAVEANEWBIRTHOFFREEDOMANDTHAT
GOVERNMENTOFTHEPEOPLEBYTHEPEOPLEFORTHEPEOPLESHALLNOT
PERISHFROMTHEEARTH'
```

—it's the famous Gettysburg Address of Abraham Lincoln.

A known-plaintext attack is trivial, since given plaintext and corresponding ciphertext the keyword can be found immediately.

3.6 The one-time pad

This is the only unbreakable cryptosystem in existence. It can be considered as a variation of the Vigenère cipher, where the keyword is as long as the plaintext. If the keyword is random, then there is no possible correspondence between plaintext and ciphertext, making cryptanalysis impossible. For example, using the same plaintext as before, and keyword

`ETJWBOGLVWGESAKYMOPFJKWGG,`

the resulting ciphertext is

`ABCDEFGHIJKLMNOPQRSTUVWXY.`

There are difficulties inherent in using this particular cryptosystem. Both persons must agree on a suitable random string, long enough to be used for all possible messages between them. If they start using a short string more than once, then effectively they are using a Vigenère cipher, and cryptanalysis is then possible. Then there is the problem of getting this string from one person to the other, without it's being intercepted by a third person.

It is important that the string be random. If the persons are using as a keyword an English text, for example, a book known to both of them, then there is the possibility of a successful cryptanalysis using frequency analysis.

3.7 Permutation ciphers

The permutation cipher is different from the ciphers discussed so far. Rather than changing just the values of plaintext letters, their position is changed instead. For example, suppose every block of five letters is permuted according to:

$$\begin{matrix} 1 & 2 & 3 & 4 & 5 \\ 2 & 4 & 5 & 3 & 1 \end{matrix}$$

Then the ciphertext

```
WITHDRAW ONE HUNDRED DOLLARS
```

is first broken up into groups of five letters:

```
WITHD RAWON EHUND REDDO LLARS
```

and then encrypted by the permutation as:

```
IHDTW AONWR HNDUE EDODR LRSAL.
```

This is not particularly secure, since letter frequencies are preserved, so cryptanalysis using frequencies is possible. However, some modern cryptosystems include permutations as part of their algorithms.

Such permutations are easy in Sage, remembering to use the permutation of numbers 0 through $n-1$ instead of from 1 to n. With the plaintext `pl`, first divide it into blocks of 5 each, permute each block and join the results:

```
sage: perm = [1,3,4,2,0]
sage: ps = [pl[5*i:5*i+5] for i in range(len(pl)//5)]; ps
['WITHD', 'RAWON', 'EHUND', 'REDDO', 'LLARS']
sage: pp = [[y[x] for x in perm] for y in ps]; pp
[['I', 'H', 'D', 'T', 'W'], ['A', 'O', 'N', 'W', 'R'],
 ['H', 'N', 'D', 'U', 'E'], ['E', 'D', 'O', 'D', 'R'],
 ['L', 'R', 'S', 'A', 'L']]
sage: ct = join(flatten(pp),''); ct
'IHDTWAONWRHNDUEEDODRLRSAL'
```

3.8 Matrix ciphers

Matrix ciphers are important not from a security point of view, but because they were one of the first examples of mathematics applied to cryptography. These ciphers were first discussed by mathematician Lester Hill in 1929 [36, 37], and for that reason this cipher is often called the *Hill cryptosystem.*

Now let M be an $n \times n$ matrix whose determinant is relatively prime to 26 (this requirement will be shown to be necessary later). The plaintext is divided into blocks of n symbols each, and encryption is performed by premultiplying each block by the matrix, and taking residues modulo 26. Thus encryption and decryption are defined as

$$\begin{aligned} c &= Mp \pmod{26} \\ p &= M^{-1}c \pmod{26} \end{aligned}$$

where p and c are blocks of size n of plaintext and corresponding ciphertext.

For example, let

$$M = \begin{bmatrix} 22 & 11 & 19 \\ 15 & 20 & 24 \\ 25 & 21 & 16 \end{bmatrix}$$

so that

$$M^{-1} = \begin{bmatrix} 14 & 25 & 6 \\ 2 & 3 & 11 \\ 21 & 9 & 25 \end{bmatrix} \pmod{26}$$

(check it!). To encrypt

```
WITHDRAW ONE HUNDRED DOLLARS.
```

first split the plaintext into blocks of three symbols each, and rather than treating each block separately, put them as columns into one big matrix, padding the plaintext at the end with X's to fill the final block. Then:

$$\begin{bmatrix} 22 & 11 & 19 \\ 15 & 20 & 24 \\ 25 & 21 & 16 \end{bmatrix} \begin{bmatrix} \mathtt{W} & \mathtt{H} & \mathtt{A} & \mathtt{N} & \mathtt{U} & \mathtt{R} & \mathtt{D} & \mathtt{L} & \mathtt{S} \\ \mathtt{I} & \mathtt{D} & \mathtt{W} & \mathtt{E} & \mathtt{N} & \mathtt{E} & \mathtt{O} & \mathtt{A} & \mathtt{X} \\ \mathtt{T} & \mathtt{R} & \mathtt{O} & \mathtt{H} & \mathtt{D} & \mathtt{D} & \mathtt{L} & \mathtt{R} & \mathtt{X} \end{bmatrix}$$
$$= \begin{bmatrix} \mathtt{X} & \mathtt{Q} & \mathtt{O} & \mathtt{V} & \mathtt{Q} & \mathtt{H} & \mathtt{N} & \mathtt{T} & \mathtt{U} \\ \mathtt{K} & \mathtt{B} & \mathtt{W} & \mathtt{B} & \mathtt{I} & \mathtt{R} & \mathtt{R} & \mathtt{B} & \mathtt{I} \\ \mathtt{I} & \mathtt{Q} & \mathtt{K} & \mathtt{B} & \mathtt{P} & \mathtt{L} & \mathtt{Z} & \mathtt{B} & \mathtt{B} \end{bmatrix}$$

and now the ciphertext can be read off as

```
XKIQBQOWKVBBQIPHRLNRZTBBUIB
```

Look at the first column to see how this is done:

$$\begin{bmatrix} 22 & 11 & 19 \\ 15 & 20 & 24 \\ 25 & 21 & 16 \end{bmatrix} \begin{bmatrix} \mathtt{W} \\ \mathtt{I} \\ \mathtt{T} \end{bmatrix} = \begin{bmatrix} 22 & 11 & 19 \\ 15 & 20 & 24 \\ 25 & 21 & 16 \end{bmatrix} \begin{bmatrix} 22 \\ 8 \\ 19 \end{bmatrix}$$
$$= \begin{bmatrix} 933 \\ 946 \\ 1022 \end{bmatrix}$$
$$= \begin{bmatrix} 23 \\ 10 \\ 8 \end{bmatrix} \pmod{26}$$
$$= \begin{bmatrix} \mathtt{X} \\ \mathtt{K} \\ \mathtt{I} \end{bmatrix}.$$

All this can be done readily in Sage. The first step is to put the plaintext into a matrix. Given an $n \times n$ matrix, first the plaintext must be padded to fill n rows:

```
sage: n = 3
sage: pl = "WITHDRAWONEHUNDREDDOLLARS"
sage: lp = str2lst(pl)
sage: ln = len(lp)
sage: for i in range((-ln)%n):
          lp.append(23)
sage: Mpl = transpose(matrix(len(lp)/n,n,lp)); Mpl

[22  7  0 13 20 17  3 11 18]
[ 8  3 22  4 13  4 14  0 23]
[19 17 14  7  3  3 11 17 23]
```

The final command produces a matrix with n rows, whose elements are the list of numbers obtained from the plaintext. This list has been padded to produce a list whose length is a multiple of n.

Matrix multiplication, modulo 26, is done automatically in Sage if the matrix is constructed over the ring of integers mod 26, $\mathbb{Z}_{26}$:

```
sage: M = matrix(Zmod(26),[[22,11,19],[15,20,24],[25,21,16]])
sage: C = M*Mpl; C

[23 16 14 21 16  7 13 19 20]
[10  1 22  1  8 17 17  1  8]
[ 8 16 10  1 15 11 25  1  1]
```

To convert back into a ciphertext string, the matrix must be converted into a list first:

```
sage:  lst2str(transpose(C).list())

'XKIQBQOWKVBBQIPHRLNRZTBBUIB'
```

These commands can very easily be placed into a small program, where n can be obtained as the numbers of rows of the enciphering matrix `M`:

```
sage: n = M.nrows()
```

Cryptanalysis

Matrix ciphers are difficult to break using a ciphertext-only attack, but are vulnerable to a known plaintext attack. Given corresponding plaintext and ciphertext, it is possible to determine the matrix used.

To see this algebraically, suppose there are n blocks of plaintext P, and corresponding blocks of ciphertext C. If the encrypting has been done with an $n \times n$ matrix M, then

$$MP = C$$

or

$$M = CP^{-1}.$$

Thus the encryption matrix M can be obtained by multiplying the inverse of P by C. This requires that P be invertible, that its determinant is relatively prime to 26. However, in a large enough plaintext, it should be possible to pick out columns that form an invertible matrix.

Consider, for example, the following plaintext and ciphertext pair:

plaintext: `SELLMYCARTOMORROW`
ciphertext: `QRSLIVSFUBQFHBKIJS`

which uses a 3×3 matrix for encryption. Look at the columns of the plaintext as numeric values:

1		2		3		4		5		6	
S	18	L	11	C	2	T	19	O	14	O	14
E	4	M	12	A	0	O	14	R	17	W	22
L	11	Y	24	R	17	M	12	R	17	X	23

and the columns of the ciphertext:

1		2		3		4		5		6	
Q	16	L	11	S	18	B	1	H	7	I	8
R	17	I	8	F	5	Q	16	B	1	J	9
S	18	V	21	U	20	F	5	K	10	S	18

The matrix consisting of the first three columns of the plaintext

$$P = \begin{bmatrix} 18 & 11 & 2 \\ 4 & 12 & 0 \\ 11 & 24 & 17 \end{bmatrix}$$

has a determinant $2852 = 18 \pmod{26}$. As 18 is not relatively prime to 26, this matrix is not invertible. Nor is the matrix consisting of columns 2, 3 and 4 (its determinant mod 26 is 4—check it!).

However, the matrix consisting of columns 3, 4 and 5

$$P = \begin{bmatrix} 2 & 19 & 14 \\ 0 & 14 & 17 \\ 17 & 12 & 17 \end{bmatrix}$$

has a determinant (mod 26) of 17, which *is* relatively prime to 26. And so its inverse exists:

$$P^{-1} = \begin{bmatrix} 2 & 23 & 9 \\ 17 & 14 & 24 \\ 12 & 13 & 20 \end{bmatrix}.$$

Then the corresponding matrix C will consist of columns 3, 4 and 5 of the ciphertext:

$$C = \begin{bmatrix} 18 & 1 & 7 \\ 5 & 16 & 1 \\ 20 & 5 & 10 \end{bmatrix}.$$

Finally, the enciphering matrix M can be obtained:

$$M = CP^{-1} = \begin{bmatrix} 137 & 519 & 326 \\ 294 & 352 & 449 \\ 245 & 660 & 660 \end{bmatrix} = \begin{bmatrix} 7 & 25 & 14 \\ 8 & 14 & 7 \\ 11 & 10 & 6 \end{bmatrix} \pmod{26}.$$

Inverting matrices modulo 26

Given a square matrix, such as the matrix A used above, how can its inverse modulo 26 be computed? Recall from basic linear algebra that the inverse of a matrix A can be defined as

$$A^{-1} = \frac{1}{\det(A)}\mathrm{adj}(A)$$

where $\mathrm{adj}(A)$ is the adjoint of A. This formula can be used directly, rewriting it slightly as

$$A^{-1} = \det(A)^{-1}\mathrm{adj}(A)$$

where the inversion is to be done modulo 26. This explains why for the matrix A to be invertible, its determinant must be relatively prime to 26. For if not, the inverse would not exist.

Using the matrix A above, first compute

$$\mathrm{adj}(A) = \begin{bmatrix} -184 & 223 & -116 \\ 360 & -123 & -243 \\ -185 & -187 & 275 \end{bmatrix} = \begin{bmatrix} 24 & 15 & 14 \\ 22 & 7 & 17 \\ 23 & 21 & 15 \end{bmatrix} \pmod{26}$$

and

$$\det(A) = -3603 = 11 \pmod{26}.$$

By using any of the methods discussed in the previous chapter, next compute

$$11^{-1} = 19 \pmod{26}.$$

Thus

$$A^{-1} = 19 \begin{bmatrix} 24 & 15 & 14 \\ 22 & 7 & 17 \\ 23 & 21 & 15 \end{bmatrix}$$

$$= \begin{bmatrix} 456 & 285 & 266 \\ 418 & 133 & 323 \\ 437 & 399 & 285 \end{bmatrix}$$

$$= \begin{bmatrix} 14 & 25 & 6 \\ 2 & 3 & 11 \\ 21 & 9 & 25 \end{bmatrix} \pmod{26}.$$

In Sage, given the type of the matrix, inversion is very straightforward:

```
sage: M.inverse()

[14 25  6]
[ 2  3 11]
[21  9 25]
```

3.9 Glossary

Frequency analysis. A method of breaking a cipher by comparing frequencies in the ciphertext with known frequencies of letters or letter blocks in (English) text. It could be also used with other alphebetical languages.

Monoalphabetic cipher. A cipher in which letters of the plaintext are encrypted to letters of the ciphertext by a single alphabetic substitution.

Polyalphabetic cipher. A cipher in which several different alphabetic substitutions are employed to create the ciphertext.

Exercises

Review Exercises

1. How can a general translation cipher be broken?

2. How can a general transposition cipher be broken?

3. How could you easily determine if a ciphertext had been obtained by using only a permutation of the plaintext?

4. What are the setup requirements to ensure full security of the one-time pad?

5. Why do these ciphers show that a known plaintext attack is far stronger than a ciphertext-only attack?

Beginning Exercises

6. Decrypt the following Caesar ciphers:

   (a) `WUDYHOQRUWKRQZHGQHVGDB`

   (b) `PHHWPHDWWKHUDLOZDBVWDWLRQRQSODWIRUPQLQH`

   (c) `VHQGDOOBRXUPRQHBWRWKLVDGGUHVV`

7. Decrypt the following shift ciphers, using the shift given:

   (a) `GTDTEZYESPHPPVPYOLYOMCTYRJZFCPWPASLYE`, 11

   (b) `DPNRIKYFXZJSZXXVIKYREPFLIJ`, 17

   (c) `XIQVBBPMEITTAZMLIVLBPMKMQTQVOJTCM`, 8

8. By using a brute-force attack, decrypt the shift ciphers:

   (a) `XPPEXPZYHPOYPDOLJ`

   (b) `LXVNCXVHYJACHCXWRPQC`

   (c) `IUBBOEKHSQHVEHJUDJXEKIQDTTEBBQHI`

9. Show how to obtain the inverse permutation so as to use the program in the previous question to decrypt a ciphertext. Using this permutation and the `transp` program, decrypt the following ciphertexts:

   (a) `GSOVQJDXJVTXCSIZFFHXSHFX`

   (b) `ODXGOVFFOXJDIXXTXXJZWZVG`

   (c) `GSOVQJDXOVGJXXISCSPIBVQYSGJXGJ`

10. A variation of the translation cipher is the *affine cipher*, which is encrypted by

    $$c = ap + b$$

    where p and c are the plaintext and corresponding ciphertext letters, and a is invertible modulo 26.

(a) Given $a = 11$ and $b = 13$, encrypt the plaintext:

`MEET AT THE AIRPORT TOMORROW.`

(b) Show how to decrypt the affine cipher.

(c) With the same a and b as above, decrypt

`YSXABNENSBFWLOLQOFN.`

11. Using the Vigenère cipher with keyword `KEY`, encrypt the message `THIS IS A SECRET MESSAGE.`

12. Encrypt the same message, but using the keyword `TRIAL`.

13. You have received the ciphertext `UQ XO FMFREL HQQOEFQA`, which has been encrypted with the keyword `OCEAN`. Decrypt this message.

14. Suppose that a plaintext has been encrypted first with a Vigenère cipher using a keyword of length n, followed by another encryption with a keyword of length m. Does this provide any further security? What is the least number of Vigenère decryptions required to obtain the plaintext?

15. The *Beaufort cipher* is a variation of the Vigenère cipher in which the correspondence between plaintext, ciphertext and key letters is given by

$$c = k - p.$$

What is the one advantage of this cipher over the Vigenère cipher?

16. Show how to change the Vigenère cipher routines to implement the Beaufort cipher, and apply it to the plaintexts and keys in questions 2 and 3. Check the decryptions in each question.

17. Vigenère himself proposed a stronger cipher than the "Vigenère cipher" discussed in Section 3.5. This stronger cipher is an *autokey cipher*, where the plaintext itself is used as a key. It works by starting with a keyword, and using plaintext characters after that. For example:

plaintext	`W I L L T R A V E L S O U T H B Y C A R N E X T F R I D A Y`
key	`C O D E W I L L T R A V E L S O U T H B Y C A R N E X T F R`
ciphertext	`Y W O P P Z L G X C S J Y E Z P S V H S L G X K S V F W F P`

(a) Check the above example.

(b) Write down the formula that provides the encryption.

(c) Show how to implement the cipher in Sage. One way of doing this is to use two loops: one for the first few plaintext characters (corresponding to keyword characters), and a second loop for the rest of the plaintext.

18. Show how to decrypt an autokey cipher using the keyword, and apply your decryption method to decrypting

    `DLGOWJPRUMOMKFNYEERLXME`

    using the key `KEYWORD`.

19. Decrypt the following autokey ciphertexts using the keywords given:

    (a) `PKBDKYUAEYKGALACPIRTQ`, `PROVERB`

    (b) `CZHNANBABUNTBZWSYMJWTMWZUDIJBMCQV`, `PLATO`

    (c) `RHSGGYWCCOHTWYHZDLVHTYISDMCD`, `PASTEUR`.

20. Another autokey cipher (also developed by Vigenère), uses the letters of the ciphertext, instead of plaintext, to form new key letters, as in

plaintext	`W I L L T R A V E L S O U T H B Y C A R N E X T F R I D A Y`
key	`C O D E Y W O P R N O K V Y G Y P R N Z N T N Q A X K J F O`
ciphertext	`Y W O P R N O K V Y G Y P R N Z N T N Q A X K J F O S M F M`

    Show that this is a much weaker cipher than the first autokey method, and is vulnerable to a brute-force attack.

21. Decrypt the following ciphertext (which has been formed using this weak autokey method):

    `NEASJFINVCMMZJPQKSQXIKXJBZXLXO`

    except for possibly the first few letters.

22. Using the permutation

$$\begin{matrix} 1 & 2 & 3 & 4 & 5 \\ 5 & 4 & 1 & 3 & 2 \end{matrix}$$

    encrypt the following plaintexts:

    (a) `SEND MONEY TO YOUR UNCLE`

    (b) `TRAVEL TO THE CITY NEXT MONDAY`.

23. Using the inverse of the permutation in the previous question, decrypt the following ciphertexts:

    (a) `OISWNHTIETEOCTMMFOEOWRDAR`

    (b) `UYOTPRENPUISOLCTETHNBESLA`.

24. Show that the transposition cipher is vulnerable to a known plaintext attack.

25. Using the matrix M given on page 67, encrypt the message THESE CODES GIVE ME A HEADACHE.

26. The matrix has been used to send you the ciphertext HBANJFJZHWPC—decrypt the message.

27. Now encrypt the plaintext SELL EVERYTHING AT HALF PRICE, this time using the matrix

$$M = \begin{bmatrix} 23 & 18 & 4 \\ 5 & 2 & 13 \\ 3 & 11 & 6 \end{bmatrix}.$$

28. Using the inverse of the matrix in the previous question, which is

$$M^{-1} = \begin{bmatrix} 9 & 4 & 20 \\ 23 & 10 & 15 \\ 1 & 23 & 6 \end{bmatrix},$$

decrypt the ciphertext MHWASITJKNANAXY

29. Using the standard formula

$$\begin{bmatrix} a & b \\ c & d \end{bmatrix}^{-1} = (ad - bc)^{-1} \begin{bmatrix} d & -b \\ -c & a \end{bmatrix}$$

find the inverse (mod 26) of

$$M = \begin{bmatrix} 23 & 12 \\ 8 & 19 \end{bmatrix}.$$

30. A 3×3 matrix has been used to encrypt the message SENDITTOMYUNCLE to the ciphertext KIDHCNREEACFHXG. Determine the matrix.

Sage Exercises

31. Here's a little Sage program to implement the transposition cipher:

```
def transp(str,p): # Transposition cipher;
#str is a string and p is a
# permutation of the numbers 0-25
    pl = str2lst(str)
    pl2 = [p[i] for i in pl]
    return lst2str(pl2)
```

Use this program to test the example given in Section 3.4.

32. Perform the following experiment to find letter frequencies:

    (a) Download a large ASCII text file and store it as `mytext.txt`. You can start at the Project Gutenberg page
    `http://www.gutenberg.org/wiki/Main_Page`
    for a collection of free downloadable texts.

    (b) Read it into sage with

```
sage: text = file('mytext.txt')
sage: t = text.read()
```

    (c) Now translate it into upper case and remove all non-alphabetical characters with the following:

```
sage: TS = ''
sage: for i in t:
....: if i.isalpha():
....:     TS += i.upper()
```

    Now `TS` contains just the letters from the original text, with no spaces.

    (d) Perform a frequency count with

```
sage: alph = "ABCDEFGHIJKLMNOPQRSTUVWXYZ"
sage: N = len(TS)
sage: for i in alph:
....:     print i,floor(TS.count(i)*1000/N)
```

    How closely does your frequency table compare from the one given in Section 3.4?

33. Edit the Vigenère cipher program so as to implement the Beaufort cipher.

34. Here is a program to implement the autokey cipher:

```
def autokey1(text,key):
    ps = str2lst(text)
    ks = str2lst(key+text[:-len(key)])
    return lst2str([(x+y)%26 for x,y in zip(ps,ks)])
```

    Check this program with the example given in question 17.

35. Use the commands given in the text to write a small program to implement the Hill cryptosystem. Given a plaintext `pl` and invertible matrix `M`, a program called `ct=hill(M,pl)` should produce ciphertext, and `hill(M.inverse(),ct)` should recover the plaintext.

36. Compare your program with Sage's `HillCryptosystem` class.

Further Exercises

37. Suppose the Hill cipher is used twice: once with an $n \times n$ matrix M_1 and then further encrypting the ciphertext with an $n \times n$ matrix M_2. Does this increase the security of the cipher? If not, why not?

38. Suppose the Hill cipher is used as follows: first the plaintext is divided into blocks of size n and an $n \times n$ matrix M_1 is used to produce the ciphertext. The ciphertext is then divided into blocks of size m, where $m \neq n$, and further encrypted with an $m \times m$ matrix. Does this system provide better security than using a single matrix?

39. Show how to break the Hill cipher quickly with a chosen-plaintext attack.

40. What does your answer to question 39 tell you about this attack as compared to a known-plaintext attack?

Chapter 4

Introduction to information theory

Recall from Chapter 3 that some cryptosystems seem to do a better job of "hiding" the plaintext and some cryptosystems are harder to cryptanalyse than others. In order to formalize these notions, and to be able to provide a formal measurement of the usefulness of a cryptosystem, it is necessary to have a basic grounding in information theory. This is the topic of this chapter. It will include:

- Entropy, or the formal quantification of uncertainty.
- Methods of quantifying the secrecy of a cryptosystem.
- The entropy and redundancy of English.
- The unicity distance, which is the amount of ciphertext required to obtain a unique plaintext.

Historical note: Information theory was the creation of one man, Claude Shannon, with two epochal papers in the late 1940s [79, 80]. He developed a complete theory which helped to provide a solid mathematical foundation to the study of the sending and secrecy of information.

4.1 Entropy and uncertainty

Intuitively, there is the notion of the uncertainty of an outcome. The result of a roll of a fair die (with 6 sides) is more uncertain than the toss of a fair coin because there are more possible outcomes in a die roll than a coin toss. And the outcome of a race with 10 entrants is more uncertain still. Suppose a loaded die is tossed with probabilities of $2/5$ of obtaining a five, $2/5$ of obtaining a six, and probability of $1/20$ each for the other four numbers; the outcome is less uncertain than for a fair die.

Entropy is the formal measurement of the uncertainty of an outcome. One immediate requirement is that entropy should be a function of probabilities alone. Suppose that a large integer is picked at random: the probability of its being even or odd is 1/2 each. So the entropy of this outcome should be the same as the entropy for the outcome of tossing a fair coin.

In order to provide a formula for entropy, start by looking at the information obtained after a random trial. Suppose for example that three fair coins are tossed. The probability of any particular outcome is 1/8; once a toss is made there are 3 bits of information obtained. In general, with n coins the probability of any particular outcome is $1/2^n$ and once a toss is made, there are n bits of information. If p is the probability of an outcome for a coin toss, then the amount of information obtained after a toss can be seen to be

$$\log_2(1/p) = -\log_2 p.$$

Suppose, however, there are k outcomes all with different probabilities p_1, p_2 and so on up to p_k, with their sum $p_1+p_2+\cdots+p_k=1$. To extend the above formula to this situation, take the weighted average of all the probabilities:

$$-p_1 \log_2 p - p_1 \log_2 p_2 - \cdots - p_k \log_2 p_k.$$

In general then, the entropy associated with a random variable X whose k outcomes have probabilities $p_1, p_2, \ldots, p_k$ is defined to be

$$H(p_1, p_2, \ldots, p_k) = -\sum_{i=1}^{k} p_i \log_2 p_i.$$

Shannon derived this formula by providing several axioms that entropy must satisfy, and showed that the above formula satisfied all his axioms. However, an intuitive approach is quite sufficient for this chapter. From the above formula, the following consequences can be shown:

1. $$H\left(\frac{1}{n}, \frac{1}{n}, \ldots, \frac{1}{n}\right) < H\left(\frac{1}{n+1}, \frac{1}{n+1}, \ldots, \frac{1}{n+1}\right).$$

   (The more outcomes there are, the more uncertain the event.)

2. $H(p_1, p_2, \ldots, p_n)$ is maximum when each $p_i = 1/n$. (If one outcome is more likely than any other, then there is less uncertainty about the outcome.)

3. The value of H is independent of the order of the variables. For example,

   $$H\left(\frac{1}{2}, \frac{1}{3}, \frac{1}{8}, \frac{1}{24}\right) = H\left(\frac{1}{24}, \frac{1}{8}, \frac{1}{2}, \frac{1}{3}\right)$$

   and equally for all other permutations.

4. $H \geq 0$ and is zero only if $p_i = 1$ for some i. (If one probability is 1, then all other probabilities are zero, and there is no uncertainty in the outcome.)

5. H is a continuous function of its variables; a small change in a probability should result in only a small change in the uncertainty.

6. Entropy is additive: if X and Y are independent variables, then

   $$H(X,Y) = H(X) + H(Y).$$

   For example, the entropy of tossing 5 coins is equal to the entropy of tossing 2 coins, plus the entropy for tossing 3 coins. If X and Y are not independent, then

   $$H(X,Y) < H(X) + H(Y).$$

One way to imagine entropy is the average number of yes/no questions required to determine the result. Suppose for example that there are four outcomes

elephant, mosquito, mountain, breadbox

each with probability 1/4. The entropy is 2, which is the number of questions required. (Is it an animal? Is it larger than a person?)

For a more formal discussion of entropy, with proofs, see Welsh [93].

Conditional entropy

Recall that the *conditional probability* of x given y is defined as

$$\Pr(x|y) = \frac{\Pr(x \cap y)}{\Pr(y)}.$$

Similarly the conditional entropy of a random variable A given B can be defined as

$$H(A|B) = \sum_{x \in A} \Pr(x|B) \log_2 \Pr(x|B).$$

In other words, whereas entropy is a function of probabilities, conditional entropy is a function of conditional probabilities. For example, suppose A is the random variable generated by rolling a fair die, and B is the variable obtained by rolling the same die and returning 1 if the result is odd and 0 if the result is even. Then a table of conditional probabilities can be easily computed:

x	1	2	3	4	5	6
$\Pr(x\|B)$	1/3	0	1/3	0	1/3	0

and so

$$\begin{aligned} H(A|B) &= -\frac{1}{3}\log_2\left(\frac{1}{3}\right) - \frac{1}{3}\log_2\left(\frac{1}{3}\right) - \frac{1}{3}\log_2\left(\frac{1}{3}\right) \\ &= \log_2 3. \end{aligned}$$

Two obvious consequences of the definition are that $H(X|X) = 0$ (if $x \in X$ then the probability of x being in X is 1, so there is no uncertainty about the outcome), and that $H(X|Y) = H(X)$ if X and Y are independent. (For independence implies that Y has no bearing on the outcome of X.)

A less obvious consequence is that conditional entropy satisfies

$$H(X,Y) = H(Y) + H(X|Y).$$

Again see Welsh [93] for a proof.

4.2 Perfect secrecy

In chapter 3 the one-time pad was said to be perfectly secure in that the ciphertext gave no information about the plaintext. This is not the case with the Caesar cipher, where the ciphertext gives away some information. For example, since letters are always encrypted to the same ciphertext character, ciphertext obtained from the Caesar cipher shows where similar letters are in the plaintext.

The notion of entropy allows a formal definition of perfect secrecy. For a simple example, suppose there are four possible plaintexts: `NORTH`, `SOUTH`, `EAST` and `WEST`, four keys a, b, c and d, and four ciphertexts 1, 2, 3, 4. Suppose a cryptosystem is defined with encryption E by the following table:

E	`NORTH`	`SOUTH`	`EAST`	`WEST`
a	1	2	3	4
b	2	3	4	1
c	3	4	1	2
d	4	1	2	3

If the messages and keys are all equally likely to be chosen, then every ciphertext is equally likely. With M being the set of message outcomes, and C the set of ciphertexts, the conditional entropy

$$H(M|C)$$

is the entropy of the message given a known ciphertext. But an examination of the above encryption table shows that for any ciphertext, each message is equally likely, and so

$$H(M|C) = 2 = H(M).$$

Suppose on the other hand an encryption function E' is defined as follows:

E'	NORTH	SOUTH	EAST	WEST
a	1	1	2	2
b	1	1	2	2
c	3	3	4	4
d	3	3	4	4

For E', each ciphertext corresponds to only two plaintexts, and so

$$H(M|C) = 1.$$

In general, a cryptosystem with messages M and ciphertexts C satisfies *perfect security* if

$$H(M|C) = H(M).$$

What this means is that knowledge of the ciphertext (without the key) gives no information about the plaintext. Look at the first example above. Since for each of the four ciphertexts, each message is equally likely, an enemy who obtains a ciphertext has no possible way of determining the plaintext. But for the second example, once a ciphertext is obtained, the messages are restricted to two possibilities only.

Although perfect secrecy was described above in terms of information and entropy, it can also be described in terms of probability. Specifically, a cryptosystem has perfect security if the conditional probability of a plaintext m given a ciphertext c is equal to the probability of obtaining c

$$\Pr(m|c) = \Pr(c)$$

for all $m \in M$ and $c \in C$.

Consider the first cryptosystem above, with $c = 2$. Since there are four possible messages, the probability is

$$\Pr(m|c) = \frac{1}{4}$$

for each m. And this is also the probability that c will be obtained. The same value will be obtained for each of the other ciphertexts.

For the second system though, assume that each ciphertext is equally likely, so that

$$\Pr(c) = \frac{1}{4}$$

for each $c \in C = \{1, 2, 3, 4\}$. Take again $c = 2$. For the four messages we have

$$\begin{aligned}
\Pr(\texttt{EAST}|c) &= \frac{1}{2} \\
\Pr(\texttt{WEST}|c) &= \frac{1}{2} \\
\Pr(\texttt{NORTH}|c) &= 0 \\
\Pr(\texttt{SOUTH}|c) &= 0.
\end{aligned}$$

For another example, define a Vigenère cipher on messages m which are eight character strings of uppercase characters (so that the set of all messages M has cardinality 26^8), with key words of length four. Then there are 26^4 possible keywords. For every plaintext then, there are 26^4 possible ciphertexts, and there are 26^8 different possible ciphertexts. Suppose each ciphertext is equally likely, so that

$$\Pr(c) = 26^{-8}.$$

For a given ciphertext c, what is $\Pr(m|c)$?

Since $m = c - k$, there are only 26^4 possible plaintexts corresponding to this ciphertext, as there are only 26^4 different keys to apply. If m is one of those messages, then

$$\Pr(m|c) = 26^{-4}.$$

If m is not one of those messages, then

$$Pr(m|c) = 0.$$

In each case,

$$\Pr(m|c) \neq \Pr(c)$$

and so this system does not have perfect security.

Suppose instead the keywords are to be of length eight (the length of each message). Then for each ciphertext, the number of possible messages is 26^8 and so

$$\Pr(m|c) = 26^{-8}$$

for all m and c. Thus this system has perfect security.

A fundamental result about perfect security was first proved by Shannon; it can be stated as follows:

> Suppose a cryptosystem is such that the numbers of plaintexts, ciphertexts and keys are all equal (to N, say). The the system has perfect security if and only if:
>
> 1. $\Pr(k) = 1/N$ for all keys k,
> 2. For each message m and ciphertext c there is exactly one key k which encrypts m to c.

For a proof see Hoffstein et al. [38], and for further discussion [80].

4.3 Estimating the entropy of English

There are various methods used to calculate the entropy H_E of English, the number of bits of information per character. To keep calculations manageable, assume there are only 27 characters (26 uppercase letters and space). If all letters were equally probable, then the entropy would be

$$\log_2(27) \approx 4.755.$$

But letters are not equally probable—for a start they differ in their frequencies; E being the most commonly occurring, and Z the least frequent.

Letters also differ in their frequencies of juxtaposition. In standard English, for example, the letter Q is almost always followed by a U (the only exceptions that I know of are transliterations of Arabic proper names).

There are various methods for calculating entropy. One goes back to Shannon, who asked people to guess the next letter in some English text, and recorded the number of guesses until the correct letter was obtained. Suppose there were N characters in the text, and it took g_i guesses to guess the i-th character. Let t_i be the number of times g_i guesses are made, with t_k being the maximum value. So for example, if seven letters required 2 guesses, then $t_2 = 7$. It follows that

$$t_1 + t_2 + \cdots + t_k = N$$

and so the probability of t_i guesses being made is

$$p_i = \frac{t_i}{N}.$$

By the standard formula for entropy, then

$$H_E \approx \sum_i p_i \log_2(p_i).$$

Shannon found that that $H_E \approx 1.6$. What this means is that although ASCII text may require 8 bits per character, English text could be compressed (say, using Huffman codes) in a lossless manner using an average of only 1.6 bits per character.

A computational experiment

This section will demonstrate how to estimate the entropy using some English text and Sage. The first step is to obtain some text, and read it into Sage as a large string with only 27 characters. One example is *The Adventures of Tom Sawyer* by Mark Twain, which is available at Project Gutenberg[1]. Suppose the text version of the book is downloaded and saved as `TomSawyer.txt`. A text editor can be used to trim off the generic Project Gutenberg material at each end of the document, but that won't make much difference to the result.

Now it can be read in, and turned into a long string:

[1]It was a favorite book of mine as a child.

```
sage: f = open("TomSawyer.txt","r")
sage: fr = f.read()
sage: fr = fr.replace('\r\n',' ')
sage: FR = fr.upper()
sage: TS=''
sage: for i in FR:
....: if (i.isalpha()) or (i.isspace()):
....:     TS+=i
```

Now, from Shannon's original paper on the entropy of printed English [81], the entropy can be calculated as

$$H_E = \lim_{n\to\infty} F_n$$

where F_n is defined as

$$\sum_{i,j} \Pr(b_i + j) \log_2(\Pr(j|b_i).$$

Here b_i ranges over all possible n-character strings (called "n-grams"), and j ranges over all single characters. So $b_i + j$ is an $n+1$ length string obtained by appending j to b_i, and $\Pr(j|b_i)$ is the probability of j following immediately after b_i, and is equal to

$$\frac{\Pr(b_i + j)}{\Pr(b_i)} = \frac{\#(b_i + j)}{\#(b_i)}$$

where $\#(x)$ is the number of times the string or character x appears in `TS`. For simplicity, restrict the experiment to considering only 4-grams. There are $27^4 = 531441$ possible such strings, but only a relatively few of them will appear in printed English text. This means that it will be more efficient to create a set of all the 4-grams in the string. This will be most easily done with the `str2num` and `num2str` programs:

```
sage: S4 = []
sage: for i in range(len(TS)-3):
....:     S4 += [str2num(TS[i:i+4])]
....:
sage: S4s = set(S4)
sage: quads = [num2str(x) for x in S4s]
sage: len(quads)
  20434
```

The list `quads` contains all 4-grams in the document; and the final number indicates that there are only 20434 different such 4-grams.

The next requirement is the number of characters in the text, to compute the probabilities:

```
sage: N = float(len(TS));N
    387895.0
```

Now for Shannon's formula:

```
sage: alph = 'ABCDEFGHIJKLMNOPQRSTUVWXYZ '
sage: s4 = 0.0
sage: for q in quads:
....:     for a in alph:
....:         qc = TS.count(q)
....:         qac = TS.count(q+a)
....:         if (qc<>0) and (qac<>0):
....:             s4 -= qac/N*log(qac/float(qc),2.0)
sage: s4
  1.67821787961163
```

This gives 1.68 as a rough approximation to the entropy of English, taking only four previous characters for each trial. This agrees surprisingly well with the empirical values found by Shannon and others. However, the generally agreed information rate of written English [81] is slightly over 1 bit per letter.

Recall that if all 27 characters were equally likely, then the entropy would be approximately 4.8. This value is called the *absolute rate* of the language and is denoted R. The entropy of English, denoted r, has been found above to be about 1.6. The difference of these two values is the *redundancy* of the language, and is denoted D, so that $D = R - r$. For English then, $D \approx 3.2$. One way of thinking about this is to think of all character strings of length n. There are $27^n = 2^{Rn}$ possible strings, of which only about 2^{rn} will be meaningful.

In fact, if the table of frequencies given on page 60 is used to determine the absolute rate of English with only 26 characters (discounting the space), then

$$\begin{aligned}D &= -0.078\log_2(0.078) - 0.009\log_2(0.009) - \dots \\ &\qquad - 0.019\log_2(0.019) - 0.001\log_2(0.001) \\ &\approx 4.157\end{aligned}$$

and so $D \approx 3$.

4.4 Unicity distance

The *unicity distance* is a formal measurement of how much ciphertext is required to obtain a unique plaintext. It can be taken as the value n for which

$$H(K|C^{(n)}) \approx 0$$

where K is a random variable whose values are the keys (which may be considered to be all equally likely), and $C^{(n)}$ is the random variable whose values are the n-length ciphertexts.

In practice, this is difficult to determine, and so

$$\frac{\log_2 |K|}{D}$$

may be taken as an approximation.

Consider the Vigenère cipher with keyword length 4. There are 26^4 possible different keys, and so the unicity distance is

$$\frac{\log_2(26^4)}{3.15} \approx 5.9688$$

which means approximately 6 characters will be required to determine the plaintext uniquely. In general, for a Vigenère cipher with keyword length m, approximately $1.5m$ ciphertext characters will be required to determine the plaintext uniquely. Since a one-time pad may be considered to be a Vigenère cipher with keyword as long as the plaintext, this indicates that to break a one-time pad requires half again as much plaintext as there is available—a clear impossibility.

For a general substitution cipher, there are 26! possible keys, and so the unicity distance is

$$\frac{\log_2(26!)}{3.15} \approx 28.06$$

which means that given a general transposition cipher, 28 characters of ciphertext are required to produce a unique plaintext.

Note that the unicity distance is not the amount of ciphertext required to break the cipher, but the amount required to obtain a *unique plaintext.*

For one last example, consider the block cipher DES, which will be discussed in Chapter 8. It uses 56-bit keys on 64-bit blocks. Note that 64 bits is 8 bytes, or 8 (extended ASCII) characters. Its unicity distance is thus

$$\frac{56}{3.15} \approx 17.8$$

or slightly over two blocks.

4.5 Glossary

Absolute rate. The base 2 logarithm of the number of symbols used for a language.

Conditional entropy. The entropy of an outcome that is conditional on another outcome.

Digram, trigram, n-gram. A group of two, three, or n letters in a text.

Entropy. A measure of the uncertainty of an outcome.

Perfect secrecy. The condition of a ciphertext providing no information about the plaintext.

Redundancy. The difference between the absolute rate and the entropy.

Unicity distance. For a cipher, the amount of ciphertext required to obtain a unique plaintext.

Exercises

Review Questions

1. List the following in order of increasing entropy: the outcome when a fair die is rolled, the sum of the toss of two fair dice, the outcome of the tosses of two fair coins, the outcome of the tosses of four fair coins.

2. How is perfect secrecy related to the probability of the key?

3. What is meant by the entropy of a natural language (such as English)?

4. Give an informal argument that entropy is maximized when all outcomes are equally probable.

5. What can be deduced about a cipher if its unicity distance is longer than the ciphertext?

Beginning Exercises

6. Show that the formula for entropy satisfies all the consequences given in section 4.1.

7. Suppose that a random variable X is defined by the number of heads obtained when a fair coin is tossed four times. What is the entropy of X?

8. Suppose the above random variable X uses a biased coin, so that $\Pr(\text{head}) = 0.4$ and $\Pr(\text{tail}) = 0.6$. What is the entropy of X?

9. Repeat the above problem, but with probabilities reversed, so that $\Pr(\text{head}) = 0.6$ and $\Pr(\text{tail}) = 0.4$.

10. Suppose a coin is tossed until a head appears, and let X be the random variable of the number of tosses required. Determine $H(X)$. Use the identities

$$\sum_{k=0}^{\infty} r^k = \frac{1}{1-r}$$

and

$$\sum_{k=0}^{\infty} kr^k = \frac{r}{(1-r)^2}.$$

11. Repeat the above exercise but with a biased coin for which

    (a) $\Pr(\text{head}) = 0.6$ and $\Pr(\text{tail}) = 0.4$,

    (b) $\Pr(\text{head}) = 0.4$ and $\Pr(\text{tail}) = 0.6$.

12. Suppose X is defined to be the absolute difference of the numbers obtained when two fair six-sided dice are tossed. What is the entropy of X?

13. Define X to take the integer values 1 to 20 with equal probability, and define Y to be equal to 1 if X is prime, and 0 if X is composite (or 1). Determine the conditional entropies $H(Y|X)$ and $H(X|Y)$.

14. Consider the following cryptosystem

$\hat{E}$	NORTH	SOUTH	EAST	WEST
a	1	2	3	4
b	2	3	4	1
c	3	4	1	2
d	4	1	2	3

but with

$$\begin{aligned}
\Pr(\texttt{NORTH}) &= 1/6 \\
\Pr(\texttt{SOUTH}) &= 1/3 \\
\Pr(\texttt{EAST}) &= 1/3 \\
\Pr(\texttt{WEST}) &= 1/6
\end{aligned}$$

and

$$\begin{aligned}
\Pr(a) &= 1/10 \\
\Pr(b) &= 1/5 \\
\Pr(c) &= 3/10 \\
\Pr(d) &= 2/5
\end{aligned}$$

(a) determine the probabilities of obtaining each of the four ciphertexts.

(b) determine the probabilities $\Pr(\texttt{NORTH}|2)$, $\Pr(\texttt{EAST}|4)$, $\Pr(\texttt{SOUTH}|3)$. Does this system have perfect security?

15. Determine the following proverbs, from which all vowels have been removed:

    (a) `T MNY CKS SPL TH BRTH`

    (b) `BRD N TH HND S WRTH TW N TH BSH`

16. Determine the following proverbs, from which alternate fourth and fifth characters (treating the space as a character) have been removed:

    (a) `A STCHN TE SES NE`

    (b) `MANHAN MA LIT WK`

17. Suppose the entropy of English was $r = 1.3$. Determine the unicity distances of the Caesar, Vigenère and general substitution ciphers.

18. Repeat the above exercise but using $r = 1.1$.

19. Determine the unicity distance of the Playfair cipher from Chapter 1. Use the fact that for this cipher the alphabet length is only 25 (the letter J is not used).

Sage Exercises

20. Consider the random variable X of the number of heads obtained when a fair coin is tossed 20 times. Determine $H(X)$.

21. Repeat the above exercise for 100 tosses.

22. The *Poisson distribution* is defined for integer values n by

$$\Pr(X = n) = \frac{\lambda^n e^{-\lambda}}{n!}$$

where λ is the expected number of occurrences of n.

Estimate $H(X)$ for (a) $\lambda = 5$, (b) $\lambda = 10$, (c) $\lambda = 20$.

23. Repeat the experiment for the entropy of English, but with a text of your own choosing.

Chapter 5

Public-key cryptosystems based on factoring

This chapter will contain investigations of

- The RSA cryptosystem: its definition, and its strengths and weaknesses.
- The Rabin cryptosystem, which counters one possible weakness of the RSA cryptosystem.
- Methods of attack on these cryptosystems.
- General notions of security on public-key cryptosystems.
- Some approaches to factoring.

5.1 Introduction

Recall that a *public-key cryptosystem* is a cryptosystem with two distinct, but related keys. Each user of the cryptosystem has a *public key*, which he or she makes public, so that anyone else can use it to encrypt messages to the user. The user also has a *private key*, known to nobody else, which is used for decryption. To make the system secure, it is important that there is no easy way to determine a private key from the corresponding public key. To do this requires some nifty mathematics, as we shall see.

5.2 The RSA cryptosystem

The RSA cryptosystem was one of the first public-key cryptosystems to be developed, and it still appears to be reasonably secure. (It gets its name from the initials of the first team to publish its details: Ronald Rivest, Adi Shamir, and Leonard Adleman.)

Historical note: Although Rivest, Shamir and Adleman were the first to publicize the algorithm, it was in fact developed before them, in England by the cryptographer Clifford Cocks. However, Cocks was unable to publish his findings because of security restrictions on his work. The idea of public-key cryptography, although "officially" first introduced by Diffie and Hellman (see Section 6.4), again seems to have been first discussed in England by James Ellis in the late 1960s.

The security of the RSA cryptosystem stems from the fact that although it is a computationally trivial job to multiply two large numbers, factoring a large number is in general very difficult. For instance, given two very large primes and their product, there is no known general way of factoring an arbitrarily large product in a reasonable time. (There are methods of factoring if the prime factors have particular properties, for example if one of them, p, has $p-1$ being a product of only small factors.) Note that it is relatively straightforward to test if a number is prime or not. However, even if a number can be shown to be composite, finding its factors is more difficult. For example, the 20th "Fermat number"

$$2^{2^{20}}+1$$

which has 315653 digits, is known to be composite, but none of its factors are known. Some methods of prime testing and factorization are given in appendix B.

Now for the cryptosystem.

To generate both public and private keys, choose two large prime numbers p and q. In the context of this cryptosystem, *large* means at least several hundred digits long each. Then compute the product

$$n = pq.$$

Now pick any number e which is relatively prime to $(p-1)(q-1)$. Then the inverse

$$d = e^{-1} \pmod{(p-1)(q-1)}$$

exists; it can be found using Euclid's algorithm. Then the values d and n are also relatively prime. The values e and n together are the public key, and the value d is the corresponding private key. The original prime numbers p and q are no longer needed, but their values should never be revealed. After all, if p and q are known, then the factorization of their product is known, which defeats the use of the cryptosystem.

To encrypt a message, break it into blocks, each of which can be given a numerical value less than n. Let m be one such block. Its encrypted value is

$$c = m^e \pmod{n}.$$

The value c can be decrypted by

$$m = c^d \pmod{n}.$$

To see why this works, distinguish two cases for m:

1. m is relatively prime to both p and q,
2. m is a multiple of one of the primes, say p.

Note that m can't be a multiple of both p and q, for then it would be a multiple of $n = pq$ and so be greater than or equal to n.

Consider the first case. Note that

$$\begin{aligned} c^d &= (m^e)^d \pmod{n} \\ &= m^{ed} \pmod{n}. \end{aligned}$$

Now recall from above that e and d are inverses modulo $(p-1)(q-1)$. That is,

$$ed = k(p-1)(q-1) + 1$$

for some k. Then

$$\begin{aligned} m^{ed} &= m^{k(p-1)(q-1)+1} \\ &= mm^{k(p-1)(q-1)}. \end{aligned}$$

Now since p and q are both primes, then

$$\begin{aligned} \phi(n) &= \phi(pq) \\ &= (p-1)(q-1) \end{aligned}$$

and by Euler's theorem, if a and n are relatively prime, then

$$a^{\phi(n)} = 1 \pmod{n}.$$

Since m is relatively prime to both p and q, it is relatively prime to n, and so this result can be applied directly. Thus

$$m^{k(p-1)(q-1)} = 1 \pmod{n}$$

and so finally

$$c^d = m \pmod{n}.$$

For the second case, first note that if $m = kp$ for some k, then

$$m^{ed} = m \pmod{p} \tag{5.1}$$

because both sides are equal to zero modulo p. Since m must be relatively prime to q, then

$$\begin{aligned} m^{ed} &= m^{k(p-1)(q-1)+1} \pmod q \\ &= m.(m^{q-1})^{k(p-1)} \pmod q. \end{aligned}$$

By Euler's theorem,

$$m^{q-1} = 1 \pmod q$$

and so

$$m^{ed} = m \pmod q. \tag{5.2}$$

Applying the Chinese remainder theorem to equations 5.1 and 5.2 shows that

$$m^{ed} = m \pmod{pq}$$

which completes the proof. The system is described in Figure 5.1.

Parameters. Two large primes p and q, and their product $n = pq$.

Key generation. The public key is (n, e), where $e < n$ and is relatively prime to $\phi(n)$. The private key is $d = e^{-1} \pmod{(p-1)(q-1)}$.

Encryption. For a message $m < n$ the ciphertext is $c = m^e \pmod n$.

Decryption. $m = c^d \pmod n$.

FIGURE 5.1: The RSA cryptosystem.

An example

Here's an example with small numbers. Pick $p = 37$ and $q = 43$. (These numbers are too small for serious cryptography, but they will do for this example.) Then

$$n = pq = 1591.$$

Now a value e must be chosen that is relatively prime to

$$(p-1)(q-1) = 36 \times 42 = 1512.$$

Choose for example $e = 17$. Then

$$d = e^{-1} \pmod{1512} = 89.$$

The public key consists of the numbers $e = 17$ and $n = 1591$; the private key is $d = 89$. Suppose somebody wants to send a message that has been encoded as

$$28562810173.$$

This can be broken into blocks of three digits each for encryption:

$$m_1 = 285$$
$$m_2 = 628$$
$$m_3 = 101$$
$$m_4 = 73.$$

To encrypt m_1, the sender evaluates

$$285^{17} \pmod{1591} = 935 = c_1$$

which is sent. The other blocks are similarly encrypted, and received ciphertexts are:

$$\begin{array}{rcrcl} 628^{17} \pmod{1591} & = & 665 & = & c_2 \\ 101^{17} \pmod{1591} & = & 1158 & = & c_3 \\ 73^{17} \pmod{1591} & = & 406 & = & c_4. \end{array}$$

Each ciphertext block c_i can be decrypted by evaluating

$$c_i^{89} \pmod{1591},$$

obtaining

$$935^{89} \pmod{1591} = 285$$
$$665^{89} \pmod{1591} = 628$$
$$1158^{89} \pmod{1591} = 101$$
$$406^{89} \pmod{1591} = 73$$

and the original message has been recovered.

Speeding up decryption

The public key e is generally chosen to be fairly small; popular choices are 3, 17 or 65537—all of the form $2^n + 1$. The binary representation of such numbers have only two ones, and the rest zeros, which makes modular exponentiation to such powers very efficient. But in general the private key d will be a number similar in size to pq. In order to speed up the decryption, exponentiation can be computed modulo p and q separately, and the results joined with the Chinese remainder theorem.

That is, to compute

$$c^d \pmod{pq},$$

the powers with smaller moduli

$$c^d \pmod{p}, \quad c^d \pmod{q}$$

can be computed first. But by Fermat's theorem, these last two exponents can be simplified to

$$c^{d \bmod (p-1)} \pmod{p}$$

and

$$c^{d \bmod (q-1)} \pmod{q}.$$

So decryption can be performed entirely using exponentiation modulo the smaller individual primes p and q. First define

$$m_p = c^{d \bmod (p-1)} \pmod{p}$$
$$m_q = c^{d \bmod (q-1)} \pmod{q}.$$

Using s, t with $sp + tq = 1$, then the plaintext will be

$$m = spm_q + tqm_p \pmod{n}.$$

For large primes p, q this method is about four times faster than attempting to compute $c^d \pmod{pq}$.

Security issues

The RSA cryptosystem is secure only if the private key d cannot be determined from the public key e, n. To obtain d requires the value of $\phi(pq) = (p-1)(q-1)$ which itself requires the value of p and q. But the only way to obtain p and q is to factor n, and this is known to be a computationally intractable task. That is, given a sufficiently large number n (say of the order of 200 digits), which is the product of two 100 digit primes, it is impossible to factor n in a reasonable time (say, the age of the universe).

There may be ways of obtaining $\phi(pq)$ without factoring pq. However, if $\phi(pq)$ can be computed, then p and q can easily be found. To see this, note that

$$\phi(n) = (p-1)(q-1) = pq - p - q + 1$$

so that

$$p + q = n - \phi(n) + 1.$$

Thus given n and $\phi(n)$ the sum and product of p and q are both known, from which p and q can be easily calculated.

It is conjectured (but not known for sure) that any method for breaking this scheme leads to an efficient factoring algorithm.

Factoring is one of the major areas of current computational number theory research, and all indications are that it remains a very difficult problem, in spite of the introduction of some very powerful and sophisticated algorithms. Although this does not prove that the RSA system is secure, it does allow it to be used with some degree of confidence. Even though it is one of the oldest public-key cryptosystems, it is still holding up as one of the best.

5.3 Attacks against RSA

Common modulus attack

One common scenario is for the same $n = pq$ to be used amongst a group of people, but different values for the exponent e. Suppose a message is encrypted twice, once using e_1, and again using e_2. This will yield two different ciphertexts:

$$\begin{aligned} c_1 &= m^{e_1} \pmod{n}, \\ c_2 &= m^{e_2} \pmod{n}. \end{aligned}$$

If n, e_1, e_2, c_1 and c_2 are all known, then m can be found without knowing either d_1 or d_2. Using the extended Euclidean algorithm, find s and t for which

$$se_1 + te_2 = 1.$$

One of s and t will be negative; the other positive. Suppose the negative one is s. Then:

$$m = ((c_1)^{-1})^{-s}(c_2)^t \pmod{n}.$$

To try this, use $p = 37$ and $q = 43$ as in the above example. Using a message $m = 500$, and the encryption exponents $e_1 = 17$ and $e_2 = 5$ produces the ciphertexts

$$\begin{aligned} c_1 &= 500^{17} \pmod{1591} = 849, \\ c_2 &= 500^5 \pmod{1591} = 22. \end{aligned}$$

The extended Euclidean algorithm finds that

$$(-2)(17) + (7)(5) = 1.$$

So $s = -2$ (this is the negative value) and $t = 7$. Then

$$\begin{aligned} m &= ((c_1)^{-1})^{-s}(c_2)^t \pmod{n} \\ &= ((849)^{-1})^2(22)^7 \pmod{1591} \\ &= 500. \end{aligned}$$

Low encryption attack

Using a low value of e can seem to simplify encryption, as it lowers the amount of work involved. And sometimes a low value, like 3 or 17, is recommended for this very purpose. But if the same message is encrypted using a low exponent, but with different values of n, the Chinese remainder theorem can be used to recover the message.

For example, here are three pairs of primes, with their products:

$$(37, 43, 1591), \quad (41, 47, 1927), \quad (31, 53, 1643).$$

Suppose the message $m = 500$ was encrypted with the exponent 3 using each of these values of n:

$$\begin{aligned} c_1 &= m^3 \pmod{1591} = 1494 \\ c_2 &= m^3 \pmod{1927} = 1291 \\ c_3 &= m^3 \pmod{1643} = 560. \end{aligned}$$

By reversing these equations, this sequence of equations for m^3 can be obtained:

$$\begin{aligned} m^3 &= 1494 \pmod{1591}, \\ m^3 &= 1291 \pmod{1927}, \\ m^3 &= 560 \pmod{1643}. \end{aligned}$$

Applying the Chinese remainder theorem produces

$$m^3 = 125000000$$

or

$$m = 500.$$

Timing attack

In 1996, Paul Kocher [51] demonstrated that it was possible to time the decryption of ciphertexts so as to recover the private key. The trick is that in any modular exponentiation algorithm, some steps will take longer than others. A careful analysis of the speeds at which an exponentiation runs allows the private key to be recovered bit by bit. This was a remarkable and unexpected ciphertext-only attack, as well as being quite practical.

One method of countering this attack is to randomize the ciphertext c before decryption, using a process known as *blinding*:

1. Generate a random number r between 0 and $n = pq$.
2. Compute $c' = cr^e \pmod{n}$ where e is part of the public key.
3. Compute $m' = (c')^d \pmod{n}$.
4. Finally recover the plaintext by $m = m'r^{-1} \pmod{n}$.

This works because $r^{ed} = r \pmod{n}$. Blinding certainly counters the timing attack, as no information about the ciphertext can be used. However, there is a performance degradation of up to 10%.

There are other attacks; for a comprehensive survey see Boneh [11].

5.4 RSA in Sage

This section will demonstrate how to perform RSA encryption and decryption in Sage, starting with an example with some small primes. First enter the primes:

```
sage: p = 67
sage: q = 101
```

and test them for primality:

```
sage: is_prime(p)
True
sage: is_prime(q)
True
```

Now compute their product n and choose a public key e:

```
sage: n = p*q; n
  6767
sage: e = 17
```

The private key d is quickly determined:

```
sage: d = inverse_mod(e,(p-1)*(q-1));d
  1553
```

For a plaintext choose an integer smaller than n, say 1000:

```
sage: m = 1000
sage: c = power_mod(m,e,n);c
  4838
```

and test the decryption:

```
sage: power_mod(c,d,n)
  1000
```

Now try RSA with some large primes and some text strings. This will require a method of converting a text string into a large digit, and the large digit back into a text string. Here's how:

```
sage: pl = "Withdraw one hundred dollars"
sage: plc = map(ord,pl)
sage: plc
[87, 105, 116, 104, 100, 114, 97, 119, 32, 111, 110, 101, 32,
104, 117, 110, 100, 114, 101,100, 32, 100, 111, 108, 108, 97,
114, 115,46]
sage: m = ZZ(plc,256)
sage: m
  1215796697534383283049438240413797624586023760948255
  6241781282072919
```

The `ord` command simply replaces every character by its ASCII index. The list of digits `plc` can be understood to be the digits of a large number of base 256. The next command simply computes this large number. To go back in the other direction use the `digits` method of the `Integer` type:

```
sage: ml = m.digits(256)
sage: ml
[87, 105, 116, 104, 100, 114, 97, 119, 32, 111, 110, 101, 32,
104, 117, 110, 100, 114, 101,100, 32, 100, 111, 108, 108, 97,
114, 115,46]
```

then convert to characters by use of the `chr` command:

```
sage: ms = map(chr,ml)
sage: ms
['W', 'i', 't', 'h', 'd', 'r', 'a', 'w', ' ', 'o', 'n', 'e',
' ', 'h', 'u', 'n', 'd', 'r', 'e', 'd', ' ', 'd', 'o', 'l',
'l', 'a', 'r', 's', '.']
```

and finally collect all together in one string by using the `join` command, with an empty delimiter between individual characters.

```
sage: join(ms,'')
  'Withdraw one hundred dollars'
```

It will be convenient to put these commands into functions `str2num` and `num2str` which translates a string into a single integer, and back again:

```
def str2num(s):
   return ZZ(map(ord,s),256)

def num2str(n):
   nl=n.digits(256)
   return join(map(chr,nl),'')
```

Now to demonstrate RSA with some non-trivially small integers first create some large primes:

```
sage: p = next_prime(2^330)
sage: p
2187250724783011924372502227117621365353169430893212436425770
6064099529991993759232235131770230539 17
sage: q = next_prime(3^210)
sage: q
1568424042913152925468569828489075118463940614573029159280267
6915731672495230992603635422093849215077
```

Now all the computations are the same as before:

```
sage: n = p*q
sage: n
3430536624628895533427375282734632051593010928406255638280036
4350689064258510927090285032214837547534301942037315246302860
0502382189378437653723315626366964446477497462389430464462110
28135643900306609
sage: e = 41
sage: d = inverse_mod(e,(p-1)*(q-1))
sage: d
2593820374719408817957283750360331551204471677575461580162954
3777350268097898505848752097528291804219740097009400533142386
7644271348176611452853512954417727732245623351955343194733709
57200282045589417
sage: c = power_mod(m,e,n)
sage: c
3360811692968287616621569069727884791254367654392756928805583
4398972815323834699479298811609238446804954966628539877923352
5991756996021636381637183758135817755010298210225228190314307
32784613078697396
```

This enormous number is the ciphertext. The plaintext can be recovered easily:

```
sage: pl = power_mod(c,d,n)
sage: pl
121579669753438328304943824041379762458602376094825562417812
2072919
sage: num2str(pl)
'Withdraw one hundred dollars'
```

Here is the computation using the speeded up approach with the Chinese remainder theorem:

```
sage: g,s,t = xgcd(p,q)
sage: s*p+t*q
  1
sage: mp = power_mod(c,d%(p-1),p)
sage: mq = power_mod(c,d%(q-1),q)
sage: m = (s*p*mq+t*q*mp)%n
sage: num2str(m)
  'Withdraw one hundred dollars'
```

5.5 Rabin's cryptosystem

Recall that the RSA cryptosystem drew its strength from the difficulty of factoring large numbers. And yet there is the possibility of a "backdoor" attack—could RSA be broken *without* factoring? Could it be possible to determine the private key d from the public key e using some other, yet undiscovered, method? It may seem that this is not worth worrying about. But there is a serious security issue here: if an attack has not been conclusively proved to be unworkable, then the possibility remains that such an attack might be mounted in the future. Although factoring is a known difficult problem, the RSA cryptosystem is not necessarily unbreakable.

Historical note: Michael Rabin published this cryptosystem only a few years after the publication of RSA. He also helped develop the Miller–Rabin primality test, which is still one of the most useful and strongest tests known.

Rabin's cryptosystem has the property that breaking it is as hard as factoring. This means that there is no possible way of breaking the system except by factoring a large number (in this case, the public key). It works as follows: the user chooses two large primes p and q, both of which are equal to 3 mod 4.

In fact the system works for *any* primes, but the decryption is easier to describe when both primes are equal to 3 mod 4. The public key is the product $N = pq$, and the corresponding private key the pair of primes (p, q).

To encrypt a message, suppose that as above it has been encoded and broken into blocks m each of which is less than N. Given such a block m, it is encrypted as

$$c = m^2 \pmod{N}.$$

Note that an alternate form of the system is sometimes given which involves an extra parameter B as

$$c = m^2 + B \pmod{N}$$

or

$$c = m(m + B) \pmod{N}.$$

However, this provides no extra security, and the simpler form, with $B = 0$, is now generally preferred. To decrypt the received message block c, the receiver has to solve the equation

$$m^2 = c \pmod{N}.$$

This can be done by the following steps:

1. Find the values s and t for which

$$sp + tq = 1.$$

2. Compute

$$\begin{aligned} c_p &= c^{(p+1)/4} \pmod{p} \\ c_q &= c^{(q+1)/4} \pmod{q}. \end{aligned}$$

3. Then the solutions m to the quadratic congruence above are

$$\pm spc_q \pm tqc_p \pmod{N}.$$

That this works is a consequence of the fact that c_p and c_q are square roots of p and q respectively, along with the Chinese remainder theorem. To show that c_p is a square root of p, note that

$$\begin{aligned} c_p^2 &= c^{(p+1)/2} \pmod{p} \\ &= c.c^{(p-1)/2} \pmod{p}. \end{aligned}$$

Now, since

$$m^2 = c \pmod{pq},$$

it follows immediately that

$$m^2 = c \pmod{p}.$$

Raising both sides of this last congruence to the power of $(p-1)/2$ produces

$$m^{p-1} = c^{(p-1)/2} \pmod{p}.$$

By Fermat's theorem, the right hand side is equal to 1 (in fact the result holds in general, no matter whether m is relatively prime to p or not), and so

$$c^{(p-1)/2} = 1 \pmod{p}$$

from which it follows from above that

$$c_p^2 = c \pmod{p}.$$

One desirable property of a cryptosystem, which is satisfied more by Rabin's system than RSA, is that encryption is very fast. The smallest exponent that can be used in RSA is $e = 3$, and this requires one squaring and one multiplication. In comparison, Rabin encryption requires a single squaring. Rabin decryption is not so fast, but is comparable to RSA decryption. Rabin's cryptosystem is given in Figure 5.2.

Parameters. Two large primes p and q each equal to 3 modulo 4, and their product $n = pq$.

Key generation. The private key consists of the two primes p and q. The public key is their product n.

Encryption. For a message $m < n$ the ciphertext is $c = m^2 \pmod{n}$.

Decryption. Solve the equation $m^2 = c \pmod{n}$ for m. This can be done using the factorization of n and the Chinese remainder theorem.

FIGURE 5.2: Rabin's cryptosystem.

Security issues

Clearly Rabin's cryptosystem is secure only if the equation

$$x^2 = c \pmod{N}$$

cannot be easily solved, given c and N. But there is no known way of efficiently solving such an equation if N is composite (as it is in Rabin's cryptosystem). However such an equation can be solved by solving the equations

$$x^2 = c \pmod{p_i}$$

for all prime factors p_i of N and then collecting the results using the Chinese remainder theorem. And of course this requires factoring N.

An example

Suppose Alice chooses $p = 31, q = 67$, so that $N = pq = 2077$ (note that both primes are equal to 3 mod 4). She then makes the product 2077 available as her public key. Then to encrypt a message block $m = 1897$, Bob computes

$$c = 1897^2 \pmod{2077} = 1245.$$

which is the value sent to Alice. To decrypt this value, Alice follows the steps below:

1. She uses the extended Euclidean algorithm to find that

   $$(13)31 + (-6)67 = 1$$

   so that $s = 13$ and $t = -6$.

2. She then computes

   $$\begin{aligned} c_p &= (1245 \bmod 31)^{(31+1)/4} \pmod{31} \\ &= 5^8 \pmod{31} \\ &= 25 \end{aligned}$$

   $$\begin{aligned} c_q &= (1245 \bmod 67)^{(67+1)/4} \pmod{67} \\ &= 39^{17} \pmod{67} \\ &= 21. \end{aligned}$$

3. Now the plaintext she requires is one of:

   $$\begin{aligned} (13)(31)(21) + (-6)(67)(25) &\pmod{2077} = 490 \\ (13)(31)(21) - (-6)(67)(25) &\pmod{2077} = 1897 \\ -(13)(31)(21) + (-6)(67)(25) &\pmod{2077} = 180 \\ -(13)(31)(21) - (-6)(67)(25) &\pmod{2077} = 1587 \end{aligned}$$

   of which the second value, 1897, is correct.

This illustrates a possible weakness in this cryptosystem: that Alice has to choose between four solutions of the decryption equation. However, in practice is it usually very easy to make that choice, as the message generally has a known property (for example, it represents English text). If this system is being used for key exchange, then there is no way for Alice to determine which of the four values represents the correct key sent to her by Bob. In such a case, either a different system should be chosen, or Bob should add a standard header to the key enabling Alice to choose it from the four solutions.

Weakness of Rabin

A major weakness of the Rabin cryptosystem is that it is completely insecure against a chosen-ciphertext attack. Suppose the four square roots s_1, s_2, s_3, s_4 corresponding to a known plaintext m are all known. Compute the values

$$\gcd(s_i - s_j, N)$$

for all possible pairs of square roots. (Recall that N is the public key, so it is known.) A value not equal to 1 will be a factor of N. For example, in the above example,

$$s_1 = 490,$$
$$s_2 = 1897,$$
$$s_3 = 180,$$
$$s_4 = 1587.$$

Now the gcds of all the differences with $N = 2077$ can be calculated:

$$\gcd(s_1 - s_2, N) = 67,$$
$$\gcd(s_1 - s_3, N) = 31,$$
$$\gcd(s_1 - s_4, N) = 1,$$
$$\gcd(s_2 - s_3, N) = 1,$$
$$\gcd(s_2 - s_4, N) = 31,$$
$$\gcd(s_3 - s_4, N) = 67.$$

Both factors of N have been determined, and the system is completely broken.

The cryptosystem is also vulnerable when the same message has been encrypted twice, with different primes. For example, here are two sets of primes with their products:

$$(37, 43, 1591), \quad (31, 53, 1643).$$

Suppose $m = 500$ is encrypted using each of these sets of primes:

$$c_1 = m^2 \pmod{1591} = 213,$$
$$c_2 = m^2 \pmod{1643} = 264.$$

These equations can be written as equations for m^2:

$$m^2 = 213 \pmod{1591},$$
$$m^2 = 264 \pmod{1643}.$$

Using the Chinese remainder theorem, the solution can be found to be

$$m^2 = 250000$$

or $m = 500$. The point is that although modular square roots are, in general, difficult to compute, integer square roots can be calculated very efficiently and easily.

5.6 Rabin's cryptosystem in Sage

Number-theoretic computations with large integers and strings have been discussed with RSA above. This section concentrates on those aspects of the Rabin cryptosystem that are unique to it. Start by finding two large primes, both equal to 3 mod 4. Here's how to do it in Sage:

```
sage: p = next_prime(2^100); p; f = mod(p,4); f
1267650600228229401496703205653
1
sage: while mod(p,4)==1:
....p = next_prime(p+1)
sage: p
1267650600228229401496703205707
sage: mod(p,4)
3
```

The prime q is computed similarly:

```
sage: q = next_prime(p+1)
sage: while mod(q,4)==1:
    q = next_prime(q+1)
....:
sage: q
1267650600228229401496703205823
sage: mod(q,4)
3
```

Choose as the plaintext the message

```
sage: pl = "Go North today!"
```

for which the corresponding integer can be found:

```
sage: pn = str2num(pl); pn
173807684108695187949871218074611527
```

With $N = pq$, encryption using Rabin is straightforward:

```
sage: ct = power_mod(pn,2,N)
sage: ct
117107168047808259987660631961310493235626209807601352554644 7
```

To decrypt, first invoke the extended Euclidean algorithm:

```
sage: x,s,t = xgcd(p,q)
sage: s;t
-1147442353654862820320291694926
1147442353654862820320291694821
```

Next, determine c_p and c_q. There's a slight subtlety here; the value `(p+1)/4` can't be used as written in an argument of `power_mod` because all arguments must be integers, and an integer division, even when the result is an integer, is classified as a member of the ring of rationals. So either use integer division, which forces the output to be an integer,

```
(p+1)//4
```

or change the type to be an integer:

```
Integer((p+1)/4)
```

Using the first method,

```
sage: cp = power_mod(ct,(p+1)//4,p)
sage: cp
110311402654710658241540124757
sage: cq = power_mod(ct,(q+1)//4,q)
sage: cq
110311402654710658241524219997
```

Finally, evaluate all four of $\pm spc_q \pm tqc_p \pmod{N}$, using the percentage sign for the modulus, to ensure that the output is of type `Integer`:

```
sage: (s*p*cq+t*q*cp)%N
173807684108695187949871218074611527
sage: (s*p*cq-t*q*cp)%N
1133527991359161024708069416767247740636887420520246623398532
sage: (-s*p*cq+t*q*cp)%N
473410052899829250833892676560147028862878047626981305833329
sage: (-s*p*cq-t*q*cp)%N
1606938044258990275541961919519710660804577518276009854620334
```

In this case, the second computation produces the correct result. And in fact this can be tested by converting each of these numbers back into strings:

```
sage: num2str((s*p*cq+t*q*cp)%N)
```

The second will produce the correct plaintext; the first and the final two will produce gibberish.

5.7 Some notes on security

In general it is impossible to prove that a public-key cryptosystem is completely secure. For example, the RSA and Rabin schemes derive their strengths from the intractability of factoring, and yet there is no proof that factoring is hard (in a formal computational sense). However, there are several definitions of security that can be applied to public-key systems, which will be mentioned here.

Semantic security

Roughly speaking, a public-key cryptosystem is *semantically secure* if no information—even partial information—can be determined about the plaintext from the ciphertext in polynomial time. A related notion is *ciphertext indistinguishability* which means that if one of two plaintexts is encrypted to a given ciphertext, it is not possible to determine which of the plaintexts was used. This can be thought of in terms of a cryptographic black box: the adversary feeds in a plaintext, and is presented with the ciphertext, and a random value. Can the adversary determine which of these two is the correct plaintext?

It can be shown that RSA in its simplest form is not semantically secure. The adversary, given a plaintext, can encrypt it first. Then it will be easy to see which of two outputs is the correct one. RSA is an example of a *deterministic cryptosystem*, where for each plaintext there is just one ciphertext, and no such cryptosystem is semantically secure.

Non-malleability

Non-malleability is the property that given a ciphertext, it is not possible to generate another ciphertext so that the respective plaintexts are somehow related. A malleable system, poorly used, may allow an attacker to somehow change the ciphertext so that the resulting plaintext is also changed. The RSA and Rabin cryptosystems are malleable. To see this, note that in RSA, encryption is performed with

$$c = m^e \pmod{n}.$$

Then for any w

$$cw^e = (mw)^e \pmod{n}$$

so that a related plaintext can be produced.

Plaintext awareness

Plaintext awareness means that it is computationally infeasible to produce a valid ciphertext without starting from a plaintext and then encrypting it. If a system is both plaintext aware and semantically secure, then a chosen ciphertext attack cannot be mounted against the system since the only way to produce a ciphertext would be to know the plaintext first. Again, in its standard form the RSA cryptosystem is not plaintext aware since any value is a possible ciphertext.

For further discussion of non-malleable and plaintext-aware cryptosystems, and methods of strengthening standard systems so that they are secure in these senses, see Menezes et al. [61].

5.8 Factoring

The problem of integer factorization has exercised some of the finest mathematicians for many hundreds of years, and recently of course this problem has obtained special significance because of cryptosystems such as RSA and Rabin. This section will discuss just two of the many methods known.

Pollard's rho method

Pollard's rho method is one of the simplest methods that is substantially faster than trial division; it is particularly suited for finding smallish prime factors (10–20 digits long) in large numbers. It is known as a "Monte Carlo" method, in that it produces a random sequence of numbers—more precisely a pseudo-random sequence—and uses that sequence to generate a possible factor.

Suppose that n is the number to be factored, and a random sequence

$$x_1, x_2, x_3, \dots$$

of residues modulo n is produced. The aim of this method is to find two values x_i and x_j in the sequence for which

$$x_i \neq x_j \pmod{n}$$

but

$$x_i = x_j \pmod{d}$$

for some divisor d of n. Then the computation $\gcd(x_i - x_j, n)$ should yield a proper factor.

The pseudo-random sequence is easily generated by starting with, say $x_0 = 2$ (although any value will do), and defining

$$x_k = f(x_{k-1})$$

for all $k \geq 1$ for some non-linear function f. In practise a simple quadratic is used, such as

$$f(x) = x^2 + 1.$$

To find the values i and j, produce a subsequence of the original sequence by taking every second term. For example, suppose $n = 17 \cdot 19 = 323$. There will be two sequences x_i and y_i:

x_0	x_1	x_2	$\overline{x_3}$	$\overline{x_4}$	$\overline{x_5}$	$\overline{x_6}$	$\overline{x_7}$	$\overline{x_8}$	x_9	x_{10}	x_{11}	x_{12}
2	5	26	31	316	50	240	107	145	31	316	50	240
y_0		y_1		y_2		y_3		y_4		y_5		y_6

with the x_i sequence starting to repeat at x_3 with a period of six; the first period is shown with an over-bar. Such a repetition in the sequence can be found by comparing x_i with y_i; in the sequence shown, $x_6 = y_6$. Pollard's rho method works by evaluating $\gcd(x_i - y_i, n)$ for every i until the period is reached, at which case, if a proper factor has not yet been found, try again with either a different value for x_0, or a different function f. For the sequence above,

i	x_i	y_i	$\gcd(x_i - y_i, n)$
1	5	26	1
2	26	316	1
3	31	240	19

and a proper factor has been found. In practice rather than producing two sequences, the values x_i and y_i are just updated each time, with $y_0 = x_0$, $x_k = f(x_{k-1})$, and

$$y_k = f(f(y_{k-1})).$$

This means that the method can be implemented with two variables, x and y, set initially to be equal, and for each step

$$x = f(x),$$
$$y = f(f(y)).$$

The method obtains its name "rho" from the similarity of a diagram illustrating the periodicity to the Greek character rho: ρ. For the sequence above,

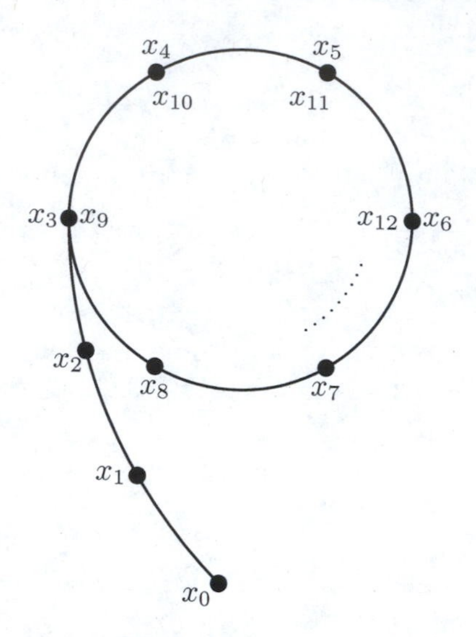

It is very easy to implement this in Sage:

```
def pollard_rho(n):
   x = Mod(2,n)
   y = x
   while true:
      x = x^2+1
      y = (y^2+1)^2+1
      g = gcd(x-y,n)
      if g>1 & g<n:
         print g;break
      if x==y:
         print "No factor found";break
```

And for a little test:

```
sage: pollard_rho(2^67-1)
  193707721
```

Historical note: The number $2^{67} - 1$ is famous in factorizing history, being one of the largest numbers to be factored "by hand," by an American mathematician, Frank Nelson Cole, who demonstrated his factorization at a meeting of the American Mathematical Society, in 1903. At the meeting, Cole walked to the blackboard, wrote all the arithmetic for computing $2^{67} - 1$, and then on a blank part of the board, carefully multiplied together his two factors. The two results agreed, and he obtained a standing ovation, without having spoken a single word. It is easy to show that $2^{67} - 1$ is composite—it is not a pseudoprime to base 2—but finding its factors had eluded the mathematical world until Cole's amazing result.

5.9 Glossary

Blinding: Randomizing the plaintext before encryption so that a timing attack won't reveal any useful information.

Chosen ciphertext attack: An attack where properties of the ciphertext or decryption process can be used to break the system.

Ciphertext indistinguishability: Being unable to determine which of two plaintexts was used to produce a particular ciphertext.

Common modulus attack: An attack on the RSA system that uses the same $n = pq$, but different exponents, for encryption.

Low encryption attack: An attack on the RSA system when a low encryption exponent is used.

Non-malleability: The property of not being able to change the ciphertext in such a way as to produce a corresponding change in the plaintext.

Plaintext awareness: Not being able to produce a valid ciphertext, except as a result of decryption of a plaintext.

Semantic security: The property of a cryptosystem providing no information about the plaintext from the ciphertext.

Timing attack: An attack on RSA based on the timings required to perform exponentiation.

Exercises

Review Questions

1. What difficult problem provides the security of the RSA cryptosystem?
2. What is a common modulus attack?
3. How can the use of a low encryption value allow RSA to be broken?

4. What is a timing attack?

5. What possible weakness of the RSA system does Rabin's system overcome?

6. How can the receiver choose from the four possible decryptions of a Rabin ciphertext?

7. What is a weakness of the Rabin system?

8. What is "non-malleability," and why does the RSA system fail this property?

9. Is the Rabin system malleable?

Beginning Exercises

10. Let $p = 11$, $q = 13$ and $e = 7$.

    (a) Find $d = e^{-1} \pmod{(p-1)(q-1)}$.

    (b) Using RSA, encrypt the message 99.

    (c) Now decrypt the result.

11. Repeat the above question, but use $p = 31$, $q = 37$, $e = 17$, and let the message to be encrypted be 700.

12. This time, use $p = 17$, $q = 23$, $e = 3$, and the message 100.

13. Pick a pair of small primes p and q, and choose a number e which is relatively prime to $(p-1)(q-1)$.

    (a) Determine your private key.

    (b) Exchange public keys with a partner.

    (c) Use your partner's public key to send a message to him or her.

    (d) Now decrypt the message you have been sent.

14. Your private key for a Rabin cryptosystem is the pair of primes 11 and 19, and your public key is their product $N = 187$.

    (a) Find the encrypted values of the messages 123 and 97.

    (b) Now follow through the decryption of the results.

15. Repeat the above question for the primes 31 and 43, and the messages 750 and 1000.

Sage Exercises

16. Create some large primes and their product:

```
rnd = lambda: randint(1,2^100)
p=next_prime(rnd())
q=next_prime(rnd())
n=p*q
```

17. Choose a value $e < n$ and ensure that it is relatively prime to your $(p-1)(q-1)$, and determine $d = e^{-1} \bmod (p-1)(q-1)$. (Use the `inv_mod` function here.)

18. Create a plaintext:

```
pl="This is my plaintext."
```

(or any plaintext you like.) Convert it to a number using the `str2num` procedure from the file above:

```
pln=str2num(pl).
```

19. Encrypt this number using the RSA method:

```
ct=power_mod(pln,e,n)
```

and decrypt the result:

```
decrypt=power_mod(ct,d,n)
num2str(decrypt)
```

20. With a friend, swap your public keys and use them to send each other a ciphertext encrypted with your friend's public key.

21. Now decrypt the ciphertext you have received using your private key.

22. Now try Rabin: create two large primes p and q and ensure that each is equal to 3 mod 4. You can use the method described in the text, or trial and error.

23. Create $N = pq$ and create a plaintext `pl`, and its numerical equivalent.

24. Determine the ciphertext c by squaring your number mod N.

25. Determine the s and t for which $sp + tq = 1$ by using the `xgcd` function.

26. Now follow through the Rabin decryption:

```
cp=power_mod(c,(p+1)//4,p)
cq=power_mod(c,(q+1)//4,q)
c1=(s*p*cq+t*q*cp)%N;num2str(c1)
c2=(s*p*cq-t*q*cp)%N;num2str(c2)
```

```
c3=(-s*p*cq+t*q*cp)%N;num2str(c3)
c4=(-s*p*cq-t*q*cp)%N;num2str(c4)
```

One of the outputs c1, c2, c3 and c4 should produce the correct plaintext; the others should be gibberish.

27. As in question 20, swap public keys with a friend, and use those public keys to encrypt a message to him or her. Now decrypt the ciphertext you have been given.

Further Exercises

28. Prove the correctness of the formula for the RSA common modulus attack.

29. Experiment with the low encryption attack: first work through the example given.

30. For another example, take these pairs of primes:

$$(11617, 69313), \qquad (46747, 34589), \qquad (55399, 19433).$$

Using $e = 3$ the following are the corresponding ciphertexts:

$$521156063, \qquad 21150682, \qquad 228705842.$$

Recover the plaintext.

31. Recall that in order to simplify RSA computation, rather than calculating $m^e \pmod{n}$, the same value could be obtained from the smaller calculations $m^e \pmod{p}$ and $m^e \pmod{q}$, followed by the Chinese remainder theorem [88]. That is, if

$$x = m^e \pmod{p}$$
$$y = m^e \pmod{q}$$

and s, t satisfy $sp + tq = 1$, then

$$c = yps + xqt \pmod{n}.$$

Suppose this method was being used, and an attacker manipulated the value of x by adding a value v relatively prime to q (for example, v could be a power of 2, or of 10). Suppose c was then calculated using $x + v$ and y, to obtain a value c'.

What is $\gcd(c - c', n)$, and how does this help the attacker?

32. Using the values in section 5.6, factor N by taking the gcd of N with each of the differences $s_i - s_j$ of the square roots.

33. Why does this method work?

Chapter 6

Public-key cryptosystems based on logarithms and knapsacks

This chapter will include discussions of

- Primitive roots.
- The El Gamal cryptosystem.
- Methods of calculating discrete logarithms.
- Cryptosystems based on integer knapsack problems.
- Methods of attacking knapsack cryptosystems.

6.1 El Gamal's cryptosystem

Recall from Chapter 2 that although it is very easy to compute a modular exponentiation:

$$b = a^x \pmod{p}$$

it is in general very difficult to go in the other direction, that is, given integers a, b and p, to find x satisfying the above equation. Finding the value of x is called the *discrete logarithm problem*, and although it can be solved if p satisfies certain properties—for example, if $p-1$ has only small factors—there is no known efficient way of solving it in general.

Before describing El Gamal's cryptosystem, it will be necessary to introduce the concept of a *primitive root*. Recall that Fermat's theorem says that if p is prime, and $a < p$, then

$$a^{p-1} = 1 \pmod{p}.$$

However, it may be that there is a divisor d of $p-1$, for which

$$a^d = 1 \pmod{p}.$$

For example, suppose $p = 11$. Then

$$\begin{aligned} 3^5 &= 1 \pmod{11} \\ 4^5 &= 1 \pmod{11} \\ 5^5 &= 1 \pmod{11} \\ 9^5 &= 1 \pmod{11} \\ 10^2 &= 1 \pmod{11} \end{aligned}$$

and for all other values $a < 11$, there is no value d smaller than 10 for which

$$a^d = 1 \pmod{11}.$$

These values a are called the *primitive roots modulo 11.* So the primitive roots modulo 11 are 2, 6, 7, and 8.

An equivalent way of defining a primitive root a is that powers of a generate all the non-zero residues of p. That is, the values of

$$1, a, a^2, a^3, \ldots, a^{p-1} \pmod{p}$$

are just the values

$$1, 2, 3, \ldots, p-1$$

in some order. For example, if $a = 6$ and $p = 11$,

n	1	2	3	4	5	6	7	8	9	10
6^n mod 11	6	3	7	9	10	5	8	4	2	1

Since the numbers on the bottom are *all* the values 1–10, it follows that 6 is a primitive root modulo 11.

If the powers of $a = 4$ are computed

n	1	2	3	4	5	6	7	8	9	10
4^n mod 11	4	5	9	3	1	4	5	9	3	1

then the bottom row contains only five distinct residues, and so 4 is not a primitive root modulo 11.

To test if a value a is a primitive root modulo p, first find all the factors d of $p-1$. Then a is a primitive root if for all factors d

$$a^{(p-1)/d} \neq 1 \pmod{p}.$$

Note that this test (which is quite fast) requires the factorization of $p-1$, and factorization is known to be difficult, so it can be quite hard to find the primitive root of a given large prime. In practice this is not really a problem because if $p-1$ is hard to factor, just choose another prime p.

In general, any prime p has $\phi(p-1)$ primitive roots, but since finding a primitive root requires factoring, in general primitive roots are hard to find.

This has led to the formation of another public-key cryptosystem, for which the security is based on the difficulty of calculating discrete logarithms.

Historical note: The system was developed by the Egyptian-American mathematician and cryptographer Taher El Gamal in 1985. The spelling "El Gamal" seems to be reasonably popular in the cryptographic literature, and hence its use in this text. But the preferred spelling now by its owner is "Elgamal."

El Gamal's cryptosystem works as follows: all users of the system are informed of a large prime p and a primitive root a modulo p. Suppose Alice is one user. Her private key is a randomly chosen integer $A < p$. Her public key is the value

$$B = a^A \pmod{p}.$$

Suppose now that Bob wishes to send a message to Alice, and suppose that (as usual) it has been encoded and subdivided into blocks each of which has a numerical value less than p. Suppose that m is one such block. To encrypt m Bob takes the following steps.

1. He chooses at random an integer $k < p$.
2. He computes the "message key":
$$K = B^k \pmod{p}.$$
3. He computes the values
$$C_1 = a^k \pmod{p}, \qquad C_2 = Km \pmod{p}.$$
4. He sends the pair (C_1, C_2) to Alice.

To decrypt, Alice first recovers K:

$$\begin{aligned} K &= B^k \pmod{p} \\ &= (a^A)^k \pmod{p} \\ &= (a^k)^A \pmod{p} \\ &= (C_1)^A \pmod{p}. \end{aligned}$$

This is easy for Alice (but not for anybody else) since only she knows the value A. Now she can find m by multiplying C_2 by $K^{-1} \pmod{p}$. (This inverse will definitely exist since p is prime.)

A few observations about this cryptosystem: first, since the message key is chosen at random for each message, the same message may well have two different encryptions each time it is sent. Also, the encryption consists of a pair of integers, thus effectively doubling the length of the message. The system is described in Figure 6.1.

Parameters. A large prime p and a primitive root $a \pmod p$.

Key generation. The private key is any $A < p$, and the public key is $B = a^A \pmod p$.

Encryption. For a message $m < n$ the ciphertext is $(C_1, C_2) = (a^k, B^k m)$ for a randomly chosen $k < p$.

Decryption. $m = C_2/C_1^A \pmod p$.

FIGURE 6.1: El Gamal's cryptosystem.

An example

Suppose the prime chosen is 71, and the primitive root is 7. Alice chooses 40 as her private key, and makes known her public key:

$$7^{40} \pmod{71} = 20.$$

Bob wishes to send the message block 62 to Alice. He chooses (randomly) the integer 30, and computes the message key:

$$20^{30} \pmod{71} = 45.$$

He then computes

$$7^{30} \pmod{71} = 32, \text{ and } 45 \times 62 \pmod{71} = 21$$

and sends the pair $(32, 21)$ to Alice.

Alice now recovers the message key by computing

$$32^{40} \pmod{71} = 45.$$

She finds the inverse $45^{-1} \pmod{71} = 30$, and obtains the message by

$$30 \times 21 \pmod{71} = 62.$$

It is clear that if the discrete logarithm problem could be easily solved, then this cryptosystem would not be secure, as a private key could be obtained from the corresponding public key. It is not known whether breaking this system is equivalent to solving the discrete logarithm problem, but it is conjectured to be the case.

6.2 El Gamal in Sage

The El Gamal system can be easily implemented in Sage. Here's the above example. Start with the prime and its primitive root:

```
sage: p = 71
sage: a = mod(2,p)
```

The reason for putting a inside a modulus is to define a as being in the residue class modulo p. This means that all further operations on a will be automatically restricted to this class, that is, in the class of residues modulo p. There will thus be no need for the `power_mod` or `inverse_mod` commands. Next choose the private key and compute the public key:

```
sage: A = 40
sage: B = a^A; B
  20
```

Now enter the message and the random k value, compute K and the ciphertext; this last can be done in one step:

```
sage: m = 62
sage: k = 30
sage: C = [a^k,B^k*m]; C
  [32,21]
```

Decryption takes one line:

```
sage: C[1]/C[0]^A
  62
```

Here's another example with larger values: starting by choosing a prime p, determining a primitive root, and creating a private/public key pair:

```
sage: p = next_prime(2^100); p
  1267650600228229401496703205653
sage: a=mod(primitive_root(p),p); a
  2
sage: A = randint(1,p); A
  300896262077666803415870714639L
sage: B = a^A; B
  676961842225380265947508047692
```

Choose a plaintext and encrypt it:

```
sage: m = 10^30
sage: m < p
  True
sage: k = randint(1,p); k
  625647259871478620045585524044L
sage: C = [a^k,B^k*m]; C
```

```
[925932179882107407189545457216,
 674060551589602342667719446404]
```

Now for the decryption:

```
sage: C[1]/C[0]^A
  1000000000000000000000000000000
```

Security issues

The El Gamal cryptosystem is semantically secure if the *decisional Diffie–Hellman problem* is assumed to hold in $\mathbb{Z}_p$. This assumption means that given a primitive root a modulo p, there is no way to decide, given $a^x \pmod{p}$, $a^y \pmod{p}$ and $a^z \pmod{p}$, whether $z = xy$. In fact, in this mathematical context of modular powers of primitive roots, the decisional Diffie–Hellman problem reduces to the discrete logarithm problem. Diffie–Hellman problems will be discussed in more detail in Section 6.4.

It is important to note that the El Gamal system is not actually *equivalent* to the discrete logarithm problem, but is based on it.

The system is an example of a *probabilistic cryptosystem*, where any plaintext can have many possible different ciphertexts. This helps to show that the cryptosystem is semantically secure. Even if you can produce one ciphertext corresponding to your plaintext, there's no way of knowing, given another ciphertext and random value, which one is the ciphertext.

The system is also malleable, for since a ciphertext corresponding to the plaintext m is

$$[C_1, C_2] = [a^k \,(\mathrm{mod}\, p), Km \,(\mathrm{mod}\, p)]$$

then for any w, the changed ciphertext

$$[C_1, wC_2]$$

represents the encryption of wm.

It is important to use a different value k for each encryption. Suppose two messages m and m' were encrypted using the same k. Then

$$C = [a^k \,(\mathrm{mod}\, p), Km \,(\mathrm{mod}\, p)]$$
$$C' = [a^k \,(\mathrm{mod}\, p), Km' \,(\mathrm{mod}\, p)].$$

But then

$$m'/m = C_1'/C_1 \pmod{p},$$

and so given one plaintext the other can be determined.

6.3 Computing discrete logarithms

There are several methods for computing discrete logarithms, none of which are efficient in all possible situations. (If such an algorithm were always efficient, then there would be no security for the El Gamal system.)

One of the easiest methods to understand is the "baby step, giant step" method developed by the American mathematician Daniel Shanks. Given a, y, n the problem is to find x so that

$$y = a^x \pmod{n}. \tag{6.1}$$

This equation can be written to make x the object, using standard log notation, with a modulus:

$$x = \log_a y \pmod{n}.$$

To explain the method, first note that if $s = \lceil\sqrt{n}\rceil$, then

$$x = is + j$$

for some $0 \le i, j \le s$. Then the log problem (6.1) can be written

$$y = a^{is+j} \pmod{n}$$

or

$$ya^{-j} = a^{is} \pmod{n}.$$

This means if values w and z between 0 and s can be found for which

$$ya^w = (a^s)^z \pmod{n} \tag{6.2}$$

then

$$x = sz - w.$$

The baby step, giant step method just computes all values $ya^w \pmod{n}$ and $(a^s)^z \pmod{n}$ until a match is found.

For example, take the problem

$$8 = 3^x \pmod{17}.$$

Here $s = \lceil\sqrt{17}\rceil = 5$, and $a^s = 3^5 = 5 \pmod{17}$, so this array of values can be computed:

k	$ya^k \bmod n$	$(a^s)^k \bmod n$
0	8	1
1	7	5
2	4	8
3	12	6
4	2	13

and the columns compared to find that

$$ya^0 = a^{2s} \pmod{n}$$

and so

$$x = 2s - 0 = 10.$$

As a slightly larger example in Sage, start with $n = 997$ and $a = 7$. The problem is to compute

$$\log_7 900 \pmod{997},$$

that is, to find the value x for which

$$900 = 7^x \pmod{997}.$$

Start by defining all the values:

```
sage: n,a,y = 997,7,900
sage: s = ceil(sqrt(n))
```

Now the computation of the "baby steps":

```
sage: B = [(y*a^k)%n for k in range(s+1)]
```

and the "giant steps":

```
sage: G = [power_mod(a,k*s,n) for k in range(s+1)]
```

Their intersection can be found by treating them as sets:

```
sage: sx = list(set(B).intersection(set(G)))[0]; sx
  401
```

The value 401 just obtained is the term common to both lists. To find the required logarithm x, use this value to find the relevant indices:

```
sage: s*G.index(sx)-B.index(sx)
  348
```

This method is certainly more efficient than a brute-force search, but it is not very efficient for very large n.

For large primes p, but for which $p - 1$ has only small factors

$$p - 1 = p_1^{a_1} p_2^{a_2} \cdots p_k^{a_k}$$

a method known as the (Silver-)Pohlig–Hellman algorithm may be used. In effect, to solve

$$y = a^x \pmod{p}$$

the algorithm uses an extension of the baby step, giant step algorithm to compute the discrete logarithm modulo $p_i^{a_i}$ for each i, and then uses the Chinese remainder theorem to find the logarithm modulo p. This method will be discussed in Appendix B.

6.4 Diffie–Hellman key exchange

Although not strictly speaking a public-key cryptosystem, the Diffie–Hellman protocol was one of the first instances of the use of a public key.

Historical note: Whitfield Diffie and Martin Hellman published this scheme in 1976, in a paper entitled "New directions in cryptography," which introduced the concept of public-key cryptography, and which is considered to be one of the most influential papers ever written (certainly in the cryptographic world!). However, as with the RSA system, it seems to have been developed earlier in England, by Malcolm Williamson, who, as with Clifford Cocks, was unable to publish his work because of security restrictions.

It works as follows: Alice and Bob between them choose a large prime p and a primitive root g. These values can be made public. Alice and Bob each then privately choose random values less than p; Alice chooses a and Bob chooses b. Then Alice computes

$$A = g^a \pmod{p}$$

and sends this result to Bob. And Bob computes

$$B = g^b \pmod{p}$$

and sends this result to Alice. Because of the difficulty of computing discrete logarithms (especially if p is chosen so that $p - 1$ has no small factors), no eavesdropper can determine the values of a and b. And even Alice and Bob can't determine each other's private values.

Then Alice computes

$$B^a \pmod{p}$$

which she can do: she knows p (it's the publicly chosen prime), she knows a because it's her private value, and she knows B, because that is what she was sent by Bob. And similarly Bob can compute

$$A^b \pmod{p}.$$

But since

$$\begin{aligned} B^a &= (g^b)^a \pmod{p} \\ &= g^{ab} \pmod{p} \\ &= (g^a)^b \pmod{p} \\ &= A^b \pmod{p} \end{aligned}$$

the values that Alice and Bob have ended with are the same! This value is their shared key, which they can now use for further communications.

The difficulty of computing discrete logarithms has given rise to the *Diffie–Hellman problems*, which exist in several forms. One version, known as the *computational Diffie–Hellman problem* can be stated as follows: if a^x and a^y are given, what is the value of a^{xy}? Note that this is a general question about *any* mathematical system in which some sort of exponentiation is defined. If the exponentiation refers to powers of a primitive root modulo p (with p being prime), then this version of the Diffie–Hellman problem appears to be very hard, hence the security of the El Gamal cryptosystem. A stronger problem is the *decisional Diffie–Hellman problem*: given a triplet $\langle a^x, a^y, a^z \rangle$, output "true" if $z = xy$ and "false" otherwise.

The importance of the Diffie–Hellman problems is that they allow a formal way of deciding on the security of some cryptosystems. If breaking the system is equivalent to one of the Diffie–Hellman problems (preferably the decisional problem) then there is good reason for deciding on the system's security.

6.5 Knapsack cryptosystems

Each public-key cryptosystem discussed so far is based on a problem known to be hard: factoring for RSA and Rabin; discrete logarithms for El Gamal. There is also a third class of hard problems: the *knapsack problems*, which has been used to develop several public-key cryptosystems.

Knapsack systems are important for several reasons:

1. they are among the earliest of workable public-key cryptosystems,
2. their construction is very elegant,
3. they have been completely broken, so are a good example of the pitfalls of cryptography.

The knapsack problem

There are a number of different "knapsack problems"; one of the most interesting for cryptography may be stated thus: given a sequence of integers

$(a_1, a_2, \ldots, a_n)$ and an integer N, is there a subset J of the list such that the sum of the elements of J is equal to N? For example, given the sequence

$$(14, 28, 56, 82, 90, 132, 197, 284, 341, 455)$$

can the number 516 be expressed as a sum of a subset of those values? (No, it can't.) However, 515 can be expressed as a sum of a subset (in three ways):

$$\begin{aligned} 515 &= 14 + 28 + 132 + 341 \\ &= 28 + 56 + 90 + 341 \\ &= 14 + 82 + 90 + 132 + 197. \end{aligned}$$

This problem, the *subset sum problem*, is known to be a very hard problem; in general there is no better way of determining whether or not we can have a subset sum than by trial and error. For very large sequences (several hundred elements), this approach becomes computationally infeasible.

However, there is one class of sequences for which the subset sum problem is easily solved: those that are "super-increasing":

> A sequence of integers $(a_1, a_2, a_3, \ldots, a_n)$ is said to be *super-increasing* if every a_i is strictly greater than the sum of all previous values.

For example, the sequence

$$(2, 3, 7, 25, 67, 179, 356, 819)$$

is super-increasing. Now to solve the subset sum problem for a super-increasing sequence is straightforward:

Step 1 Let a be the largest value in the sequence less than or equal to N. If $a = N$ then stop with a success.

Step 2 Let a' be the largest value less than $N - a$. If $a + a' = N$, then stop with a success.

Step 3 Let a'' be the largest value less than $N - a - a'$. If $a + a' + a'' = N$, then stop with a success.

Step n ...and so on.

With the super-increasing sequence given above, and the value $N = 207$, the algorithm produces $a = 179$, $a' = 25$, $a'' = 3$. Since $207 = 179 + 25 + 3$, the algorithm stops successfully. If the starting number is $N = 400$, then $a = 356$, $a' = 25$, $a'' = 7$, $a''' = 3$, $a'''' = 2$. At this stage a sum of 400 has not been obtained, and there are no more values in the set to use. Thus 400 can't be expressed as a subset sum.

This is easy to implement in Sage:

```
def si_solve(lst,N):
   M=N;res=[]
   for i in reversed(lst):
      if i<=M:
         res=[i]+res
         M=M-i
   if M<>0:
      print "Unsolvable"
   else:
      return res
```

And for example:

```
sage: se=[2, 3, 7, 25, 67, 179, 356, 819]
sage: si_solve(se,207)
  [3, 25, 179]
sage: si_solve(se,400)
  Unsolvable
```

The Merkle–Hellman additive knapsack cryptosystem

Alice chooses a super-increasing sequence $(a_1, a_2, a_3, \ldots, a_n)$, an integer N greater than $a_1 + a_2 + a_3 + \cdots + a_n$, and an integer w relatively prime to N. She then creates a new sequence $(b_1, b_2, b_3, \ldots, b_n)$ by

$$b_i = wa_i \pmod{N}.$$

The sequence $(b_1, b_2, b_3, \ldots, b_n)$ is her public key. Note that even though the original sequence was super-increasing, this new public key sequence will in general not be super-increasing.

Now suppose Bob wishes to send a message to Alice. He encodes it into a binary string, and breaks this up into blocks of size n (where n is the size of Alice's public key set). Suppose that a particular message block m has binary digits $m_1 m_2 m_3 \ldots m_n$. Then m is encrypted as

$$c = b_1 m_1 + b_2 m_2 + b_3 m_3 + \cdots + b_n m_n$$

and this value is sent to Alice.

To decrypt, Alice computes

$$cw^{-1} \pmod{N}$$

and solves the resulting subset sum using her super-increasing sequence. The numbers used correspond to the ones in the binary block. The system is described in Figure 6.2.

Parameters. The size of the message block n should be at least 200.

Key generation. The private key consists of a super-increasing sequence $(a_1, a_2, \ldots, a_n)$, an integer $N > \sum_{k=1}^{n} a_k$ and an integer w relatively prime to N. The public key is the sequence $(b_1, b_2, \ldots, b_n)$ where $b_i = wa_i \pmod{N}$.

Encryption. For a message $m = (m_1, m_2, \ldots, m_n)$ of n bits, the ciphertext is $c = \sum_{k=1}^{n} m_k b_k$.

Decryption. Solve the subset sum problem for $c' = cw^{-1} \pmod{N}$ using the super-increasing sequence a_i.

FIGURE 6.2: Merkle–Hellman additive knapsack cryptosystem.

An example

Alice chooses

$$(2, 3, 7, 25, 67, 179, 356, 819)$$

as her super-increasing sequence, and $N = 1600$, $w = 57$. She then multiplies every value of her sequence by 57, and finds the residue modulo 1600. The result, her public key, is

$$(114, 171, 399, 1425, 619, 603, 1092, 283).$$

Now Bob wants to send a message to Alice; he breaks it up into 8-bit blocks. Suppose one of these blocks is 10110010. He encrypts this as $114 + 399 + 1425 + 1092 = 3030$.

Now Alice computes

$$57^{-1} \pmod{1600} = 393.$$

She then computes

$$3030 \times 393 \pmod{1600} = 390.$$

Now she solves the subset sum problem for 390 using her original super-increasing sequence. She finds that $390 = 2 + 7 + 25 + 356$, and deduces that the message block is 10110010.

Here's how this can be done in Sage. First, creating the public key requires mapping the function

$$f(x) = wx \pmod{N}$$

to all elements of the super-increasing private key:

```
sage: a = [2, 3, 7, 25, 67, 179, 356, 819]
sage: w = 57; N = 1600
sage: b = [(x*w)%N for x in a]; b
  [114, 171, 399, 1425, 619, 603, 1092, 283]
```

And given the plaintext, obtaining the ciphertext is also straightforward:

```
sage: m = [1, 0, 1, 1, 0, 0, 1, 0]
sage: c = sum(x*y for x,y in zip(m,b)); c
  3030
```

Decryption requires multiplying by w^{-1} (mod N) and solving the result with the super-increasing sequence:

```
sage: s = si_solve(a,mod(c/w,N)); s
  [2, 7, 25, 356]
sage: [s.count(i) for i in a]
  [1, 0, 1, 1, 0, 0, 1, 0]
```

This list is the recovered plaintext.

The Graham–Shamir version

This is a version of the above knapsack which "hides" the super-increasing property of the original sequence. To create her public key, Alice first generates a sequence a'_k, for which each element is a binary string consisting of three parts:

$$a'_k = (R_k, I_k, S_k).$$

Here R_k and S_k are long random strings, and I_k is a string of length n whose k-th bit (counting from the left) is one and all other bits are zero. Each string S_k has $\log_2(n)$ zeros in its higher bit positions so that the adding up does not cause overflows into the area of the I_k's.

For example, here is a possible sequence of length $n = 8$, where for simplicity each R and S value will also have length 8:

k	R_k	I_k	S_k	a'_k
1	10110010	10000000	00001010	11698186
2	11001011	01000000	00011101	13320221
3	01110100	00100000	00001111	7610383
4	10111111	00010000	00011001	12521497
5	11101001	00001000	00001101	15271949
6	10100110	00000100	00000101	10880005
7	01111110	00000010	00010110	8258070
8	11011000	00000001	00001100	14156044

The a'_k values are just the decimal versions of the binary numbers created by concatenating R_k, I_k and S_k.

Now the sequence a'_i is used to create b_i by the same method as above:

$$b_i = wa'_i \pmod{N}.$$

Bob sends a message to Alice by adding the relevant elements of b_i to create the ciphertext c. Alice computes

$$cw^{-1} \pmod{N}$$

and then extracts from the middle the bits of m.

An example

Suppose a_k is as above, and Alice chooses $N = 123456789$ and $w = 1234567$. Then the sequence b is

95761453, 14070929, 95695894, 108608953, 46901792, 119946424,

79070070, 41722108.

Bob sends the message 10110010 to Alice by adding the appropriate b elements to obtain

$$c := 379136370.$$

Now Alice computes

$$379136370(1234567)^{-1} \pmod{123456789} = 40088136.$$

Converting this to binary and taking the second left-most block of eight bits

$$1001100011\boxed{10110010}01001000$$

reveals the message.

A multiplicative system

Merkle and Hellman also devised a knapsack based on multiplying and discrete logarithms rather than adding. The private key is a list of numbers $(a_1, a_2, \ldots, a_n)$ that are relatively prime. Creation of the public key requires a prime p that satisfies

1. $p > a_1 a_2 \cdots a_n$,

2. $p - 1$ has only small factors.

The system also requires a primitive root r modulo p. The public key consists of $(b_1, b_2, \ldots, b_n)$, where

$$a_i = r^{b_i} \pmod{p}.$$

That is, the public key values are the discrete logarithms of the private key values (mod p). Although discrete logarithms are in general hard to compute, there is an algorithm that is efficient if $p-1$ has small factors. But this is guaranteed by condition 2 above.

Given a message m that is a binary block of length n, encryption is the same as for the additive system; the ciphertext c is

$$c = b_1 m_1 + b_2 m_2 + b_3 m_3 + \cdots + b_n m_n.$$

To decrypt, compute

$$c' = r^c \pmod{p}$$

and solve the equation

$$a_1^{x_1} a_2^{x_2} \cdots a_n^{x_n} = c'$$

for x_i. The system is described in Figure 6.3.

Key generation. The private key consists of a pairwise relatively prime sequence $(a_1, a_2, \ldots, a_n)$, a prime $p > \prod_{k=1}^{n} a_k$ and a primitive root $r \pmod{p}$. The public key is the sequence $(b_1, b_2, \ldots, b_n)$ where $b_i = \log_r a_i \pmod{p}$.

Encryption. For a message $m = (m_1, m_2, \ldots, m_n)$ of n bits, the ciphertext is $c = \sum_{k=1}^{n} m_k b_k$.

Decryption. Factorize $c' = r^c \pmod{p}$. The factors a_j will be the positions of the 1s in m.

FIGURE 6.3: Merkle–Hellman multiplicative knapsack cryptosystem.

An example

Suppose Alice chooses as her private key the first eight prime numbers:

$$(2, 3, 5, 7, 11, 13, 17, 19).$$

She needs to choose a prime p that is greater than their product

$$p > 9699690$$

and for which $p-1$ has only small factors. Such a prime is $p = 10125001$ for which

$$p - 1 = 2^3 3^4 5^6.$$

A primitive root for p is $r = 7$. Since this is one of the a_i values, another primitive root $r = 23$ will be used. Now the discrete logarithms for the private key values can be found to be

$$(7887744, 1807458, 4132406, 4412137, 6599031, 6140658, 4831563, 5414293).$$

This can be checked for a few of them:

$$\begin{aligned} 23^{788744} \pmod{10125001} &= 2 \\ 23^{1807458} \pmod{10125001} &= 3 \\ 23^{4132406} \pmod{10125001} &= 5 \end{aligned}$$

and so on. This list of discrete logarithms is the public key. To send the message $m = (10110010)$ Bob multiplies each public key value by the corresponding message bit and adds them:

$$\begin{aligned} & 7887744(1) + 1807458(0) + 4132406(1) + 4412137(1) \\ + \; & 6599031(0) + 6140658(0) + 4831563(1) + 5414293(0) \\ = \; & 21263850, \end{aligned}$$

thus obtaining the ciphertext c.

To decrypt, Alice computes

$$\begin{aligned} c' &= 23^{21263850} \pmod{10125001} \\ &= 1190. \end{aligned}$$

She now has to find which of the a_i values, when multiplied together, produce c'. In this case it is easy; simply factor c' to obtain

$$c' = (2)(5)(7)(17)$$

and the 1s in the message correspond to these factors:

private key:	2	3	5	7	11	13	17	19
factors of c':	2		5	7			17	
message:	1	0	1	1	0	0	1	0

This system can be easily implemented in Sage; first set up the private key of length 10, and the prime

```
sage: a=primes_first_n(10); a
  [2, 3, 5, 7, 11, 13, 17, 19, 23, 29]
sage: while true:
```

```
....:     e=[randint(0,3) for i in range(10)]
....:     p = prod(x^y for (x,y) in zip(a,e))+1
....:     if is_prime(p):
....:         print p, e
....:         break
....:
10999238251 [1, 1, 3, 3, 1, 2, 0, 0, 1, 0]
```

The next step is to find a primitive root that is not one of the factors of $p-1$:

```
sage: r = 2
sage: f = [x for x,e in factor(p-1)]
sage: while true:
....:     ps = [mod(r,p)^((p-1)/d) for d in f]
....:     if ps.count(1)==0 and f.count(r)==0:
....:         print r
....:         break
....:     else:
....:         r+=1
10
```

Now the creation of the public key is easy with the discrete log function:

```
sage: b = [discrete_log(x,mod(r,p)) for x in a]; b
  [6650271039, 30221027, 4348967212, 8521197821,
   1510363225, 4830189306, 3811470080, 5406152869,
   4034081373, 721824093]
```

Now the ciphertext can be obtained as for the additive knapsack system:

```
sage: m = [randint(0,1) for i in range(10)]; m
  [1, 1, 1, 0, 1, 1, 1, 1, 0, 0]
sage: c = sum(x*y for x,y in zip(m,b)); c
  26587634758
```

Finally decryption

```
sage: c1 = mod(r,p)^c; c1
  1385670
sage: cf = [x for x,e in factor(c1.lift())]; cf
  [2, 3, 5, 11, 13, 17, 19]
sage: [cf.count(i) for i in a]
  [1, 1, 1, 0, 1, 1, 1, 1, 0, 0]
```

which is the correct plaintext.

6.6 Breaking the knapsack

Suppose $V = \{v_1, v_2, \ldots, v_k\}$ is a set of linearly independent vectors in $\mathbb{R}^n$, with $n \geq k$. Define the *lattice* generated by V to be the set of points L for which

$$L = \{a_1v_1 + a_2v_2 + \cdots + a_kv_k, a_i \in \mathbb{Z}\}.$$

That is, the a_i values range over the set of integers. Any set of vectors that generates L, and that is linearly independent, forms a *basis* for the lattice. The set of vectors is said to V *span* L. For example, consider

$$v_1 = [5,1], \quad v_2 = [7,2]$$

in $\mathbb{R}^2$, as shown in Figure 6.4.

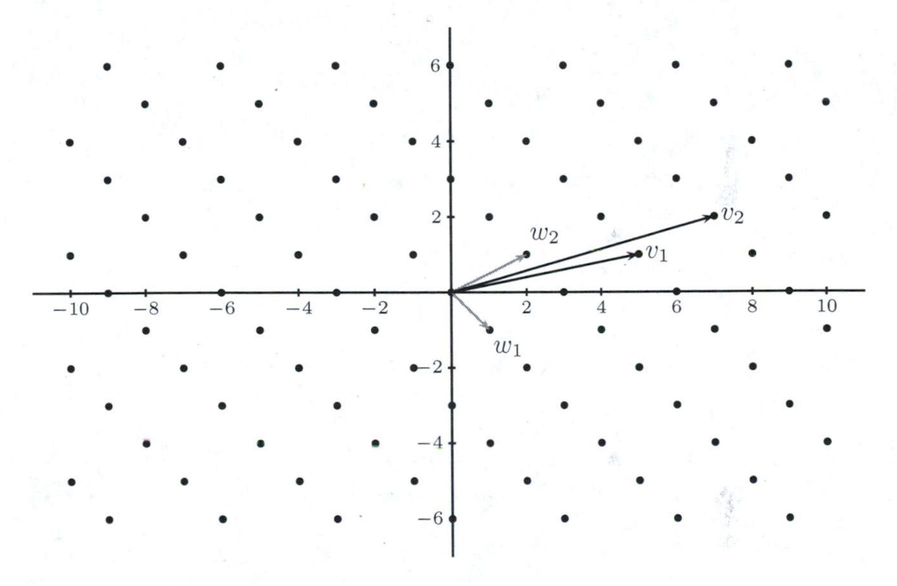

FIGURE 6.4: A lattice and its basis vectors in $\mathbb{R}^2$

The lattice can be plotted in Sage using the following:

```
sage: v1=vector([5,1])
sage: v2=vector([7,2])
sage: pts=[]
sage: for i in range(-15,15):
    for j in range(-15,15):
        tmp=i*v1+j*v2
        if abs(tmp[0])<=10 and abs(tmp[1])<=6:
```

```
            pts.append(tmp)
sage: plot(points(pts))
```

By inspection, it can be seen that the lattice can just as easily be generated by simpler, smaller, vectors, for example the vectors $w_1 = [1, -1]$ and $w_2 = [2, 1]$.

Given a lattice, the idea of *lattice reduction* is to find such small vectors in the lattice. Before discussing this, see how this could be useful in solving the subset sum problem, in particular the previous example, with the public key

$$(114, 171, 399, 1425, 619, 603, 1092, 283).$$

and ciphertext 3030. Start by forming the matrix

$$M = \begin{bmatrix} 2 & 0 & 0 & 0 & 0 & 0 & 0 & 0 & 114 \\ 0 & 2 & 0 & 0 & 0 & 0 & 0 & 0 & 171 \\ 0 & 0 & 2 & 0 & 0 & 0 & 0 & 0 & 399 \\ 0 & 0 & 0 & 2 & 0 & 0 & 0 & 0 & 1425 \\ 0 & 0 & 0 & 0 & 2 & 0 & 0 & 0 & 619 \\ 0 & 0 & 0 & 0 & 0 & 2 & 0 & 0 & 603 \\ 0 & 0 & 0 & 0 & 0 & 0 & 2 & 0 & 1092 \\ 0 & 0 & 0 & 0 & 0 & 0 & 0 & 2 & 283 \\ 1 & 1 & 1 & 1 & 1 & 1 & 1 & 1 & 3030 \end{bmatrix}$$

The rows are the vectors which generate the lattice; thus

$$\begin{aligned} v_1 &= [2, 0, 0, 0, 0, 0, 0, 0, 114] \\ v_2 &= [0, 2, 0, 0, 0, 0, 0, 0, 171] \\ &\vdots \\ v_8 &= [0, 0, 0, 0, 0, 0, 0, 2, 283] \\ v_9 &= [1, 1, 1, 1, 1, 1, 1, 1, 3030]. \end{aligned}$$

The set of integer linear combinations of these v_i form a lattice. Now we know that $114 + 399 + 1425 + 1092 = 3030$, and so

$$v_9 - (v_1 + v_3 + v_4 + v_7) = [-1, 1, -1, -1, 1, 1, -1, 1, 0].$$

Because of the nature of the rows of the original matrix, subtracting the final row from any other row will produce only ± 1 values. And since we are dealing with a subset sum, subtracting the sum of the other rows from the final row will produce a vector, the final value of which will be zero. This is a small vector in the lattice.

Conversely, if such a small vector can be found in the lattice, it is easy to find which rows of the original matrix sum to it since by the construction it will precisely be those rows whose indices are given by the positions of the -1s in the short row.

This can all be done easily in Sage. First enter the original public key and sum, and turn it into a matrix M:

```
sage: s = matrix([[114, 171, 399, 1425, 619, 603, 1092, 283]])
sage: S = 3030
sage: n = len(s.row(0))
sage: M = block_matrix([2*identity_matrix(n),transpose(s),\
      matrix([[1 for i in range(n)]]),matrix([[S]])],\
      subdivide=false)
```

Now a lattice reduction algorithm can be applied to this matrix; a standard algorithm is that developed by Lenstra, Lenstra and Lovász [55], and hence known as the *LLL lattice reduction algorithm* that can be called in Sage as

```
sage: ML = M.LLL(); ML
```

and the reduced matrix is

$$\begin{bmatrix} -1 & 1 & -1 & -1 & 1 & 1 & -1 & 1 & 0 \\ 2 & 0 & -2 & 0 & 0 & 0 & 0 & 2 & -2 \\ -1 & -1 & 1 & 1 & -1 & 3 & -1 & -1 & 0 \\ -2 & -2 & 0 & 0 & 0 & 0 & 0 & 2 & -2 \\ 1 & -1 & -1 & -1 & 1 & 3 & 1 & -1 & 1 \\ 1 & -1 & 1 & -1 & 1 & -1 & -3 & 1 & 1 \\ 0 & 0 & -2 & -2 & 0 & -2 & 0 & -2 & -2 \\ -2 & 0 & 0 & 0 & -4 & 0 & 0 & 0 & 2 \\ -2 & -2 & -4 & 2 & 0 & -2 & -2 & 0 & 1 \end{bmatrix}.$$

There is one short row here, the first one, and the positions of the -1s are the required positions.

6.7 Glossary

Diffie–Hellman problems: Determining g^{xy} from g^x and g^y, or given g^z to determine whether or not $z = xy$.

Discrete logarithm problem: Determining the value x for which $a^x = b \pmod{p}$ given a primitive root a, and b and p.

Knapsack cryptosystem: Any cryptosystem whose security is based on the difficulty of solving a knapsack problem.

Lattice: The set of all vectors $a\mathbf{u}+b\mathbf{v}$ where a and b range across the integers $\mathbb{Z}$, and $\mathbf{u}$ and $\mathbf{v}$ are linearly independent.

Primitive root: A number a whose powers (modulo p) generate all non-zero residues of a prime p.

Probabilistic encryption: An encryption scheme where the same plaintext can have multiple different ciphertexts.

Subset sum problem: A particular knapsack problem which requires finding a subset of a given set of integers, whose sum is equal to a given value.

Exercises

Review Questions

1. What is a primitive root of a prime p?
2. What tests are available to determine if a number is a primitive root?
3. How many primitive roots modulo p are there?
4. What is a probabilistic cryptosystem?
5. Why is the El Gamal cryptosystem considered to be semantically secure?
6. What difficult problem provides the security for the El Gamal cryptosystem?
7. What are the decisional and computational Diffie–Hellman problems?
8. What is the subset sum problem?
9. What is a super-increasing sequence?
10. What is *lattice reduction* and how can it be used to break the additive knapsack cryptosystem?
11. What important cryptographic lesson does the study of knapsack cryptosystems teach?

Beginning Exercises

12. By multiplication, show that 2 is a primitive root modulo 13, but that 3 is not.
13. By multiplication, show that 3 is a primitive root modulo 17, but that 2 is not.
14. Using the results of the last two questions, determine

    (a) $\log_2 8 \pmod{13}$, (b) $\log_2 10 \pmod{13}$, (c) $\log_3 5 \pmod{17}$, (d) $\log_3 2 \pmod{17}$.

15. Suppose p is a prime and a is a primitive root modulo p.

    (a) Show that if $0 < x, y < p$ then $\log_a x + \log_a y = \log_a xy \pmod{p}$.

    (b) Show for any $0 < x < p$ that $k \log_a x = \log_a x^k \pmod{p}$.

16. Produce a table of all powers of 5 modulo 23. Use this table to solve

    $$7^x = 12 \pmod{23}.$$

    Use the logarithm laws from the previous question.

17. Use the baby step, giant step method to compute the logarithms (a) $\log_2 10 \pmod{59}$, (b) $\log_{11} 44 \pmod{83}$, (c) $\log_{31} 84 \pmod{101}$.

18. For the prime $p = 43$, find all the prime factors d of $p - 1$. Put $a = 2$ and determine the values of $a^{(p-1)/d} (\bmod\ p)$ for all d. What does this tell you about whether 2 is a primitive root modulo 43?

    Repeat but with $a = 3$.

19. By applying the above test for $a = 2, 3, 4, 5$ for $p = 97$, show that 5 is the smallest primitive root modulo 97.

20. A system is set up for the use of an el Gamal cryptosystem using the prime number 17 and a primitive root 3. Your private key is the number 10.

    (a) Determine your public key.

    (b) Using your public key, somebody encrypts the message 7, choosing the random value 4. What is the result of this encryption?

    (c) Decrypt the result.

21. Now suppose that the prime is 23, with primitive root 5. Let your private key be 15.

    (a) Determine your public key.

    (b) Encrypt the message 20, using the random value $k = 10$.

    (c) Now decrypt the result.

22. For an El Gamal cryptosystem choose the prime $p = 43$ and primitive root $a = 3$, and pick $A = 20$ as your private key.

    (a) Determine your public key.

    (b) Encrypt the message $m = 10$ using the random key $k = 35$.

    (c) Decrypt the resulting ciphertext.

23. Repeat the above question, but with $p = 97$, $a = 5$, $A = 50$, and with the message $m = 75$ and random key $k = 37$.

24. A poor programmer has set up an El Gamal cryptosystem with $p = 199$, $a = 3$, and to always use the same value of k. You discover that $m = 100$ is encrypted as $(17, 39)$. Decrypt the ciphertexts

    (a) $(17, 198)$, (b) $(17, 119)$.

25. Show that the Diffie–Hellman key exchange protocol in the form described in Section 6.4 is vulnerable to a man-in-the-middle attack.

26. Solve the subset sum problem for each of the following super-increasing sequences and numbers:

    (a) $(7, 9, 21, 44, 85, 169, 337, 676)$, $N = 1041$,
    (b) Same sequence as above, $N = 397$,
    (c) Same sequence again, $N = 957$,
    (d) $(3, 5, 13, 28, 53, 105, 209, 420)$, $N = 715$,
    (e) Same sequence as in (d), $N = 438$, $N = 513$.

27. Let $(2, 7, 11, 34)$ be a super-increasing sequence. Let $N = 60$, and $w = 47$.

    (a) Determine your public key.
    (b) Determine the value of $w^{-1} \pmod N$.
    (c) Somebody sends you the message 100101011101. What are the numbers you receive?
    (d) Decrypt these numbers.

28. Repeat the above question using the super-increasing sequence $(2, 5, 8, 19, 42, 79)$, and choose $N = 200$, $w = 23$. Suppose that the message to be sent to you is 101001011011. Determine the numbers you will receive, and decrypt them.

29. Suppose that the array

k	R_k	I_k	S_k	a'_k
1	111	1000	0010	1922
2	101	0100	0011	1347
3	100	0010	0001	1057
4	110	0001	0011	1555

contains the parameters for a Graham–Shamir system.

    (a) Using $N = 6000$ and $w = 799$ determine the public key b.
    (b) Using the public key, send the message $m = 0110$.
    (c) Using $w^{-1} = 3199 \pmod N$, decrypt the ciphertext.
    (d) Repeat the last two questions for messages 1110 and 1011.

30. A multiplicative knapsack has the private key $a = (2, 3, 5, 7)$, and uses the prime $p = 257$ and primitive root $m = 12$. The public key is $b = (48, 161, 151, 117)$.

    (a) Check the public key; show that $a_i = m^{b_i} \pmod{p}$ for $i = 1, 2, 3, 4$.

    (b) Suppose that the message 0101 is to be sent. Determine the ciphertext.

    (c) Decrypt the ciphertext.

31. Repeat the above question using messages 1100 and 1011.

Sage Exercises

32. The primes $q = 1456501$ and $r = 1470149$ satisfy the fact that $p = 2qr+1$ is prime. Show that 2 is a primitive root modulo p.

33. Suppose that $q = 1206055751377$ and $r = 2160341914259$. Again $p = 2qr + 1$ is prime. Show that 5 is the smallest primitive root modulo p.

34. Use $a = 5$ and p from the previous question as the basis for an El Gamal cryptosystem. Create a public and private key pair. Encrypt the message $m = 10^{20}$ and decrypt the result.

35. The numbers $q = 3^{235}+2$ and $r = 5^{144}+2$ are both prime, and $p = 6qr+1$ is also prime. Find the smallest primitive root a modulo p.

36. Use a and p from the previous question as the basis for an El Gàmal cryptosystem. Encrypt the message

    `What a difficult problem this is!`

    (using the `str2num` procedure) and decrypt the result.

37. The following brute-force program will attempt to solve a subset sum problem with a non-super-increasing sequence:

```
def subsetsum(lst,n):
    for i in powerset(lst):
    if add(i)==n:
        return i
        break
    print "Unsolvable"
```

A simple random list and sum of elements can be created with:

```
sage: n = 10
sage: s = [randrange(10,1000) for i in range(n)]
sage: N = add(random_sublist(s,0.4))
sage: %time subsetsum(s,N)
```

Experiment with longer lists, 15, 20, 25, 30, and more elements. When do you start to see a significant wait in execution time?

38. Now do the same thing with random super-increasing lists:

```
sage: s=[]
sage: for i in range(n):
....:       s+=[sum(s)+randrange(1,10)]
sage: N = add(random_sublist(s,0.4))
sage: si_solve(s,N)
```

Try very large values of n: 100, 200. Does this affect the execution time?

39. Use Sage to follow through the first Merkle–Hellman multiplicative example given on page 133.

40. Use lattice reduction to break the knapsack example given on page 131.

Further Exercises

41. Suppose that a is a primitive root modulo p, and that $\gcd(x, p-1) = 1$. Show that $a^x \pmod{p}$ is another primitive root.

42. Find a primitive root modulo $p = 31$ and use the previous question to find all primitive roots.

Chapter 7

Digital signatures

As mentioned in Chapter 1, along with encryption of messages and other data, digital signatures are fundamental to much modern security. This chapter introduces:

- The basic requirements of a digital signature.
- Digital signature schemes based on the RSA, Rabin and El Gamal cryptosystems.
- The Digital Signature Standard.
- Security issues with the above schemes.

7.1 Introduction

Digital signatures are one of the most important and practical uses of modern cryptography. So much so, that digital signatures have been described as the "killer app" of cryptography. As well as the obvious application of signing documents, digital signatures can also be used elsewhere, for example, for digital cash.

Before starting to discuss digital signatures, it's helpful to first review the concept of signatures. Suppose Alice has decided to sell Bob her car, and she sends him a letter which tells him that he can buy her car for $15,000. To make sure he knows that her offer is genuine, she signs her letter with her name:

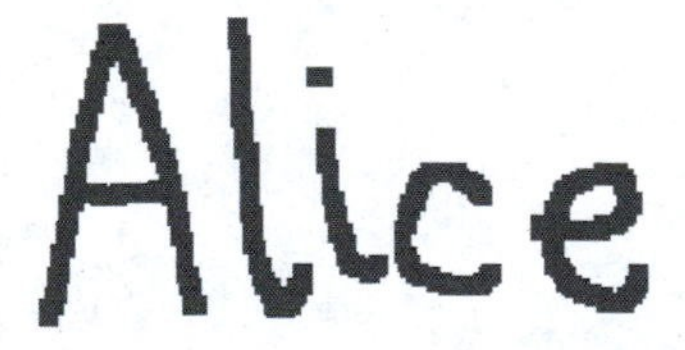

Now when Bob reads the letter he knows that it is indeed from Alice, that is, as long as it's not somebody else forging Alice's signature. How can Bob be absolutely sure that this is the genuine Alice signing her name, and not a forgery? Maybe she really wants to sell her car for $25,000 and this is a forged letter. And suppose this is the genuine letter—what is to stop the bad Mallory from opening the letter in transit, and carefully copying Alice's signature onto other documents?

Here's what might happen. Mallory copies Alice's signature onto another document, say a letter to Charlie saying that he can have her house very cheaply. When Charlie comes to see her with the letter, Alice denies ever signing such a letter. Charlie gets very cross, and demands that Alice fulfill her promise to him, and may even threaten legal action.

This leads to the following requirements for a signature:

1. *authentic*—it convinces the receiver that the message is indeed from Alice;

2. *unforgeable*—nobody else but Alice could have signed the message;

3. *not reusable*—even if the message is sent on to a third party, the signature can't travel with it.

4. The signed message is *unalterable.*

5. The signature can't be *repudiated*—Alice can't later deny that she signed the document.

As discussed above, these really aren't satisfied by a pen-and-ink signature. Look at requirement 4: suppose Alice comes to visit Bob about selling her car, and while Bob's back is turned, Alice quickly changes the "5" in $15,000 to "8." She then claims that Bob must have poor eyesight: misreading $18,000 as $15,000. Or suppose, after sending Bob the letter, Alice changes her mind about either selling her car, or its price. So when Bob comes to see her with the letter, she claims that it must be a forgery: she never signed such a letter.

The use of cryptography allows for some signature schemes that do indeed satisfy the above requirements. One of the consequences of the above requirements is that a signature will in fact change with each message: it will be a function both of the message and of the sender. This is different from pen-and-ink signatures, where a signature (such as Alice's above) does not depend in any way on the message itself. This is what makes forgery possible: a signature can be copied from one message to another.

Some definitions. A *digital signature* is data that is associated both with a given message and with the sender or originator of the message. The process of creating such a digital signature of a message is called *signing.* The signature may be a number, or a string of bits. To *verify* a signature is to determine that the signature is authentic: that it belongs to both the message and the sender. A *digital signature algorithm* is a formal method for producing a digital signature.

7.2 RSA signature scheme

Many digital signature algorithms are based on public-key cryptosystems. And in fact any public-key system can be used as the basis of a signature scheme. In particular, the RSA scheme can be used for a very simple signature system.

Suppose Alice wants to sign a message m to be sent to Bob. She has set up all the necessary information for the RSA cryptosystem: two primes p and q and their product $n = pq$, an integer $e < \phi(n) = (p-1)(q-1)$, and the corresponding $d = e^{-1} \pmod{\phi(n)}$. So Alice's public key is (n, e), and her private key is d.

Signing the message. Alice computes

$$s = m^d \pmod{n}$$

and sends the pair (m, s) to Bob. Note that the message m itself is not encrypted (although it can be if need be). Then the value s is Alice's signature for the message.

Verifying the signature. Bob receives the pair (m, s). He uses Alice's public key to compute

$$m' = s^e \pmod{n}$$

and then compares m with m'. If they are equal, he accepts the signature.

This works because

$$\begin{aligned} s^e &= (m^d)^e \pmod{n} \\ &= m^{de} \pmod{n} \\ &= m \pmod{n}. \end{aligned}$$

Note that this signature scheme is really RSA "in reverse": Alice uses her *private key* to form the signature, and Bob uses her public key to verify it. If Alice wanted to encrypt the message to Bob, she would use Bob's public key to send him the message, and he would use his private key to decrypt it.

An example

Suppose Alice's primes are $p = 853$ and $q = 929$, with $n = pq = 792437$. She chooses $e = 17$ and so $d = 511601$. To sign the message $m = 500000$ she computes

$$\begin{aligned} s &= m^d \pmod{n} \\ &= 500000^{511601} \pmod{792437} \\ &= 659911. \end{aligned}$$

This will be the signature of the message. She sends $(500000, 659911)$ to Bob.

Bob now computes

$$\begin{aligned} m' &= s^e \pmod{n} \\ &= 659911^{17} \pmod{792437} \\ &= 500000. \end{aligned}$$

Since this is equal to the message, Bob accepts the signature and concludes that the message is indeed from Alice.

Checking the signature requirements

The mathematics of the scheme has been shown to be sound, but does this scheme actually satisfy the requirements of a digital signature scheme? Consider the five requirements discussed previously.

Authentic. Only Alice has the private key corresponding to her public key. Any other value of d would produce a signature s for which the public key would not work. So Bob may be sure that the message is only from Alice.

Unforgeable. Since signing the message requires Alice's private key, nobody else can sign the message.

Not reusable. The signature can't be attached to another message, for a different message would require a different signature to be verifiable.

Unalterable. Once a message has been signed, it can't be altered. As the signature s is generated from the message m itself, any different message would have a different signature. Consequently, if Alice tries to change the message m to M after it has been signed with s, Bob will find that $m' \neq M$ and so will not verify the signature.

Non-repudiation. Since only Alice can sign a message with her private key, she can't deny signing the message. The best she can do is to conveniently "lose" her private key. But then nobody else can have signed the message.

Problems with this scheme

The RSA signature scheme is vulnerable to *existential forgery*, where it is possible to create a legitimate message and signature without knowing the private key. Suppose that (m_1, s_1) and (m_2, s_2) are two pairs of signed messages. Then the pair

$$(m', s') = (m_1 m_2, s_1 s_2) \pmod{p}$$

will be a legitimately signed message, by the multiplicativity of the modulus operation.

However, assuming that p and q are chosen so that $n = pq$ is hard to factor, then this scheme is secure against forgery of signatures to individual messages.

Another problem is that the signature is going to be roughly the same size as the message. And in fact if m is much smaller than n, it may be that the signature is larger than m. Using the numbers above for example, if $m = 10$, then $s = 422453$. This means doubling the amount of information Alice must send to Bob. Another problem is that modular powers of a large number like the message are slow to compute. One solution is to sign, not the message itself, but a sort of "fingerprint" of the message. Such fingerprints are provided by *cryptographic hash functions*, which produce a fixed size block of bits from an arbitrary length input. Hash functions will be discussed in more detail in Chapter 11.

Here is how the RSA scheme can be used with a hash function. First Alice picks p and q large enough to be cryptographically secure, but not as large as may be required for encrypting an entire message.

Signing the message. Alice computes a hash $h = H(m)$ of the message. This will be only a few bytes (say 20). She then computes

$$s = h^d \pmod{n}$$

and sends the pair (m, s) to Bob. As above, the message may remain as plaintext, and s is the signature.

Signing also protects against existential forgery. If

$$s_1 = (H(m_1))^d \pmod{n}$$
$$s_2 = (H(m_2))^d \pmod{n}$$

then it is not possible to construct a signed message $(m', s') = (m_1 m_2, s_1 s_2) \pmod{p}$ as above because in general

$$H(m_1 m_2) \neq H(m_1)H(m_2).$$

Verifying the signature. Bob receives the pair (m, s). He first computes the hash $h = H(m)$, then uses Alice's public key to compute

$$h' = s^e \pmod{n}$$

and then compares h with h'. If they are equal, he accepts the signature. This method will produce smaller signatures, and will be fast.

It is convenient to choose e to be small in relation to d. The signature will be computed only once for each message, but it may need verifying many times.

7.3 Rabin digital signatures

The Rabin signature scheme starts with similar parameters as in the RSA scheme: two primes p and q, and their product $N = pq$. Like the RSA scheme, the Rabin scheme is encryption "worked backwards."

Signing the message. Alice computes a square root s of the message m (recall that this is easiest if both p and q are congruent to 3 mod 4):

$$s^2 = m \pmod{N}$$

and sends the pair (m, s) to Bob.

Verifying the signature. Bob receives the pair (m, s). He squares s to compute

$$m' = s^2 \pmod{N}$$

and then compares m with m'. If they are equal, he accepts the signature.

Bob may be sure that the message is indeed from Alice, as a square root of N is computationally infeasible to compute unless the factorization $N = pq$ is known. But the factorization is Alice's private key.

As with the RSA scheme, the signature s is the same size as the message. This size can be reduced by signing a hash of the message, and if necessary using smaller parameters.

An example

Alice chooses primes $p = 859$ and $q = 947$ (both of which are congruent to 3 mod 4), and so $N = pq = 813473$ is her public key. Given the message $m = 500001$ she computes a square root by the following steps.

1. Using the extended Euclidean algorithm she finds x and y such that

   $$xp + yq = 1.$$

   These values are $x = -226$ and $y = 205$.

2. She then computes

   $$\begin{aligned} m_p &= m^{(p+1)/4} \pmod{p} \\ &= 500001^{215} \pmod{859} \\ &= 731 \end{aligned}$$

   $$\begin{aligned} m_q &= m^{(q+1)/4} \pmod{q} \\ &= 500001^{237} \pmod{947} \\ &= 224. \end{aligned}$$

3. Then a square root is given by

$$s = xpm_q + yqm_p \pmod{N} = 809909.$$

Bob verifies this by computing

$$\begin{aligned} s^2 \pmod{N} &= 809909^2 \pmod{813473} \\ &= 500001. \end{aligned}$$

Since this is the message, the signature is verified.

Note that this scheme is restricted to messages m for which square roots (mod N) exist. The message 500000, for example, could not be signed with the parameters above because 500000 does not have a square root mod $N = 813473$. As with the RSA scheme, efficiency will be improved by signing a hash of the message, rather than the message itself. However, with this scheme, signature verification is very fast, as it involves only a single modular squaring. Even though the generation of the signature takes some work, it will in general need to be done only once.

If $x^2 = m \bmod N$ has a solution, then m is said to be a *quadratic residue* of N which were introduced in chapter 2. Recall that one way to check if a number is a quadratic residue is by using *Euler's criterion*, which states that if p is an odd prime and a is relatively prime to p, then

$$\begin{aligned} a^{(p-1)/2} &= 1 \pmod{p} \quad \Rightarrow \quad a \text{ is a quadratic residue of } p \\ a^{(p-1)/2} &= -1 \pmod{p} \quad \Rightarrow \quad a \text{ is quadratic non-residue of } p. \end{aligned}$$

An important result is that if $N = pq$ with both p and q prime, then m is a quadratic residue of N if and only if it is a quadratic residue of both p and of q.

Recall that the *Legendre symbol* it is defined for a prime p and any number a as

$$\left(\frac{a}{p}\right) = \begin{cases} 1 & \text{if } a \text{ is a quadratic residue of } p, \\ -1 & \text{if } a \text{ is a quadratic non-residue of } p, \\ 0 & \text{if } a \text{ is a multiple of } p. \end{cases}$$

Euler's criterion is one way of computing the Legendre symbol, but it is quite slow and inefficient, and there are much faster methods.

For example, given the values $a = 500000$, $p = 859$, $q = 947$ from above, then

$$\left(\frac{500000}{859}\right) = -1$$

$$\left(\frac{500000}{947}\right) = -1$$

which shows that 500000 is not a quadratic residue of either p or q, and so not a quadratic residue of their product. But if $a = 500001$:

$$\left(\frac{500001}{859}\right) = 1$$
$$\left(\frac{500001}{947}\right) = 1$$

which shows that 500001 is a quadratic residue of both p and q, and so a quadratic residue of their product.

7.4 The El Gamal digital signature scheme

The El Gamal scheme is important in its own right, but it has also been used as the basis of the *Digital Signature Algorithm*, the DSA, which has been adopted as a standard.

Recall that El Gamal encryption requires a prime p, a primitive root a of p, and as a private key Alice chooses an integer $A < p$. Then her public key is $B = a^A \pmod{p}$. To encrypt a message m to Alice, Bob chooses an integer $k < p$ and sends the pair $(C_1, C_2) = (a^k \bmod p, B^k m \bmod p)$. Then decryption is performed by using $C_1^A = (a^k)^A = (a^A)^k = B^k$ (all mod p). Alice can do this since she knows her private key A. Then $m = C_2(B^k)^{-1} \pmod{p}$.

These ideas can be worked into a signature scheme.

Signing the message. Alice chooses $k < p-1$ and for which $\gcd(k, p-1) = 1$. She then computes

$$r = a^k \pmod{p}$$
$$s = k^{-1}(m - Ar) \pmod{p-1}.$$

She then sends (m, r, s) to Bob.

Verifying the signature. Bob receives the triple (m, r, s). He first checks that $r < p$; if not, he does not accept the signature. He then uses Alice's public key (p, a, B) to compute

$$v_1 = B^r r^s \pmod{p}$$
$$v_2 = a^m \pmod{p}.$$

If $v_1 = v_2$, he accepts the signature. As with RSA and Rabin, it is more efficient to sign a hash $H(m)$ rather than the message itself.

An example

The prime is $p = 859$, with primitive root $a = 2$. As a private key Alice chooses $A = 100$ and so her public key is $B = 2^{100} \pmod{859} = 712$. Suppose the message is $m = 500$. Alice picks $k = 199$ and confirms that $\gcd(k, p-1) = \gcd(199, 858) = 1$. Then she computes

$$\begin{aligned} r &= a^k \pmod{p} \\ &= 2^{199} \pmod{859} \\ &= 67 \end{aligned}$$

$$\begin{aligned} s &= k^{-1}(m - Ar) \pmod{p-1} \\ &= 199^{-1}(500 - 100 \times 67) \pmod{858} \\ &= 400. \end{aligned}$$

She sends $(500, 67, 400)$ to Bob as the message with signature. Bob now computes

$$\begin{aligned} v_1 &= B^r r^s \pmod{p} \\ &= 712^{67} 67^{400} \pmod{859} \\ &= 175 \end{aligned}$$

$$\begin{aligned} v_2 &= a^m \pmod{p} \\ &= 2^{500} \pmod{859} \\ &= 175. \end{aligned}$$

Since $v_1 = v_2$ Bob accepts the signature.

Why it works

Since from Fermat's theorem $x^{p-1} = 1 \bmod p$, then in general

$$x^m = x^n \pmod{p}$$

if $m = n \bmod p - 1$. Since $s = k^{-1}(m - Ar) \bmod p - 1$, then

$$m = sk + Ar \bmod p - 1$$

which means that

$$\begin{aligned} a^m &= a^{sk+Ar} \bmod p \\ &= (a^k)^s (a^A)^r \bmod p \\ &= r^s B^r \bmod p \end{aligned}$$

so that $v_2 = v_1$.

For security with this system, Alice must keep her private key A safe. Anybody who knows the value of A can sign a message "from Alice."

It is also important that a different k is used for signing each message. Suppose that Alice signs two messages m_1 and m_2 using the same value of k. The r values will be the same for both messages (because it depends on only a and k), but the s values will be different:

$$s_1 = k^{-1}(m_1 - Ar) \pmod{p-1},$$
$$s_2 = k^{-1}(m_2 - Ar) \pmod{p-1}.$$

These congruences can be written in the form:

$$s_1k = m_1 - Ar \pmod{p-1},$$
$$s_2k = m_2 - Ar \pmod{p-1}.$$

Subtracting these last congruences produces

$$(s_1 - s_2)k = m_1 - m_2 \pmod{p-1}. \tag{7.1}$$

To solve this congruence for k, note that in general the equation

$$\alpha k = \beta \pmod{n}$$

has solutions if and only if $g = \gcd(\alpha, n)$ is a divisor of β. In this case there are g solutions

$$\frac{\beta}{g}x_0 + t\frac{n}{g}, \quad t = 0, 1, \ldots, g-1$$

where

$$x_0 = (\alpha/g)^{-1} \pmod{n/g}.$$

So there will be

$$g = \gcd(s_1 - s_2, p-1)$$

solutions to equation 7.1. In general g will be a small value, and so there will be only a small number of possible values for k. The correct value of k can now be found by computing

$$a^k \pmod{p}$$

until the value r is reached.

Having found k, next solve

$$Ar = m_1 - ks_1 \pmod{p-1}$$

for A; there will be $\gcd(r, p-1)$ solutions. Now a^A can be computed for each solution until B is obtained. The system has been completely broken, and Alice's private key A is now public.

One of the verification steps is to check that $r < p$. If Bob accepts signatures for which $r \geq p$, then a signature can be forged: r and s values can be created for which the corresponding v_1 and v_2 are equal. This can be done by starting with a message m for which $\gcd(m, p-1) = 1$ and a legitimate signature (r, s) for m. Then given any other message m' a signature (r', s') can be forged by first defining

$$u = m'm^{-1} \pmod{p-1}$$

and then setting

$$s' = su \pmod{p-1}$$

and choosing r' to satisfy both

$$\begin{aligned} r' &= ru \pmod{p-1}, \\ r' &= r \pmod{p}. \end{aligned}$$

The value of r' can be computed using the Chinese remainder theorem. But this r' will be greater than p.

To see that this forgery works, note that by definition of u that

$$m' = um \pmod{p-1}.$$

Then

$$\begin{aligned} a^{m'} &= a^{mu} \pmod{p} \\ &= (a^m)^u \pmod{p} \\ &= (B^r r^s)^u \pmod{p} \\ &= B^{ru} r^{su} \pmod{p}. \end{aligned}$$

In order for r' and s' to be used for verifying the signature of m', they must satisfy the equations given above. For more information on forging El Gamal signatures see Bleichenbacher [7].

An example using Sage

Using Sage's types, the El Gamal signature scheme can be implemented very neatly. Here's the above example:

```
sage: p = 859
sage: a = mod(2,p)
```

Defining a to be of type modulo p means that all further calculations with a will automatically return a value modulo p. Now to create a private key and compute the public key:

```
sage: A = 100
sage: B = a^A;B
  712
```

Because of the definition of a, there is no need to use the `power_mod` command. Now create a message and sign it:

```
sage: m = 499
sage: k = mod(199,p-1)
sage: gcd(k,p-1)
  1
sage: r = a^k;r
  67
sage: s = (m-A*r.lift())/k;s
  663
```

Here k is defined to be of type modulo $p-1$, and because r is already of type modulo p, the "`lift()`" operation "lifts" r out of that integer ring.

Now verification is very easy:

```
sage: v1 = a^m;v1
  517
sage: v2 = B^r*r^s;v2
  517
sage: v1==v2
  True
```

Note that the verification could be done in one simple command

```
sage: a^m==B^r*r^s
  True
```

Here's how to attempt a forgery, with the message $m = 600$:

```
sage: m1 = mod(600,p-1)
sage: u = (m1/m).lift(); u
  744
sage: s1 = s*u; s1
  780
sage: r = r.lift()
sage: r1 = crt([r*u,r],[p-1,p]); r1
  14670
```

So $(r', s') = (14670, 780)$ is the forged signature for $m' = 600$. Assuming that $r' < p$ is not checked, then:

```
sage: a^m1==B^r1*r1^s1
  True
```

7.5 The Digital Signature Standard

The *Digital Signature Standard* (DSS), also known as the *Digital Signature Algorithm* (DSA) has (as its name suggests) been adopted as a standard by the American National Institute of Standards and Technology (NIST). It is based in part on the El Gamal scheme, and is similar to it.

To set up the parameters for this scheme, Alice must do the following.

1. Choose primes p and q for which p is at least 512 bits (about 154 digits), and q is 160 bits (about 48 digits) and a factor of $p-1$.

2. Find a value $h < p$ for which

$$h^{(p-1)/q} \neq 1 \pmod{p}$$

   and set

$$g = h^{(p-1)/q} \pmod{p}.$$

   It follows that $g^q = 1 \pmod{p}$. If a primitive root of p is known, then h can be chosen to be this primitive root.

3. Choose a random value A for which $1 \leq A < q-1$ and calculate $B = g^A \pmod{p}$.

4. Then her public key consists of the four values (p, q, g, B), and her private key is A.

Signing the message. This is very similar to El Gamal. Given a message m, Alice first chooses (at random) a value k for which $0 < k < q-1$. She then computes:

$$r = (g^k \pmod{p}) \pmod{q}$$
$$s = k^{-1}(m + Ar) \pmod{q}$$

and sends (m, r, s) to Bob.

Verifying the signature. Using Alice's public key, Bob computes

$$
\begin{aligned}
x &= s^{-1}m \pmod{q},\\
y &= s^{-1}r \pmod{q},\\
v &= (g^x B^y \pmod{p}) \pmod{q}.
\end{aligned}
$$

If $v = r$ then he accepts the signature.

DSS is stronger than the El Gamal system, as k is harder to determine from r because of the extra reduction mod$q$. It is also quicker and more efficient because the modular arithmetic is done modulo q, which leads to shorter signatures.

An example

Suppose $p = 1031$ and $q = 103$ (these don't satisfy the length criteria, but q is a factor of $p - 1$). The smallest primitive root of p is $a = 14$, from which $g = a^{(p-1)/q} \pmod{p} = 320$. Choose $A = 70$ so $B = g^A \pmod{p} = 48$.

Suppose the message is $m = 500$, and $k = 25$ is chosen at random. Then

$$
\begin{aligned}
r &= (g^k \pmod{p}) \pmod{q}\\
&= (320^{25} \pmod{1031}) \pmod{103}\\
&= 198 \pmod{103}\\
&= 95
\end{aligned}
$$

$$
\begin{aligned}
s &= k^{-1}(m + Ar) \pmod{q}\\
&= 25^{-1}(500 + 70 \times 95) \pmod{103}\\
&= 80.
\end{aligned}
$$

So the triple $(500, 95, 80)$ is sent to Bob.

Bob now computes

$$\begin{aligned} x &= s^{-1}m \pmod{q} \\ &= 80^{-1}500 \pmod{103} \\ &= 32 \end{aligned}$$

$$\begin{aligned} y &= s^{-1}r \pmod{q} \\ &= 80^{-1}95 \pmod{103} \\ &= 72 \end{aligned}$$

$$\begin{aligned} v &= (g^x B^y \pmod{p}) \pmod{q} \\ &= (320^{32}48^{72} \pmod{1031}) \pmod{103} \\ &= 198 \pmod{103} \\ &= 95. \end{aligned}$$

Since $v = r$, he accepts the signature.

Why it works

From the definition of s:

$$m = (-Ar + ks) \pmod{q}.$$

Multiplying through by s^{-1} produces

$$s^{-1}m = (-Ars^{-1} + k) \pmod{q}.$$

This last equation can be written

$$\begin{aligned} k &= s^{-1}m + Ars^{-1} \pmod{q} \\ &= x + Ay \pmod{q}. \end{aligned}$$

Then

$$\begin{aligned} r &= (g^k \pmod{p}) \pmod{q} \\ &= (g^{x+Ay} \pmod{p}) \pmod{q} \\ &= (g^x (g^A)^y \pmod{p}) \pmod{q} \\ &= (g^x B^y \pmod{p}) \pmod{q} \\ &= v. \end{aligned}$$

DSS in Sage

As with the El Gamal system, the DSS is very easily implemented in Sage. Here's the above example, starting with the setting up:

```
sage: p = 1031
sage: q = 103
sage: a = mod(14,p)
sage: g = a^((p-1)/q); g
  320
sage: A = 70
sage: B = g^A; B
  48
```

Now create the message and sign it:

```
sage: m = 500
sage: k = 25
sage: r = mod(g^k,q); r
  95
sage: s = (m+A*r)/k; s
  80
```

Now the signature is easily verified:

```
sage: x = m/s; x
  32
sage: y = r/s; y
  72
sage: v = mod(g^x*B^y,q)
  95
sage: v==r
  True
```

Choosing values for use in the DSS

To be secure, the primes p and q need to be large. It is too inefficient to choose a large prime p first, and then try to factor $p-1$ in the hopes of obtaining a large enough factor q. It is better to first choose a prime q, and then use a random multiplier c of large enough size so that $cq+1$ is prime.

For example, start by finding p and q of about 1024 and 160 bits respectively. First q:

```
sage: q = next_prime(randint(2^160,2^161)); q
  1758790707451389480029046050128144234163788061207
```

(Your value may well be different.) Now, p must have $1024-160=864$ more bits than q, which means the multiplier c needs to be comparable to 2^{864}. So just multiply by random values until a prime is obtained:

```
sage: while true:
....:     p=randint(2^864,2^865)*q+1
....:     if is_prime(p):
....:         print p
....:         break
```

The value p (which is not necessary to display) has 309 digits, and hence is large enough:

```
sage: len(bin(p))
  1027
```

The next requirement for the DSS is the value g. All that is required is that

$$h^{(p-1)/q} \neq 1 \pmod{p}.$$

```
sage: h = mod(randrange(1,p),p)
sage: h^((p-1)/q)<>1
  True
```

So choose h randomly until one is found for which the modular power is not equal to one. Having found an appropriate h, g can be computed, and the public and private keys produced:

```
sage: g = h^((p-1)/q)
sage: A = randrange(1,p)
sage: B = g^A
```

Since g, A, B are all very large numbers, comparable in size to p, we won't display them. But at this stage there is all the necessary material for signing a message, and for providing the information needed to verify a signature.

For more information on choosing values, as well as a great deal more information about the DSS, see the official Standard [66].

7.6 Glossary

Digital signature. An attachment to a message that, among other things, verifies the sender of the message.

Digital Signature Scheme. Any scheme that includes both signing and verification processes.

Signing. The process of creating a digital signature.

Verification. The process of verifying the sender.

Exercises

Review Questions

1. What are the five requirements that a digital signature should satisfy?

2. Which of these requirements are satisfied by a pencil and paper signature?

3. What is *existential forgery*?

4. What does it mean to digitally "sign" a message?

5. What does it mean to "verify" a signature?

6. What are some of the reasons for using a message hash in a signature scheme?

7. Why is the DSA considered to be a more powerful signature scheme than the El Gamal scheme?

8. Is the Rabin scheme vulnerable to existential forgery?

Beginning Exercises

9. Suppose that an RSA system is used for which $p = 11$ and $q = 13$. Alice chooses $e = 7$ as her public key, so that $d = 103$ is her private key. Verify that the following are valid message and signature pairs:

   (a) $(99, 44)$, (b) $(50, 106)$, (c) $(75, 36)$,
   (d) $(2, 63)$, (e) $(120, 120)$, (f) $(15, 141)$.

10. Suppose now that the values are $p = 41$, $q = 47$, and Alice's public key is $e = 3$, so that her private key is $d = 1227$. Verify that the following are valid message and signature pairs:

    (a) $(1000, 10)$, (b) $(750, 1577)$, (c) $(2, 538)$,
    (d) $(1500, 546)$, (e) $(33, 1638)$, (f) $(972, 1211)$.

11. For a Rabin cryptosystem, Alice chooses $p = 43$ and $q = 47$ (both of which are equal to 3 mod 4). Verify that the following are valid message and signature pairs:

    (a) $(100, 225)$, (b) $(1999, 1368)$, (c) $(6, 1541)$,
    (d) $(900, 1991)$, (e) $(64, 1841)$, (f) $(572, 625)$.

12. Show that with the values of p and q in the previous question, the message $m = 2$ cannot be signed by this method.

13. Show that the Rabin system is vulnerable to existential forgery, and that hashing protects against it.

14. For an El Gamal system, the prime $p = 71$ is chosen with primitive root $a = 7$. Alice chooses $A = 10$ as her private key, so that $B = 45 = a^A \bmod p$ is her public key. Using $k = 3$ each time, show that the following triples (m, r, s) are valid messages and signatures:

    (a) $(20, 59, 20)$ (b) $(50, 59, 30)$ (c) $(2, 59, 14)$

15. Repeat the above question, but with values $p = 1009$, $a = 11$, $A = 500$, $B = 96$ and $k = 5$:

    (a) $(2, 620, 698)$, (b) $(1000, 620, 696)$, (c) $(591, 620, 211)$.

16. With the El Gamal values $p = 859$, $a = 2$, Alice uses her private key A and the same value of k to sign $m_1 = 500$ as $(r, s_1) = (67, 400)$, and $m_2 = 600$ as $(r, s_2) = (67, 698)$. She makes available her public key $B = 712$.

    (a) Determine $\gcd(s_1 - s_2, p - 1)$.
    (b) Find all the solutions k of the equation $(s_1 - s_2)k = (m_1 - m_2) \pmod{p - 1}$.
    (c) For each value of k, determine $A^k \pmod{p}$. Which value of k is the correct one (which produces r)?
    (d) Determine $\gcd(r, p - 1)$.
    (e) Find all the solutions A to the equation $Ar = m_1 - ks_a \pmod{p-1}$.
    (f) For each value of A, compute $a^A \pmod{p}$. Which value is the correct one, which produces B?

17. Suppose the DSS is to be applied with the same parameters as in the example given in section 7.5. Sign the message $m = 1000$.

    Note that $r = 95$ (since g and k haven't changed), and use $k^{-1} \bmod q = 33$.

18. Using the values of r and s from the previous question, verify the signature.

19. Using the same parameters, verify the following messages and signatures:

    (a) $(2, 95, 23)$, (b) $(750, 95, 90)$, (c) $(591, 95, 96)$.

Sage Exercises

20. For an El Gamal signature system, the prime $p = 322871$ and primitive root $a = 17$ are chosen. Suppose Alice's public key is $B = 392$. Verify the following messages and signatures:

    (a) $(100000, 28762, 98402)$, (b) $(99999, 220177, 155911)$

21. Using the same values as in the previous question, Alice uses the same value of k to produce the following signatures:

    $(44444, 193988, 186082)$, $(77777, 193988, 277985)$.

    Using the method described in question 16, find the values of k and the private key A.

22. A DSS Scheme is set up with $p = 13295704797979873$, $q = 576749$, $h = 10$ so $g = h^{(p-1)/q} = 10445699122117129 \pmod{p}$.

    Suppose Alice chooses $A = 3^{30}$ as her private key, so that $B = g^A = 7103123715399196 \pmod{p}$.

    (a) Sign the message $m = 10^{16}$ using the "random" value $k = 123456789$.

    (b) Verify the resulting signature.

23. A simple hash function can be set up as follows: let p, q be two large primes, and a a number with large order in the ring $\mathbb{Z}/\mathbb{Z}_{pq}$. If m is the numerical value of a message, then its hash is defined as

    $$h(m) = a^m \pmod{pq}.$$

    (a) Let $p = 330877$, $q = 732799$, and $a = 17$. Convert the plaintext "`Please send me all your money by next Tuesday.`" to a numerical value using the `str2num` command, and compute its hash.

    (b) Sign the hash using all the DSS parameters given in question 22.

    (c) Verify the signature.

24. For an RSA signature scheme, I provide the public key (n, e), where

    $$n = 2^{137} - 1, \quad e = 17.$$

    (This value n has two large prime factors.) Use my public key to verify my signature of the following message:

    ```
    This is my text.
    59813426890666400174073214601409067068836
    ```

    Do this by using the `str2num` command to turn the message into an integer m, and so show that $s^e = m \pmod{n}$.

25. Suppose you intercept the following message and RSA signature, with the parameters from the previous question:

```
150 dollars
```

$$56840167481593713576269217621738921774093$$

Show how to use the `factor` command to break this system, and send on the amended message

```
250 dollars
```

with a digitally forged signature.

26. Now try with the public key

$$n = (6^{67} - 1)/5, \quad e = 17$$

to verify the RSA signature:

```
Please feed my dog!
```

$$62685679920085174734947219739407165890608700377299 2$$

27. For a Rabin signature scheme, I provide the public key $N = (7^{47} - 1)/6$, which I know can be factorized into two large primes.

Check the following message and signature:

```
Arrive Thursday.
```

$$77982630048255907681798126464740755348$$

28. For an El Gamal signature scheme, I choose the next prime after 2^{150} which has a primitive root $a = 2$. My public key is

$$B = 3839098709688075864806587944986962459792644 24$$

Verify the signature

```
Leave AT ONCE!
```

$$40975520084777535329086035006149371171948219 0$$
$$62258521080960506724947009299222876537609154 2$$

29. For a DSS signature, choose p to be the next prime after 2^{170}, and $q = 1434415054685906 96209$. Verify that q is a divisor of $p - 1$.

 (a) A primitive root of p is $a = 3$. Use this primitive root to determine

$$g = a^{(p-1)/q} \bmod p.$$

(b) The public key value is

$$B = 262500069131803851447159605077073651476979504412692$$

Now using these values, verify this signature:

```
Now's your chance!
138670423972596752241
126414342335575050310
```

Further Exercises

30. Show that the El Gamal system is vulnerable to existential forgery; if y satisfies $\gcd(y, p-1) = 1$ and (m, r, s) is a signed message, then if

$$r' = B^y a^x \pmod{p}$$
$$s' = -ry^{-1} \pmod{p-1}$$
$$m' = su \pmod{p-1}$$

the pair (r', s') is a legitimate signature for m'.

31. Show how the use of a hash function protects against existential forgery.

32. A digital signature algorithm based on the Diffie–Hellman key exchange protocol is given by both Stallings [83] and Wagstaff [90]. This is a very slightly simplified version of it. All users share a prime p and a primitive root a of p. Alice chooses $A < p$ as her private key and publishes $B = a^A \pmod{p}$ as her public key.

Messages $m < p$ can be signed only if $\gcd(m, p-1) = 1$. A message m is signed by computing

$$s = a^{Am^{-1} (\bmod p-1)} \pmod{p}$$

and the signature is verified by showing that

$$B = s^m \pmod{p}.$$

(a) Prove the validity of this method.

(b) Show that for an arbitrary signable message m' a user can easily forge Alice's signature.

33. Suppose that Alice and Bob wish to both encrypt and sign messages using RSA. Alice creates primes p_A and q_A and publishes their product n_A and public key e_A; her private key being d_A. Similarly Bob produces primes p_B and q_B with product n_B and public key e_B; his private key being d_B.

(a) How does Alice encrypt and sign a message to Bob in a way that the signature can be verified and the message recovered if $n_A > n_B$?

(b) What if $n_A < n_B$?

Chapter 8

Block ciphers and the data encryption standard

Block ciphers, which operate on an n-bit block of plaintext (the length of which is a function of the cipher), are some of the most powerful, fastest, and most used cryptosystems in existence. Unlike public-key systems, which obtain their security from a well-known hard mathematical problem, block ciphers are secret key systems based on bit operations, and obtain their strengths from a mixture of non-linear operations, such as substitutions, and permutations.

This chapter introduces the block ciphers, and investigates one of the historically most important such ciphers. In particular, the following topics will be discussed:

- The basic concept of a block cipher.
- Rounds and key schedules.
- Modes of encryption, which are ways in which a block cipher can be used to encrypt a ciphertext longer than a single block.
- The Feistel construction, which underlies many modern block ciphers.
- The Data Encryption Standard (DES) which was in use for about 30 years up to 2000, and has been intensively researched and analyzed.
- A simplified DES which has been designed for hand calculation.
- Methods of investigating the security of block ciphers, such as the avalanche criteria and measures of strength against differential cryptanalysis.
- Some "lightweight" ciphers, designed to run in restricted hardware or memory.

8.1 Block ciphers

A *block cipher* may be considered as two related functions, encryption and decryption, each of which takes two inputs. The inputs to the encryption function are a plaintext block and a key; the inputs to the decryption function are the ciphertext block, and the same key. Both plaintext and ciphertext blocks will have the same length.

Denote the encryption function by

$$E(p, k)$$

and the decryption function by

$$D(c, k).$$

A block cipher should satisfy the following properties:

1. The plaintext should be recoverable from the ciphertext; thus:

   $$D(E(p, k), k) = p.$$

   In other words, decryption and encryption are inverse functions.

2. Given a ciphertext and a complete working knowledge of both functions, there should be no feasible way of recovering the plaintext. (If the plaintext were easy to recover, the cipher would not be secure.)

3. The functions should be fast and efficient in both hardware and software.

Block ciphers may be distinguished by the length of the blocks of plaintext and ciphertext (this is called the *block size*), and the length of the key. One of the basic aspects of block cipher security is that the key be at least 128 bits (but see below); ciphers with smaller keys are vulnerable to brute force attacks.

An example

A simple block cipher previously investigated is the matrix or Hill cipher discussed in Chapter 3. In this case the block size was the number of rows and columns of the matrix used. For this cipher

$$E(p, k) = Mp$$

and

$$D(c, k) = M^{-1}c$$

where $M = k$ is the (matrix) key. Recall that this cipher is not particularly strong in that it is vulnerable to a known plaintext attack, so it does not satisfy condition 2 above. However, there is no limit on key size or block size: the key matrix can be of any size.

Note also that for the matrix cipher the plaintext had sometimes to be padded to make sure the last block was of the appropriate size. Also, partial blocks can't be encrypted: before an encryption, an entire block is required.

8.2 Some definitions

To discuss block ciphers and their security and use, some basic terminology will be needed.

Rounds and round keys. Almost all block ciphers work by applying some sort of mixing function to the plaintext and key, and then applying that function to the output, and so on. Each application of the mixing function is called a *round.* The input to each round consists of the block obtained from the previous round, and a sequence of bits obtained in some way from the original key. These are the *round keys.* The method by which the round keys are determined from the original key is called the *key schedule.* A general schema is shown in Figure 8.1. It is necessary to choose the number of rounds to obtain a good level of security, but not so many as to slow down the encryption. It is also necessary for the round keys all to be different: some block ciphers define their round keys in such a way that for certain keys, some or all of the round keys are the same. This produces a loss of security.

Confusion and diffusion. These terms describe how well a cipher mixes the bits from the plaintext and the key. Informally, diffusion describes how a change in the plaintext affects the ciphertext. A small change in plaintext resulting in a large change in ciphertext shows a high level of *diffusion.* For example, with the classical cryptosystems discussed in Chapter 3, neither the translation or Vigenère ciphers provide much diffusion: a single change in the plaintext will result in only that particular character of the ciphertext changing.

For example:

```
plaintext:   WITHDRAWONEHUNDREDDOLLARS
key:         CODE
ciphertext:  YWWLFFDAQBHLWBGVGRGSNZDVU
```

But suppose just one character of the plaintext is changed:

```
plaintext:   WITHDREWONEHUNDREDDOLLARS
key:         CODE
ciphertext:  YWWLFFHAQBHLWBGVGRGSNZDVU
                   ↑
```

As you see, only the corresponding character in the ciphertext changes.

The Hill cipher provides more diffusion, in that the change of a single character will affect the entire ciphertext block. For example, with the plaintext

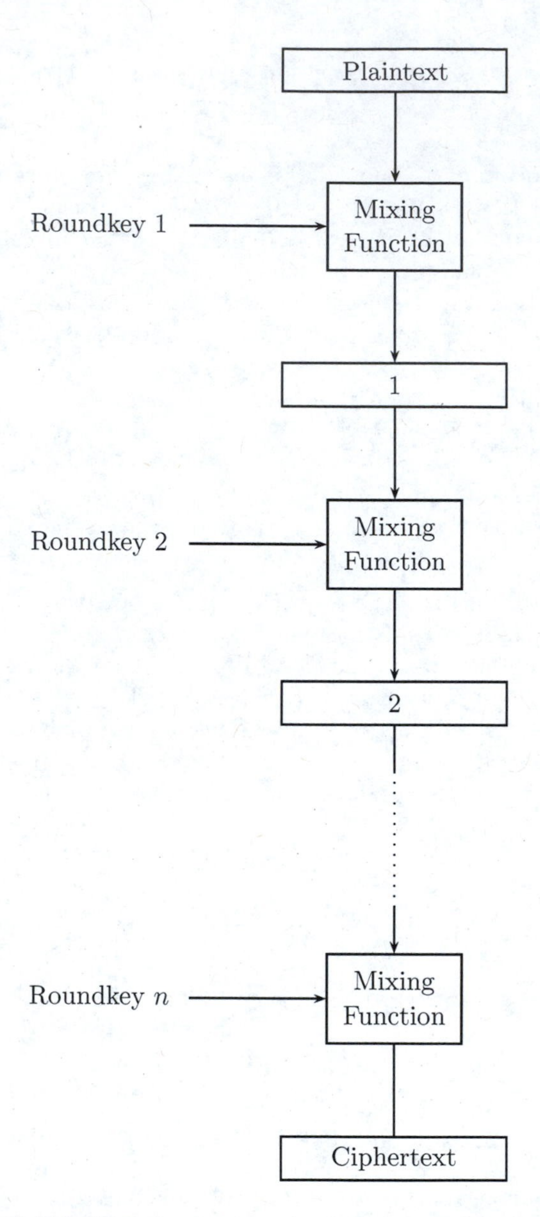

FIGURE 8.1: A general schema for a block cipher.

above, and the matrix

$$M = \begin{bmatrix} 22 & 11 & 19 \\ 15 & 20 & 24 \\ 25 & 21 & 16 \end{bmatrix}$$

the corresponding ciphertext is

```
XKIQBQOWKVBBQIPHRLNRZTBBUIB
```

If one character of the plaintext is changed as above, then the new ciphertext is

XKIQBQ[YEG]VBBQIPHRLNRZTBBUIB

where the changed block has been boxed.

Confusion is the property that the key and the ciphertext are not easily related; in particular that each character of the ciphertext should depend on many parts of the key. So if a single character of the key is changed, the ciphertext should change. In the Vigenère cipher, if a single key character is changed, all the ciphertext characters at that place will change. In an ideal cipher, changing one character of the key will change *all* characters of the ciphertext.

A good cipher will exhibit both confusion and diffusion. This is done with combination of substitution (for confusion), where strings of bits are replaced by other strings of bits as defined by substitution tables, and permutations, which spread the result of the substitutions around, and so add diffusion.

Stream ciphers. Block ciphers are used almost everywhere, but another very useful class of ciphers are the *stream ciphers* where the characters are encrypted one at a time. These ciphers have the advantage of not having to wait for an entire block to become available before encryption. It can sometimes be useful to make a block cipher act like a stream cipher.

8.3 Substitution/permutation ciphers

As mentioned above, the design of a good modern cipher will include two main components: substitution, where a group of bits is replaced by another group, and permutation, where the bits are switched around. This ensures that after enough rounds, a small change in either the key or the plaintext will propagate changes over the entire ciphertext. Figure 8.2 illustrates the setup of such a scheme.

Figure 8.3 shows how the change in a single bit will propagate through the system. In a well-designed cipher a single bit change in the input to any

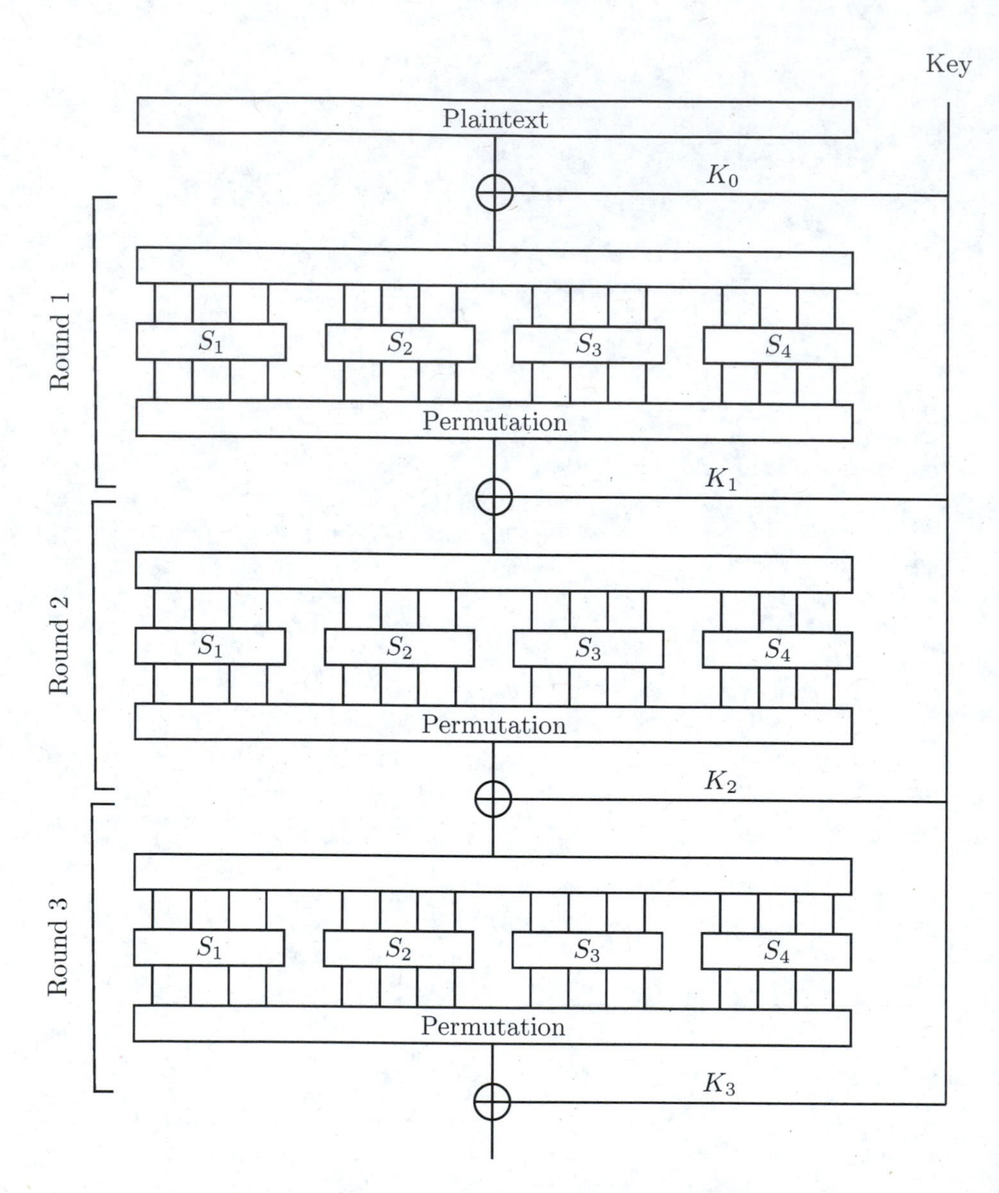

FIGURE 8.2: A substitution/permutation cipher.

substitution will result in at least half of the output bits changing. This is in fact not merely desirable, but one of the design characteristics of a good modern cryptosystem.

Notice how the number of bits that differ increases as the rounds advance. The number of rounds required for a complete mixing must be enough to ensure full bit propagation across the entire block, but not so many as to slow the operation of the system.

8.4 Modes of encryption

All block ciphers use a fixed block size. Thus when encrypting plaintext longer than the block size the cipher will need to be used many times. There are many ways to do this; these are called *modes of encryption.*

Electronic codebook mode (ECB)

ECB is the simplest mode, and in many ways the obvious one: each block of plaintext is encrypted independently of the others. If the blocks of plaintext are p_1, p_2, p_3 and so on, then

$$c_1 = E(p_1, k), \quad c_2 = E(p_2, k), \quad c_3 = E(p_3, k), \quad \dots$$

Clearly decryption is very easy, in that each ciphertext block can be decrypted on its own. ECB mode is shown in Figure 8.4.

The trouble with ECB is that it is very insecure. Equivalent blocks of plaintext will be encrypted to equivalent blocks of ciphertext. And since many files, and emails, start with standard headers, the first ciphertext block may be the same for all encryptions. This gives the enemy a huge advantage in allowing the use of a known plaintext attack.

Cipher block chaining (CBC)

Instead of encrypting a plaintext block directly, the previous ciphertext block is first added to it (with XOR). Thus

$$c_1 = E(p_1 \oplus c_0, k), \quad c_2 = E(p_2 \oplus c_1, k), \quad c_3 = E(p_3 \oplus c_2, k), \quad \dots$$

Decryption is also straightforward:

$$p_1 = D(c_1, k) \oplus c_0, \quad p_2 = D(c_2, k) \oplus c_1, \quad p_3 = D(c_3, k) \oplus c_2, \quad \dots$$

To make this mode work, start by choosing a suitable block to use for c_0; this block is known as the *initialization vector*, or IV. This block should not be fixed, as then CBC would exhibit the same problem as ECB for the first block

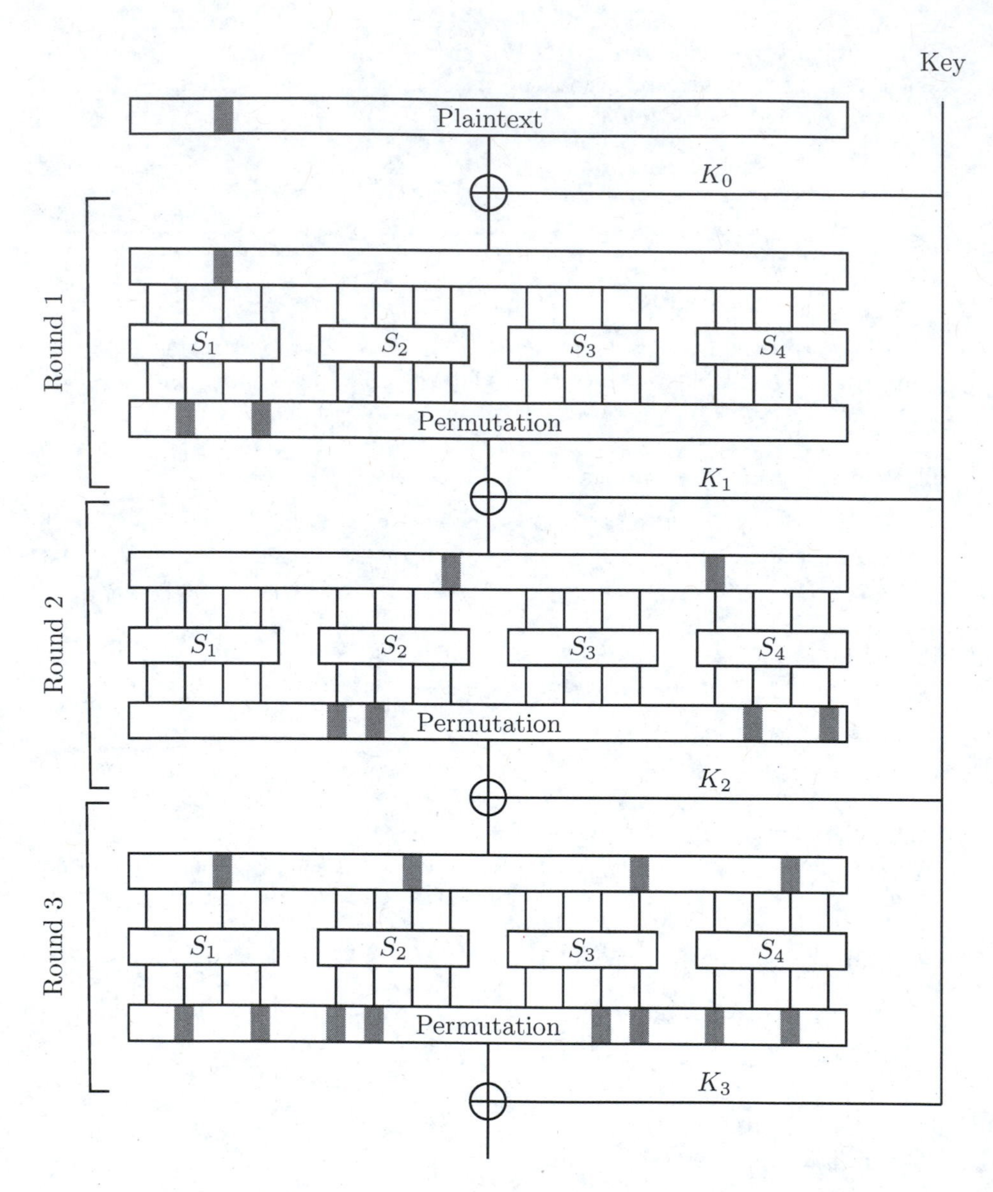

FIGURE 8.3: Propagation of a single bit change.

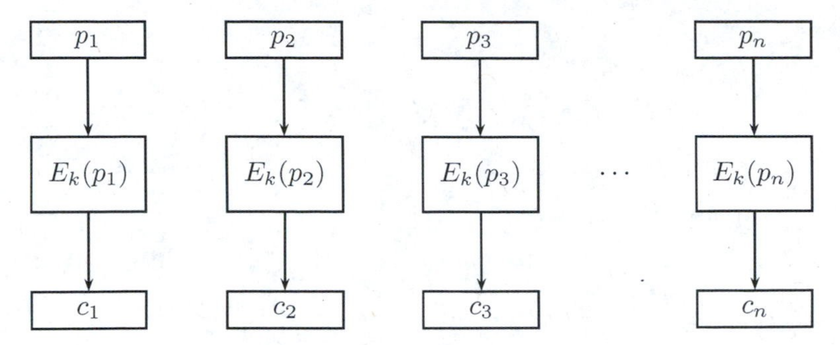

FIGURE 8.4: Electronic codebook mode.

of the message. A better idea is to use a random block, which then has to be sent along with the ciphertext. This means that the ciphertext is one block longer than the plaintext, which may be unacceptable for short messages. There is also the problem here of choosing a *random* block, which means access to a good random generator. The best solution is to use a combination of message numbers (for example, the number of the message in an email list) and any other information that uniquely characterizes the message, put it all into a single block and encrypt it to form the IV. This method means that every IV will be unique. A unique number that identifies the message is called a *nonce* value (short for **n**umber used **once**).

CBC mode is illustrated in Figure 8.5.

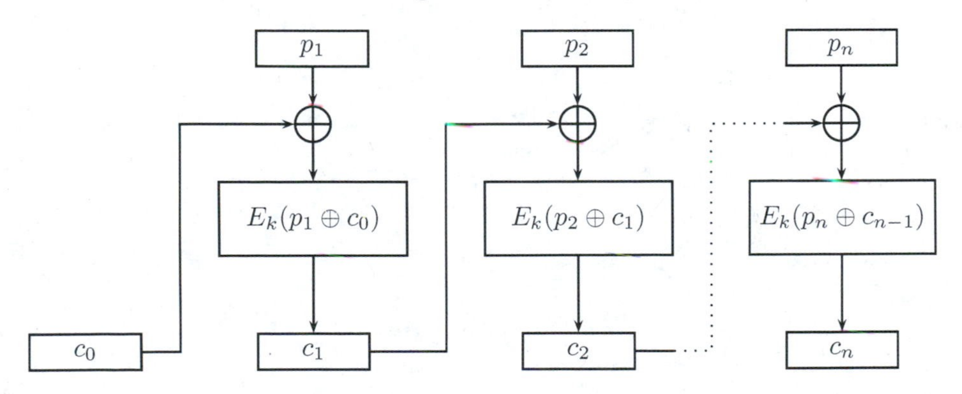

FIGURE 8.5: Cipher block chaining mode.

Output feedback mode (OFB)

Here the plaintext is never itself used as the input to the encryption function. Rather, the cipher is used to generate a pseudo-random string of bits (which is called the *key stream*), which is then XOR-ed with the plaintext. The result is a stream cipher similar in style to the one-time pad.

Start by choosing an IV in the style of CBC above, and then the key stream can be defined by

$$k_0 = \text{IV}, \quad k_1 = E(k_0, k), \quad k_2 = E(k_1, k), \quad k_3 = E(k_2, k), \quad \ldots$$

and then for each plaintext block

$$c_1 = p_1 \oplus k_1, \quad c_2 = p_2 \oplus k_2, \quad c_3 = p_3 \oplus k_3, \quad \ldots$$

There are other more general ways of defining OFB, which involve shifting the bits of the previously computed ciphertexts to obtain different numbers of new bits in the key stream each time. However, these are considered less secure than using a complete block each time.

OFB is illustrated in Figure 8.6.

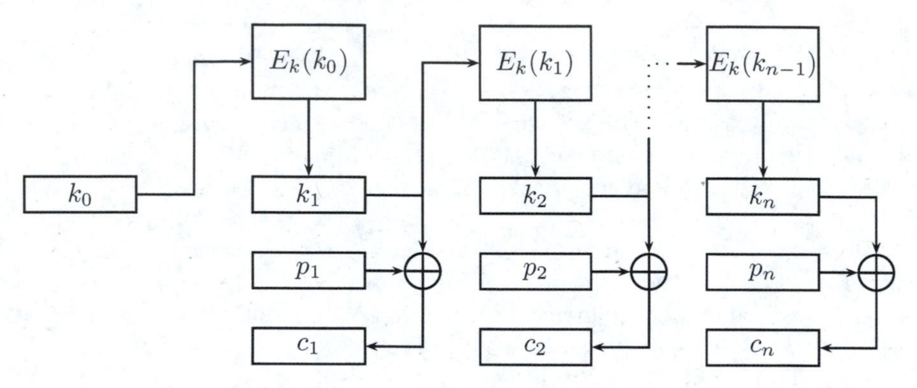

FIGURE 8.6: Output feedback mode.

One of the attractive properties of OFB is that the key stream can be generated independently of the message. Also notice that the decryption function is never used. Since for each block, encryption is defined as

$$c_i = p_i \oplus k_i$$

then using the self-inverse property of XOR

$$p_i = c_i \oplus k_i.$$

Cipher feedback mode (CFB)

This is another stream cipher mode. Its definition is similar to CBC, except that the XOR occurs *after* the encryption, rather than before:

$$c_1 = E(c_0, k) \oplus p_1, \quad c_2 = E(c_1, k) \oplus p_2, \quad c_3 = E(c_2, k) \oplus p_3, \quad \ldots$$

Going backwards,

$$p_1 = c_1 \oplus E(c_0, k), \quad p_2 = c_2 \oplus E(c_1, k), \quad p_3 = c_3 \oplus E(c_2, k), \quad \ldots$$

Notice that as with OFB above, the decryption function is never used. This mode is illustrated in Figure 8.7.

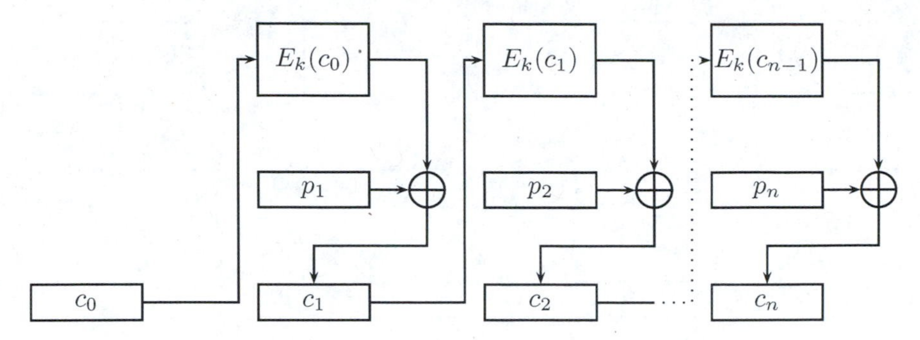

FIGURE 8.7: Cipher feedback mode.

Counter mode (CTR)

This mode is less well known than the others, as it wasn't accepted as a standard mode until more recently. However, it is both very easy to implement and very secure. To start, a nonce value such as was needed for the other modes is needed. The nonce needs to be expanded to a full block. This is done by concatenating with the nonce a *counter* for the message blocks. For a 128 block size, standard sizes are 48 bits for the message number, 16 bits for the nonce data, and the remaining 64 bits for the counter value (which is initialized to one). Then

$$k_1 = E(\text{Nonce}\|1, k), \quad k_2 = E(\text{Nonce}\|2, k), \ldots, k_i = E(\text{Nonce}\|i, k),$$

where $\|$ means concatenation, and then

$$c_1 = p_1 \oplus k_1, \quad c_2 = p_2 \oplus k_2, \quad \ldots \quad c_i = p_i \oplus k_i.$$

For security, it is vital that the counter values are never re-used. This means that the largest message that can be safely encrypted with this method is 2^{64} blocks. However, if a larger block size is available (which is possible with some of the newer block ciphers), then more space is available for the numbers, and so larger messages can be encrypted securely. As with OFB and CFB, CTR mode is a stream cipher mode, where the key stream is produced and then XOR-ed with the message to produce the ciphertext. This means that the key stream can be generated in advance. Also, the decryption function is never used.

CTR mode is illustrated in Figure 8.8.

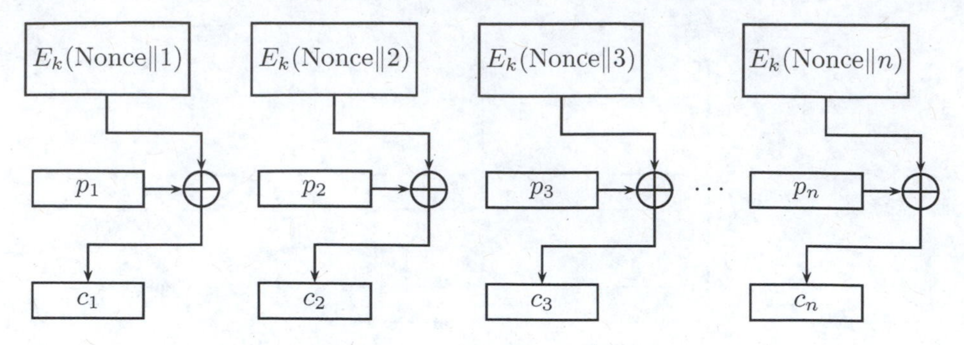

FIGURE 8.8: Counter mode.

8.5 Exploring modes of encryption

All the modes can be investigated using the Hill (matrix) cipher as the block cipher. Here the key is the encryption matrix M, and encryption is done by multiplying the matrix by a block of plaintext p, so that

$$E(p, k)$$

becomes

$$Mp.$$

Replacing exclusive or with addition modulo 26 produces the following encryption equations for each of the modes:

ECB: $c_i = Mp_i$.

CBC: $c_0 = IV$, $c_i = M(c_{i-1} + p_i)$.

OFB: $k_0 = IV$, $k_i = Mk_{i-1}$, $c_i = p_i + k_i \pmod{26}$.

CFB: $c_0 = IV$, $c_i = Mc_{i-1} + p_i \pmod{26}$.

CTR: $c_i = M(\text{Nonce}||i) + p_i \pmod{26}$ where the Nonce is a vector of size one less than the number of rows of the matrix. For this interpretation with the Hill system, only 26 columns can be encrypted, after which the numbers start repeating.

In order to see the results of the various modes, plaintexts and ciphertexts will be given as blocks of letters.

Although the ECB mode can be implemented by a single matrix product, it will be easier to describe it as a column by column procedure. The ciphertext will be a list of vectors, produced one vector at a time, each vector being appended to the list. Using the same matrix and plaintext as in Chapter 3:

```
sage: M = matrix(Zmod(26),[[22,11,19],[15,20,24],[25,21,16]])
sage: pl = "WITHDRAWONEHUNDREDDOLLARSXX"
sage: P = transpose(matrix(Zmod(26),9,3,str2lst(pl))).columns()
sage: P
[(22, 8, 19), (7, 3, 17), (0, 22, 14), (13, 4, 7), (20, 13, 3),
(17, 4, 3), (3, 14, 11), (11, 0, 17), (18, 23, 23)]
```

Now for the encryption. First obtain the number of columns in the plaintext, and set up a ciphertext matrix with an empty column:

```
sage: nc = len(P)
sage: nc
  9
```

The actual encryption will be done in a loop. The matrix product of M with the current column is the encryption itself, and that result is appended to the current ciphertext matrix C:

```
sage: C = []
sage: for i in range(nc):
....:     C += [M*P[i]]
....:
sage: map(lst2str,C)
['XKI', 'QBQ', 'OWK', 'VBB', 'QIP', 'HRL', 'NRZ', 'TBB', 'UIB']
```

Note that this results is the same ciphertext as obtained in chapter 3, but broken up into blocks.

To see why this is not a good scheme, suppose a single value in the plaintext is changed:

```
sage: pl2 = "WITHDREWONEHUNDREDDOLLARSXX"
sage: P2 = transpose(matrix(Zmod(26),9,3,str2lst(pl2))).columns()
sage: P2
[(22, 8, 19), (7, 3, 17), (4, 22, 14), (13, 4, 7), (20, 13, 3),
(17, 4, 3), (3, 14, 11), (11, 0, 17), (18, 23, 23)]
```

Now encrypt `pl2` is according to the same ECB scheme:

```
sage: C2 = []
sage: for i in range(nc):
....:     C2 += [M*P2[i]]
....:
sage: map(lst2str,C2)
['XKI', 'QBQ', 'YEG', 'VBB', 'QIP', 'HRL', 'NRZ', 'TBB', 'UIB']
```

Notice that the only block in the ciphertext that changes is the single block where the plaintext change was made.

CBC mode can be performed similarly, but the starting point is a "random" vector for c_0:

```
sage: IV = vector(Zmod(26),[7,13,19])
sage: C = [IV]
```

In order to keep the indices in the Sage loop the same as those in the definition, also add a block at the front P:

```
sage: P = [vector([0,0,0])]+P
```

Then the encryption is performed as for ECB in a loop, with the actual encryption using the $c_i = M(c_{i-1} + p_i)$ scheme:

```
sage: C = [IV]
sage: for i in range(1,nc+1):
....:     C += [M*(P[i]+C[i-1])]
....:
sage: map(lst2str,C[1:nc+1])
['FZG', 'VSI', 'UFB', 'PJY', 'RBB', 'VEF', 'QMM', 'DPN', 'EPB']
```

Suppose that as before one single change is made to the plaintext, to produce pl2. Using the same initial vector:

```
sage: C2 = [IV]]
sage: P2 = [vector([0,0,0])]+P2
sage: for i in range(1,nc+1):
....:     C2 += [M*(P2[i]+C2[i-1])]
....:
sage: map(lst2str,C2[1:nc+1])
['YQO', 'GGW', 'NLF', 'YXB', 'BDS', 'UEL', 'KVH', 'GSY', 'FZG']
```

Note now that every block of plaintext from the place of the change onwards differs from the original ciphertext.

Decryption is straightforward:

```
sage: D = []
sage: for i in range(nc,0,-1):
....:     D = [MI*C[i]-C[i-1]]+D
....:
sage: map(lst2str,D)
['WIT', 'HDR', 'AWO', 'NEH', 'UND', 'RED', 'DOL', 'LAR', 'SXX']
```

This last can be joined to make a single string:

```
sage: join(_,'')
 'WITHDRAWONEHUNDREDDOLLARSXX'
```

Output feedback mode is also straightforward: first the key-stream, using the same IV as above:

```
sage: IV = vector(Zmod(26),[7,13,19])
sage: K = [IV]
sage: for i in range(nc):
....:     K += [M*K[i]]
....:
sage: K
[(7, 13, 19), (8, 15, 24), (17, 8, 15), (19, 21, 1),
(18, 1, 22), (19, 12, 17), (15, 23, 11), (12, 13, 20),
(7, 10, 9), (19, 1, 9)]
```

Next the encryption:

```
sage: C = [x+y for (x,y) in zip(P,K)]
sage: map(lst2str,C[1:nc+1])
['EXR', 'YLG', 'TRP', 'FFD', 'NZU', 'GBO', 'PBF', 'SKA', 'LYG']
```

Decryption:

```
sage: D = [x-y for (x,y) in zip(C,K)]
sage: map(lst2str,D[1:nc+1])
['WIT', 'HDR', 'AWO', 'NEH', 'UND', 'RED', 'DOL', 'LAR', 'SXX']
```

CTR mode is also straightforward to implement; a counter stream can be created as follows, with the vector $[11, 21]$ as the nonce:

```
sage: v = [11,21]
sage: CTR = [vector(v+[i+1]) for i in range(nc)]
sage: CTR
[(11, 21, 1), (11, 21, 2), (11, 21, 3), (11, 21, 4),
(11, 21, 5), (11, 21, 6), (11, 21, 7), (11, 21, 8),
(11, 21, 9)]
```

Encryption is straightforward (this requires removing the added vector $[0, 0, 0]$ from the plaintext list `P` first):

```
sage: C = []
sage: for i in range(nc):
....:     C += [M*CTR[i]+P[i]]
....:
sage: map(lst2str,C)
['UTX', 'YML', 'KDY', 'QJH', 'QQT', 'GFJ', 'LNH', 'MXD', 'MSZ']
```

8.6 The Data Encryption Standard

The *Data Encryption Standard* (DES) was designed by IBM and after evaluation by the NSA was adopted in 1976 as an encryption standard, and authorized for use on all non-classified information. The DES algorithm is without doubt the most intensively researched algorithm in the history of cryptography. Although it is beginning to show its age, it is still much used, and there are ways of using it that increase its strength. DES is an example of a general class of secret-key cryptosystems called *Feistel ciphers*.

First note that DES, and all similar ciphers, are designed to work on strings of 0s and 1s, that is, binary data. This means that all arithmetic will be mod 2. In practice, addition mod 2 is implemented with use of the logical exclusive-or (XOR) operation, which is fast and supported in hardware. DES is a secret-key cipher, which means that for its operation the plaintext must be somehow mixed with the key. This same key is required for decryption.

8.7 Feistel ciphers

These are named after Horst Feistel, a cryptographer working for IBM (and who was one of the principal designers of DES).

Historical note: Feistel was born in Germany in 1915, and he became an American citizen in 1994. His design of DES, and in particular his elegant construction—which is still extensively used in block ciphers—has led him to be considered the father of modern block cipher design.

A Feistel cipher consists of a number of iterations, or *rounds.* At round i, the current text is broken up into a left half (L_i) and a right half (R_i), and the key k is modified to a subkey k_i. The values of k_i and R_i are the inputs to a function f, the result of which, $f(k_i, R_i)$, is added to L_i to become the new right half R_{i+1}. The original R_i becomes the new left half L_{i+1}.

That is,

$$\begin{aligned} R_{i+1} &= L_i \oplus f(k_i, R_i) \\ L_{i+1} &= R_i. \end{aligned}$$

Figure 8.9 shows one such round.

The beauty of a Feistel cipher is that it can be decrypted by the same algorithm. Assuming that the key k_i can be determined for any value of i, then

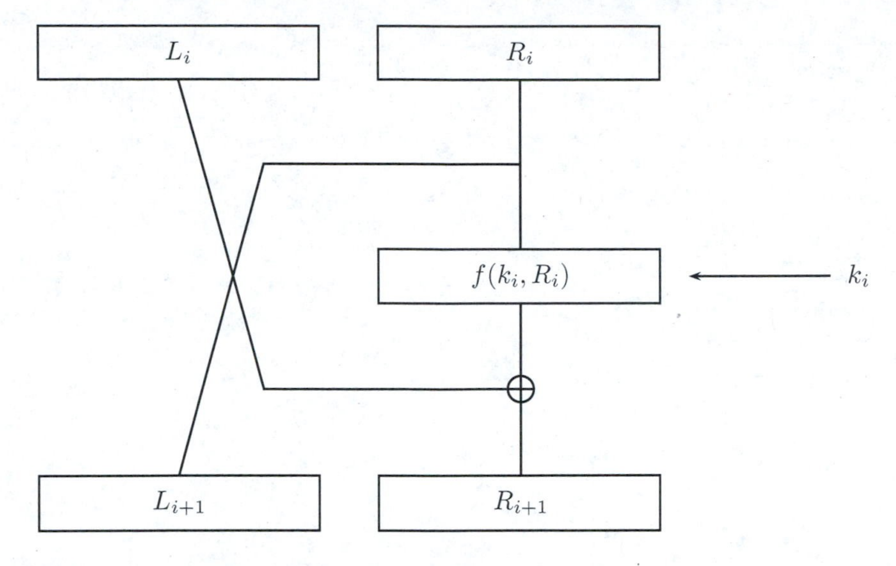

FIGURE 8.9: One round of a Feistel cipher.

the above equations can be written as

$$\begin{aligned} R_i &= L_{i+1} \\ L_i &= R_{i+1} \oplus f(k_i, R_i) \\ &= R_{i+1} \oplus f(k_i, L_{i+1}). \end{aligned}$$

The function f does not need to have any special properties; it doesn't have to be invertible. It does, however, have to be non-linear. Feistel ciphers get their cryptographic strength from these functions, so it is vital that the function does a good "mixing" job of the bits from k_i and R_i.

8.8 Simplified DES: sDES

DES is a complex algorithm, and operates on 64-bit blocks, with a key of 56 bits. This makes it almost impossible to work through the algorithm by hand to get a feeling of how it works. To help understand the intricacies and inner working of DES, a simplified version has been developed by Edward Schaefer [76]. Encryption and decryption of this simplified cipher can be performed by hand, giving a valuable insight into the workings and design of DES.

Simplified DES uses 8-bit plaintext and a 10-bit key to produce an 8-bit ciphertext. It uses an initial permutation, subkeys, a Feistel-style operation, and two rounds—all of which are described below.

Initial permutation. This starts off by permuting the bits of the plaintext:

$$b_0b_1b_2b_3b_4b_5b_6b_7 \rightarrow b_1b_5b_2b_0b_3b_7b_4b_6.$$

That is, bit 0 ends up in the third place, bit 1 ends up in the zeroth place, bit 2 ends up in the second place, and so on (for all discussion of sDES, lists and arrays will be indexed starting at zero, as in Python and Sage). This permutation can be expressed more simply as

$$(1, 5, 2, 0, 3, 7, 4, 6).$$

Corresponding to this initial permutation is its inverse, which simply puts the bits back where they started from:

$$b_0b_1b_2b_3b_4b_5b_6b_7 \rightarrow b_3b_0b_2b_4b_6b_1b_7b_5.$$

Again, this can be simply expressed as

$$(3, 0, 2, 4, 6, 1, 7, 5).$$

To see how this works, consider bit 1 of the plaintext. According to the initial permutation, it gets placed in position 5. Now look at the inverse permutation. Bit 5 is placed in position 1—exactly where it started from.

Figure 8.10 shows this.

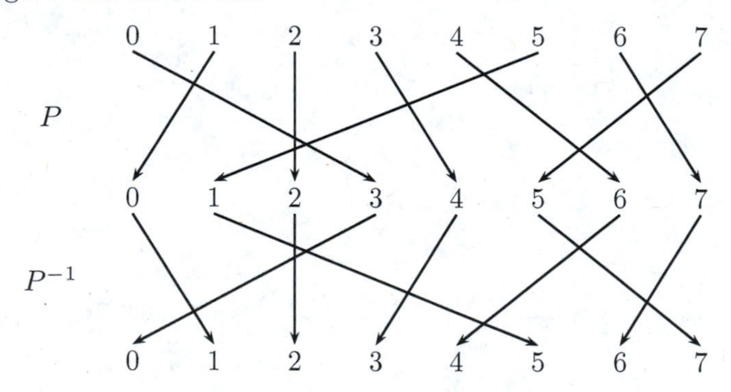

FIGURE 8.10: The permutations in sDES.

Subkeys. Simplified DES uses a 10-bit key, of which *subkeys* of 8 bits each are used in each round. The subkeys are obtained as follows:

1. The ten bits of the key are first permuted according to the permutation:

   $$(2, 4, 1, 6, 3, 9, 0, 8, 7, 5).$$

2. The result is then split into two halves of five bits each.

3. For round one, each half is cyclically shifted one bit to the left:

$$(a, b, c, d, e), (f, g, h, i, j) \rightarrow (b, c, d, e, a), (g, h, i, j, f).$$

4. Subkey one is obtained by taking these bits out of the result of the last operation:

$$(5, 2, 6, 3, 7, 4, 9, 8).$$

5. For round two, the result of step 3 is shifted two bits to the left:

$$(b, c, d, e, a), (g, h, i, j, f) \rightarrow (d, e, a, b, c), (i, j, f, g, h).$$

6. Then subkey two is obtained by taking the same bits out as were used for subkey one.

If the key is

$$k_0k_1k_2k_3k_4k_5k_6k_7k_8k_9$$

then the two subkeys are

$$\begin{aligned} K_1 &= k_0k_6k_8k_3k_7k_2k_9k_5, \\ K_2 &= k_7k_2k_5k_4k_9k_1k_8k_0. \end{aligned}$$

The Feistel step. At the heart of sDES is a Feistel-type function, as shown in Figure 8.11.

Note that the left and right halves are not switched around, as they are in the Feistel structure described in section 8.7.

The mixing function. The next part of sDES is the function $f(K_i, R_i)$ which mixes the bits in the structure shown in Figure 8.11. Inputs are an 8-bit subkey, and a 4-bit block, which is first expanded into 8-bits:

$$b_0b_1b_2b_3 \rightarrow b_3b_0b_1b_2b_1b_2b_3b_0$$

or as shown in Figure 8.12.

The subkey is XOR-ed with this expansion, and the result broken into two halves of 4 bits each. These halves are then turned into 2 bits each by means of two *S-boxes* S_0 and S_1. These are then put together to make 4 bits, and permuted according to

$$x_0x_1x_2x_3 \rightarrow x_1x_3x_2x_0.$$

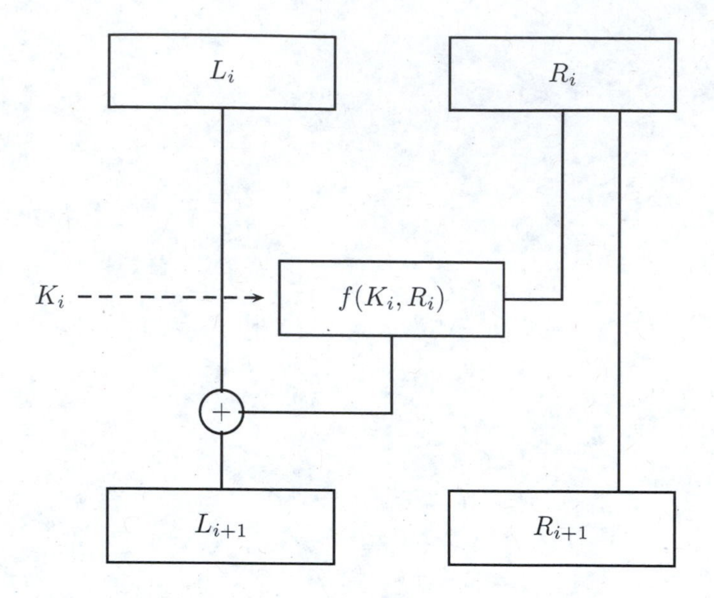

FIGURE 8.11: The Feistel-type function in sDES.

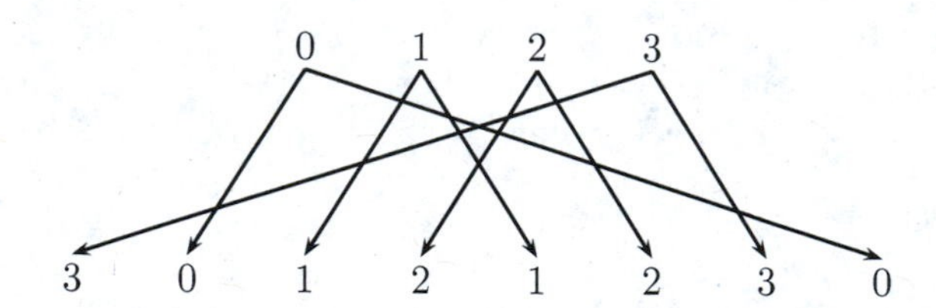

FIGURE 8.12: The expansion permutation of sDES.

The S-boxes ("S" stands for "substitution") are defined by the tables:

S-box S_0

	0	1	2	3
0	1	0	3	2
1	3	2	1	0
2	0	2	1	3
3	3	1	3	2

S-box S_1

	0	1	2	3
0	0	1	2	3
1	2	0	1	3
2	3	0	1	0
3	2	1	0	3

To apply an S-box, say S_0, suppose the 4 input bits are $x_0x_1x_2x_3$. The outer two bits give the row number; the inner two bits the column number. So for example, 1011 has row 11, or 3, and column 01, or 1. At row 3 and column 1 is the value 1, or 01 in binary. Another example: 0001. The row value is 01, or 1, and the column value 00, or 0. At row 1 and column 0 of S_0 is the value 3, or 11 in binary.

The entire mixing function is shown in figure 8.13.

The rounds. Simplified DES has two rounds. The first round involves switching the two halves of the Feistel output; the second round keeps them as they are. So a complete encryption using sDES is

1. Initial permutation.
2. Feistel operation using subkey K_1.
3. Switch left and right halves.
4. Feistel operation using subkey K_2.
5. Inverse permutation.

An example

The plaintext will be 01010101 and key 0000011111. The subkeys can be obtained immediately as

$$\begin{aligned} K_1 &= 01101011, \\ K_2 &= 10101010. \end{aligned}$$

Below are all the steps for the encryption.

Step 1. Initial permutation:

$$01010101 \to 11001100.$$

Step 2. The left and right halves of the last step are 1100 and 1100 respectively. The right half has to be mixed with the subkey K_1. This is done using the function shown in Figure 8.13.

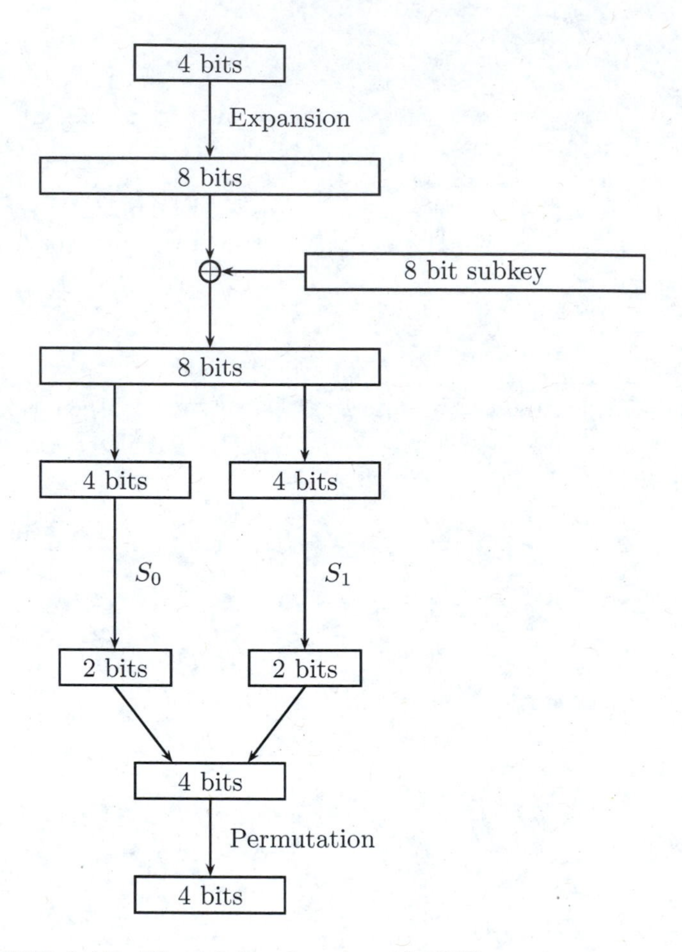

FIGURE 8.13: The mixing function of sDES.

First the 4 bits 1100 are expanded to 01101001, and added to the subkey K_1 with XOR:

$$\begin{array}{rcccccccc} & 0 & 1 & 1 & 0 & 1 & 0 & 0 & 1 \\ \oplus & 0 & 1 & 1 & 0 & 1 & 0 & 1 & 1 \\ \hline & 0 & 0 & 0 & 0 & 0 & 0 & 1 & 0 \end{array}.$$

The first four bits are used as input to S_0, and the second four bits used as input to S_1:

$$\begin{aligned} 0000 &\xrightarrow{S_0} 01 \\ 0010 &\xrightarrow{S_1} 01. \end{aligned}$$

Then the outputs are concatenated, 0101, and permuted, 1100.

This output is now XOR-ed with the left half,

$$1100 \oplus 1100 = 0000$$

and concatenated with the right half 1100 to produce

$$00001100$$

as the output of this step.

Step 3. Interchange the left and right halves: 11000000.

Step 4. The left half is now 1100 and right half is 0000, with subkey $K_2 =$ 10101010. Expand the right half

$$0000 \rightarrow 00000000$$

and XOR with the subkey:

$$00000000 \oplus 10101010 = 10101010.$$

Put the two halves of this through the S-boxes

$$\begin{aligned} 1010 &\xrightarrow{S_0} 10 \\ 1010 &\xrightarrow{S_1} 00 \end{aligned}$$

and concatenate, 1000 and permute: 0001.

Again this is XOR-ed with the left half

$$0001 \oplus 1100 = 1101$$

and concatenated with the right half 0000 to produce 11010000 as the output of this step.

Step 5. Now the final inverse permutation:

$$11010000 \rightarrow 11000100.$$

This last string is the ciphertext.

Decryption

Given that XOR is self inverse, the decryption of sDES is the same as encryption, except that the keys are used in reverse order: K_2 first, and K_1 second:

1. Initial permutation.
2. Feistel operation using subkey K_2.
3. Switch left and right halves.
4. Feistel operation using subkey K_1.
5. Inverse permutation.

This means that essentially the same program can be used for both encryption and decryption.

Although sDES has been designed specifically for pencil and paper use, so as to enable the learner to understand the sorts of functions that go into making up this cipher, it has also been added to Sage.

```
sage: from sage.crypto.block_cipher.sdes import SimplifiedDES
sage: sdes = SimplifiedDES()
sage: P = [0,1,0,1,0,1,0,1]
sage: K = [0,0,0,0,0,1,1,1,1,1]
sage: sdes.encrypt(P,K)
  [1, 1, 0, 0, 0, 1, 0, 0]
```

8.9 The DES algorithm

The DES algorithm operates on 64 bit blocks. It uses a 64 bit key, of which 8 bits are removed and used as parity check bits. Thus the effective length of the key is 56 bits, and so there are only $2^{56} = 72057594037927936$ different possible keys for use with DES.

The DES algorithm consists of a number of extra steps over and above a straightforward Feistel cipher. Although DES is complicated to describe, all the individual component steps are reasonably straightforward, and more importantly, can be implemented very efficiently in hardware. Although DES can be, and is, implemented in software, the greatest speeds and efficiencies come from using purpose-built chips.

The extra steps are:

- An initial permutation IP, which simply reorders the bits in the plaintext. There is a corresponding final permutation FP = IP^{-1}, to "undo" the effects of the initial permutation.
- Rather than just one function f that mixes the current subkey and right half, both the right half and the key are independently changed, and the results added.
- After this addition, the result is then put through two more steps to further mix up the bits of the text and subkey.

The steps of round i are more fully described as:

- The right half R_i, which consists of 32 bits, is expanded to 48 bits and these bits reordered by an operation called an *expansion permutation*, and denoted E.
- The 56 key bits are themselves divided into two halves, and each half is cyclically shifted to the left by a number of bits whose value depends on the value i of the round. The result is then compressed and permuted by a *compression permutation*, known in the DES standard as "permuted choice 2" or PC-2.
- The results of the expansion permutation of R_i and the compression permutation of the i-th subkey (both consisting of 48 bits) are added together.
- The result of this addition is then passed through a highly non-linear function to produce an output of 32 bits. This function is called an *S-box substitution*.
- These 32 bits are permuted by a *P-box permutation*, before being added to L_i to produce R_{i+1}.

Figure 8.14 illustrates all these operations which form one round of DES. Each of these steps will now be described in detail.

The initial permutation

This can be described by the following table, where the number in position k shows which bit is shifted to that position:

58	50	42	34	26	18	10	2
60	52	44	36	28	20	12	4
62	54	46	38	30	22	14	6
64	56	48	40	32	24	16	8
57	49	41	33	25	17	9	1
59	51	43	35	27	19	11	3
61	53	45	37	29	21	13	5
63	55	47	39	31	23	15	7

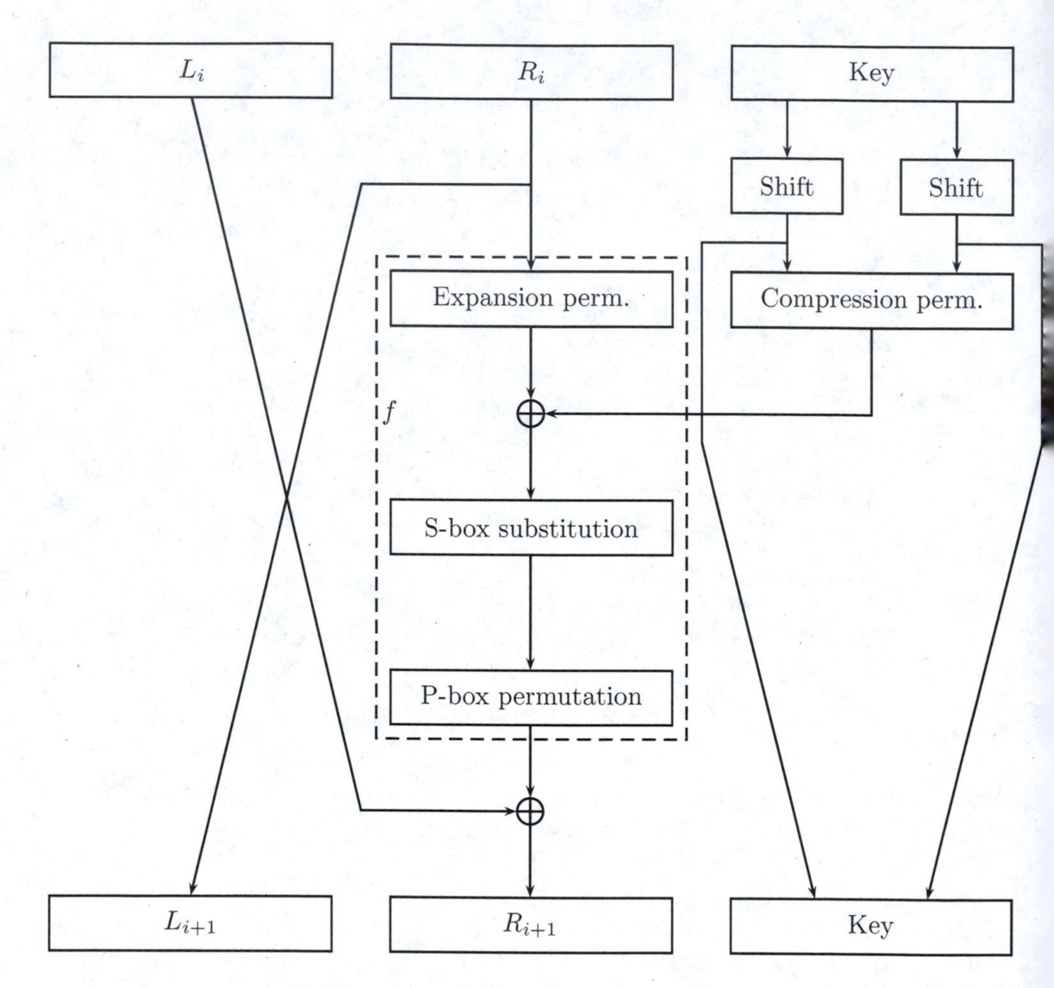

FIGURE 8.14: One round of DES.

For example: bit 58 in the plaintext is shifted to position 1; bit 50 in the plaintext is shifted to position 2; bit 42 is shifted to position 3, and so on; and conversely: bit 1 in the plaintext is shifted to position 40; bit 2 in the plaintext is shifted to position 8; bit 3 is shifted to position 48, and so on.

The final permutation is simply the inverse of this; so bit 1 goes to position 58; bit 2 goes to position 50; bit 3 goes to position 42, and so on.

Obtaining the subkeys

The bits 8, 16, 24, 32, 40, 48, 56 and 64 are removed from the key. If the remaining bits are stored as 7 bits per byte, then there is one bit left over to be used as an error-detection bit; thus each each byte can be assured to have an odd number of ones. The remaining 56 bits are permuted as follows:

57	49	41	33	25	17	9
1	58	50	42	34	26	18
10	2	59	51	43	35	27
19	11	3	60	52	44	36
63	55	47	39	31	23	15
7	62	54	46	38	30	22
14	6	61	53	45	37	29
21	13	5	28	20	12	4

There are 16 rounds in DES. In rounds 1, 2, 9, and 16, each 28-bit half key is cyclically shifted left by 1 bit. In all other rounds, the key halves are shifted by 2 bits. After this shift, the halves are joined together and the 56 bits compressed to 48 according to this table (again where the number in position k gives the bit that is shifted to that position):

14	17	11	24	1	5
3	28	15	6	21	10
23	19	12	4	26	8
16	7	27	20	13	2
41	52	31	37	47	55
30	40	51	45	33	48
44	49	39	56	34	53
46	42	50	36	29	32

So for example, bit 1 of the shifted key is moved to position 5, bit 2 is moved to position 24, bit 9 is ignored, and so on.

The expansion permutation

Before the right half can be added to the current compressed and permuted subkey, it must itself be expanded from 32 to 48 bits. The following table describes this:

32	1	2	3	4	5
4	5	6	7	8	9
8	9	10	11	12	13
12	13	14	15	16	17
16	17	18	19	20	21
20	21	22	23	24	25
24	25	26	27	28	29
28	29	30	31	32	1

Note that the expansion permutation has a simple structure: the inner four columns are just the full 32 bits written out in rows four at a time. The left hand column is the right most column of bits cyclically shifted down by one, and the right hand column is the left most column of bits cyclically shifted up by one.

Bit 4, for example, gets copied to both bits 5 and 7 in the output. Bit 5 gets copied to both bits 6 and 8. In general, bit $4k$ gets copied to both bits $6k+1$ and $6k-1$ (mod 32), and bit $4k+1$ gets copied to bits $6k$ and $6k+2$ (mod 32).

The S-box substitution

At this stage the right half has been permuted and expanded from 32 to 48 bits, and the key has been shifted and compressed from 56 to 48 bits. These two strings can now be added together, and the resulting sum put through the S-box substitutions. It is this particular step which gives DES its cryptographic strength. All the steps so far described are linear, and therefore easy to cryptanalyse. But the S-box substitutions are non-linear, and it is because of this non-linearity that DES is so difficult to cryptanalyse.

The 48 bits which are the result of the addition are broken up into eight 6-bit blocks. Each block is operated on by one of eight S-boxes; specifically block i is operated on by S-box i. Each S-box produces a 4-bit output, so the result of this step will be a total of 32 bits.

Each S-box may be considered a 4×16 array, each row of which consists of a permutation of the numbers 0 to 15. Given a 6-bit block and a corresponding S-box, the outer two bits give the row number (which is to be indexed as 0, 1, 2, 3), and the inner four bits the column number (indexed from 0 to 15). The value at this position in the S-box is the output.

And now, here are the S-boxes:

S-box 1

	0	1	2	3	4	5	6	7	8	9	10	11	12	13	14	15
0	14	4	13	1	2	15	11	8	3	10	6	12	5	9	0	7
1	0	15	7	4	14	2	13	1	10	6	12	11	9	5	3	8
2	4	1	14	8	13	6	2	11	15	12	9	7	3	10	5	0
3	15	12	8	2	4	9	1	7	5	11	3	14	10	0	6	13

S-box 2

	0	1	2	3	4	5	6	7	8	9	10	11	12	13	14	15
0	15	1	8	14	6	11	3	4	9	7	2	13	12	0	5	10
1	3	13	4	7	15	2	8	14	12	0	1	10	6	9	11	5
2	0	14	7	11	10	4	13	1	5	8	12	6	9	3	2	15
3	13	8	10	1	3	15	4	2	11	6	7	12	0	5	14	9

S-box 3

	0	1	2	3	4	5	6	7	8	9	10	11	12	13	14	15
0	10	0	9	14	6	3	15	5	1	13	12	7	11	4	2	8
1	13	7	0	9	3	4	6	10	2	8	5	14	12	11	15	1
2	13	6	4	9	8	15	3	0	11	1	2	12	5	10	14	7
3	1	10	13	0	6	9	8	7	4	15	14	3	11	5	2	12

S-box 4

	0	1	2	3	4	5	6	7	8	9	10	11	12	13	14	15
0	7	13	14	3	0	6	9	10	1	2	8	5	11	12	4	15
1	13	8	11	5	6	15	0	3	4	7	2	12	1	10	14	9
2	10	6	9	0	12	11	7	13	15	1	3	14	5	2	8	4
3	3	15	0	6	10	1	13	8	9	4	5	11	12	7	2	14

S-box 5

	0	1	2	3	4	5	6	7	8	9	10	11	12	13	14	15
0	2	12	4	1	7	10	11	6	8	5	3	15	13	0	14	9
1	14	11	2	12	4	7	13	1	5	0	15	10	3	9	8	6
2	4	2	1	11	10	13	7	8	15	9	12	5	6	3	0	14
3	11	8	12	7	1	14	2	13	6	15	0	9	10	4	5	3

S-box 6

	0	1	2	3	4	5	6	7	8	9	10	11	12	13	14	15
0	12	1	10	15	9	2	6	8	0	13	3	4	14	7	5	11
1	10	15	4	2	7	12	9	5	6	1	13	14	0	11	3	8
2	9	14	15	5	2	8	12	3	7	0	4	10	1	13	11	6
3	4	3	2	12	9	5	15	10	11	14	1	7	6	0	8	13

S-box 7

	0	1	2	3	4	5	6	7	8	9	10	11	12	13	14	15
0	4	11	2	14	15	0	8	13	3	12	9	7	5	10	6	1
1	13	0	11	7	4	9	1	10	14	3	5	12	2	15	8	6
2	1	4	11	13	12	3	7	14	10	15	6	8	0	5	9	2
3	6	11	13	8	1	4	10	7	9	5	0	15	14	2	3	12

S-box 8

	0	1	2	3	4	5	6	7	8	9	10	11	12	13	14	15
0	13	2	8	4	6	15	11	1	10	9	3	14	5	0	12	7
1	1	15	13	8	10	3	7	4	12	5	6	11	0	14	9	2
2	7	11	4	1	9	12	14	2	0	6	10	13	15	3	5	8
3	2	1	14	7	4	10	8	13	15	12	9	0	3	5	6	11

As an example, suppose that the third 6-bit block is 110110. It will thus be operated on by the third S-box. The outer two bits of the block form the binary number 10 (or 2); the inner four bits 1011 correspond to the decimal number 11. The output will be the value at row 2 (that is, the third row), and column 11 (that is, the twelfth column) of S-box 3 above; this value is 12, which in binary is 1100.

The P-box permutation

The final part of the round is a straight permutation of the 32 bits produced by the S-box substitutions. It is described as

16	7	20	21	29	12	28	17
1	15	23	26	5	18	31	10
2	8	24	14	32	27	3	9
19	13	30	6	22	11	4	25

As before, the number at position k indicates which bit moves to that position. So for example, bit 16 moves to position 1, bit 7 moves to position 2, bit 1 moves to position 9, bit 2 moves to position 17.

8.10 Security of S-boxes

S-boxes are a feature not only of DES, but of many other cryptosystems. Sometimes they are fixed, as in DES, and also in AES; for some other systems they are a function of the plaintext or key. Because the security of the system is so intimately bound up with the design of the S-boxes, there is considerable literature in the design and analysis of S-boxes. In general, an S-box that takes an n-bit input and produces an s-bit output is called an $n \times s$ S-box.

The following have been published by Coppersmith [21], who was one of the original designers of DES, as the design criteria for S-boxes:

1. Each S-box must take a 6-bit input and return a 4-bit output. (This was the maximum size available for 1974 hardware.)

2. No S-box operation on a bit is a linear or affine function.

3. If the outer two bits are fixed, then as the inner four bits vary over all their 16 possibilities, every possible output will be returned.

4. A change to a single bit of input results in at least two bit changes in the output.

5. If two inputs x and y differ in their middle bits only (so that $x \oplus y =$ `001100`) then their outputs must differ in at least two bits.

6. If two inputs x and y differ in their first two bits and are identical in their last two bits (so that $x \oplus y = \texttt{11} * *\texttt{00}$) then their outputs must differ.

7. Suppose r is a 6-bit string. Of the 32 possible pairs of inputs x and y whose exclusive-or is r, no more than 8 pairs may have the same value for $S(x) \oplus S(y)$.

8. The probability that a non-zero input difference to three adjacent S-boxes yields a zero output difference should be minimized.

The avalanche criterion

One important property of an S-box is that the bits in the output should be a function of all bits in the input. In particular, a single change in the input should affect at least half the bits in the output. This will be guaranteed for the DES S-boxes by criterion 3 above. For example, take S-box 1, and consider the input $[0, 1, 1, 0, 0, 0]$. The row number is 00, or 0, and the column number 1100, or 12. The output is 5, or 0101. Suppose one bit is changed so that the input is $[0, 1, 1, 0, 1, 0]$. The row is unchanged, but the column is now 1101, or 13. The output is 9, or 1001. Comparing the two outputs shows that 2 bits have changed.

This can be formalized as follows: suppose that $S : \mathbb{Z}_2^n \to \mathbb{Z}_2^m$, where $\mathbb{Z}_2$ is the finite field of two elements. Then S satisfies the *avalanche effect* if for all $i \leq n$

$$\sum_{x \in \mathbb{Z}_2^n} \mathrm{wt}(S(x) \oplus S(x \oplus 2^i)) = m2^{n-1}$$

where $\mathrm{wt}(x)$ is the number of ones in the binary expansion of x.

Sage provides an S-box class that enables the easy analysis of S-boxes. Here's how the avalanche effect might be tested for S-box 1. First create a single list that produces the output for every binary string between 0 and 63:

```
sage: S1 = \
[[14, 4, 13, 1, 2, 15, 11, 8, 3, 10, 6, 12, 5, 9, 0, 7],\
[0, 15, 7, 4, 14, 2, 13, 1, 10, 6, 12, 11, 9, 5, 3, 8],\
[4, 1, 14, 8, 13, 6, 2, 11, 15, 12, 9, 7, 3, 10, 5, 0],\
[15, 12, 8, 2, 4, 9, 1, 7, 5, 11, 3, 14, 10, 0, 6, 13]]
sage: B = []
sage: for i in range(64):
....:     j = ZZ(i).binary().zfill(6)
....:     B += [S1[ZZ('0b'+j[0]+j[5])][ZZ('0b'+j[1:5])]]
sage: Sb1 = mq.SBox(B)
```

This can be tested using the inputs above:

```
sage: Sb1([0,1,1,0,0,0])
  [0, 1, 0, 1]
sage: Sb1([0,1,1,0,1,0])
  [1, 0, 0, 1]
```

Note that such an S-box can use integers as input and output:

```
sage: Sb1(24)
  5
sage: Sb1(26)
  9
```

The next requirement is a function that can test the avalanche effect:

```
def testAE(X,b): # tests the avalanche effect test
  wts = 0        # on an S-box created with mq.SBox
  for i in range(64):
      s = X(i)^^X(i^^b)
      wts += sum(s.bits())
  return wts
```

This can now be applied for all values of 2^i:

```
sage: for i in [1,2,4,8,16,32]:
....:     testAE(Sb1,i)
....:
  152
  152
  156
  152
```

```
180
160
```

Since the sum of all weights is at least $m2^{n-1} = 4.2^5 = 128$, the first S-box of DES is shown to satisfy the avalanche effect.

The *strict avalanche criterion* (SAC) requires that on average half the bits will change for each bit change of the input. That is:

$$\sum_{x \in \mathbb{Z}_2^n} S(x) \oplus S(x \oplus 2^i) = (2^{n-1}, 2^{n-1}, \ldots, 2^{n-1}).$$

This can be tested to an entire DES S-box by using a function that tests the SAC:

```
def testSAC(X,b): # tests the SAC for a complete DES S-box
  m = [0,0,0,0]
  for i in range(64):
      s = X(i)^^X(i^^b)
      m = [x+y for x,y in zip(m,s.digits(2,padto=4))]
  return m
```

Now the program can be run for all values of 2^i:

```
sage: for i in [1,2,4,8,16,32]:
....:     testSAC(Sb1,i)
....:
[36, 36, 32, 48]
[36, 36, 36, 44]
[40, 40, 44, 32]
[44, 40, 36, 32]
[44, 48, 40, 48]
[48, 40, 40, 32]
```

Since each value is greater than $2^{n-1} = 2^5 = 32$, the first S-box is confirmed to satisfy the SAC.

These simple tests show that the S-boxes of DES are more than merely random permutations. As an experiment, construct an S-box with random permutations and test it:

```
sage: def r16():
....:     return [i-1 for i in random_permutation(16)]
....:
sage: Sx = mq.SBox(flatten([r16(),r16(),r16(),r16()]))
sage: for i in [1,2,4,8,16,32]:
  testAE(Sx,i)
....:
```

```
    128
    120
    128
    144
    112
    124
sage: for i in [1,2,4,8,16,32]:
....:     testSAC(Sx,i)
....:
    [32, 24, 44, 28]
    [44, 20, 28, 28]
    [36, 32, 28, 32]
    [40, 36, 32, 36]
    [20, 28, 28, 36]
    [24, 28, 36, 36]
```

Note that in the testing of the avalanche effect, some values are less than 128, and in the testing of the SAC, some values are less than 32. This shows that the randomly constructed S-box `Sx` is not a good S-box.

Resistance to differential cryptanalysis

Differential cryptanalysis is a powerful technique in which pairs of plaintexts with given differences are passed through the cryptosystem and the differences in the ciphertexts examined. S-boxes can be tested for their vulnerability to differential cryptanalysis, and the first step is the construction of a *differential distribution table* or DDT. For a DES S-box, there are 64 possible inputs, and 16 possible outputs. There are thus $64^2 = 4096$ possible pairs of inputs, and $16^2 = 256$ pairs of outputs. Suppose that x_1 and x_2 are two inputs, and $y_1 = S(x_1)$, $y_2 = S(x_2)$ are the corresponding outputs. Writing

$$x' = x_1 \oplus x_2$$

and

$$y' = y_1 \oplus y_2 = S(x_1) \oplus S(x_2)$$

the DDT shows the number of times each y' value occurs for each difference x'.

As an example, consider the S-box S_0 of sDES. Here's a little program to implement it:

```
sage: S0 = [[1,0,3,2],[3,2,1,0],[0,2,1,3],[3,1,3,2]]
sage: B = []
sage: for i in range(16):
....:     j = ZZ(i).binary().zfill(4)
```

```
....:        B += [S0[ZZ('0b'+j[0]+j[3])][ZZ('0b'+j[1]+j[2])]]
sage: Sb0 = mq.SBox(B)
```

Now it is straightforward to construct a DDT:

```
sage: Sb0.difference_distribution_matrix()
[16, 0, 0, 0]
[0, 2, 10, 4]
[0, 10, 6, 0]
[2, 4, 0, 10]
[2, 4, 8, 2]
[10, 0, 4, 2]
[0, 2, 2, 12]
[4, 10, 2, 0]
[2, 4, 8, 2]
[8, 2, 2, 4]
[4, 2, 2, 8]
[2, 8, 4, 2]
[8, 2, 2, 4]
[2, 4, 8, 2]
[2, 8, 4, 2]
[4, 2, 2, 8]
```

Note that every row in the DDT sums to 16: this is to be expected, as there are exactly 16 possible differences between outputs. Also notice that the first row consists of a single 16 followed by zeros. Again, this is expected: the first row is the row of zero difference, and the only possible difference in the S-box outputs will also be zero.

Given the first DES S-box `S1` above its DDT can also be produced:

```
sage: ddt = Sb1.difference_distribution_matrix()
```

and some rows displayed:

```
sage: for i in [0,1,2,3,4,12,62,63]:
  print i, ddt[i]
....:
  0 (64, 0, 0, 0, 0, 0, 0, 0, 0, 0, 0, 0, 0, 0, 0, 0)
  1 (0, 0, 0, 6, 0, 2, 4, 4, 0, 10, 12, 4, 10, 6, 2, 4)
  2 (0, 0, 0, 8, 0, 4, 4, 4, 0, 6, 8, 6, 12, 6, 4, 2)
  3 (14, 4, 2, 2, 10, 6, 4, 2, 6, 4, 4, 0, 2, 2, 2, 0)
  4 (0, 0, 0, 6, 0, 10, 10, 6, 0, 4, 6, 4, 2, 8, 6, 2)
  12 (0, 0, 0, 8, 0, 6, 6, 0, 0, 6, 6, 4, 6, 6, 14, 2)
  62 (4, 8, 2, 2, 2, 4, 4, 14, 4, 2, 0, 2, 0, 8, 4, 4)
  63 (4, 8, 4, 2, 4, 0, 2, 4, 4, 2, 4, 8, 8, 6, 2, 2)
```

To get a better understanding of the DDT, consider the element in row 62, column 5, which is 4. This means that there are four different tuples

$$(x_1, x_2, y_1, y_2)$$

for which $x_1 \oplus x_2 = x' = 62 =$ `0b111110`, $y_1 \oplus y_2 = y' = 5 =$ `0b0101`, and $y_1 = S(x_1)$, $y_2 = S(x_2)$. These can be found very easily:

```
sage: for x1 in range(64):
....:     for x2 in range(64):
....:         y1 = Sb1(x1)
....:         y2 = Sb1(x2)
....:         if x1^^x2 == 62 and y1^^y2 == 5:
....:             print x1,x2,y1,y2
....:
8 54 2 7
21 43 12 9
43 21 9 12
54 8 7 2
```

Note that if (x_1, x_2, y_1, y_2) is a tuple satisfying the requirements, then so is (x_2, x_1, y_2, y_1) by the commutativity of the exclusive-or (XOR) operation. So all elements in the DDT must be even.

In that same row, the first zero value corresponds to $y' = 10$, which means that if 5 in the above loop is replaced with 10, there should be no outputs. This should also provide some insight into the first row. The top left element is the number of tuples (x_1, x_2, y_1, y_2) for which $x_1 \oplus x_2 = y_1 \oplus y_2 = 0$, and $y_1 = S(x_1)$, $y_2 = S(x_2)$. But $x_1 \oplus x_2 = 0$ means that $x_1 = x_2$, and similarly $y_1 = y_2$. So the 64 tuples are all values of $(x, x, S(x), S(x))$. Also there can be no tuples (x_1, x_2, y_1, y_2) for which $x_1 = x_2$ but $y_1 \neq y_2$ and with $y_1 = S(x_1)$, $y_2 = S(x_2)$. Hence all other values in the top row are zero.

Note that from Figure 8.14 the right half of the current state is XOR-ed to the compressed key before the S-boxes are applied. However, the addition of the key does not affect the value of the DDT, since

$$(x_1 \oplus k) \oplus (x_2 \oplus k) = x_1 \oplus x_2.$$

Hence the DDT depends only on the S-box, and is independent of the key.

Now suppose that x_1 and x_2 are known, as well as the value y'. This can be used to give some information about the key k. First of all, both $x_1 \oplus k$ and $x_2 \oplus k$ must occur in a tuple corresponding to the value in row x', column y' of the DDT. This means that all the input values can be determined. For example, suppose $x_1 = 23 =$ `0b010111`, $x_2 = 57 =$ `0b111001`, and $y' = 7 =$ `0b0111`. All possible inputs can be determined using the same loop as above, noting first that $x_1 \oplus x_2 =$ `0b101110` $= 46$:

```
sage: for x1 in range(64):
```

```
....:       for x2 in range(64):
....:           y1 = Sb1(x1)
....:           y2 = Sb1(x2)
....:           if x1^^x2 == 46 and y1^^y2 == 7:
....:               print x1,x2,y1,y2
....:
1 47 0 7
17 63 10 13
25 55 9 14
47 1 7 0
55 25 14 9
63 17 13 10
```

This produces the set

$$X = \{1, 17, 25, 47, 55, 63\}$$

as all possible inputs, which means that the key k must be an element of the set

$$K = X \oplus x_1 = X \oplus x_2 = \{6, 14, 22, 32, 40, 56\}.$$

If another input pair (x_1, x_2) is given, then there will be another possible set K' of keys, and so the key k will be an element of $K \cap K'$. Further observations will narrow down the possible choices for k.

Further examples, applied to DES, are given by Pieprzyk et al. [70].

The DDT is central to differential cryptanalysis, and the starting place is finding entries in the table (aside from the top left value), which are large. A large entry means that there will be many pairs with the same difference, whose S-box values also have the same difference. Such pairs can be used to great advantage in the cryptanalysis. So a "good" S-box will have a DDT not containing large values. It can be shown (see Seberry et al. [78] and references there) that a uniform DDT is neither possible nor desirable. Instead the robustness of the S-box against differential cryptanalysis can be measured by two values: L, the largest value in the DDT (aside from the top left value), and R, the number of non-zero entries in the first column (again aside from the top left value). For the DES S-box 1, these can be calculated as:

```
sage: L = max([max(ddt.row(i)) for i in range(1,64)]); L
  16
sage: R = 64-list(ddt.column(0)).count(0)-1; R
  37
```

The value R is a smoothness measure, and as with L, a large value is undesirable. Seberry introduces the following robustness measure for an $n \times s$ S-box:

$$\epsilon = \left(1 - \frac{L}{2^n}\right)\left(1 - \frac{R}{2^n}\right).$$

A large value of ϵ is a measure of the S-box's strength to withstand differential cryptanalysis. For 6×4 DES S-box 1, this value is

```
sage: (1-R/2^6)*(1-L/2^6).n()
  0.316406250000000
```

which is considered fairly small. This shows that differential cryptanalysis can be used successfully against DES, albeit with some difficulty.

8.11 Security of DES

Given the importance of DES to the world of cryptography (it was one of the first algorithms made public by the NSA), it has been the topic of intense scrutiny and analysis. Some aspects of its security will now be briefly examined.

Key length

The original design specified a key length of 128 bits, but after the NSA evaluation of DES, it was reduced to 56 bits, which vastly reduces the available number of keys. This makes possible a brute-force attack on DES: simply try every possible key! In 1998, the Electronic Frontier Foundation produced a machine called "Deep Crack" which consisted of 1856 custom-built DES chips mounted on circuit boards, placed in a chassis and controlled by a single PC [33]. At a relatively low cost of $250,000, this machine could run through all 2^{56} keys in less than 9 days, but on average the key could be found in half that time. Later, in 2006, another machine comparable in power was constructed [52] but on account of improvements in hardware technology, at a much lower cost.

Design of the algorithm

Almost every aspect of DES has been questioned and examined, from the number of rounds to the design of the S-boxes. Recent research has shown that with any number of rounds fewer than 16, DES can be broken with a known-plaintext attack more efficiently than with a brute-force attack.

As for the S-boxes, initially there was much speculation that the NSA included a "trapdoor," a means by which a special algorithm (known only to the NSA) could be applied, by which DES could be easily broken. However, this was eventually shown not to be the case. Moreover, it has been shown that the particular design of the S-boxes provides some security from differential cryptanalysis: one of the most powerful attacks known.

Given the method in which subkeys are produced for each round of DES, some keys are *weak keys*; that is, they do not produce different enough keys at each round. For example, the keys consisting of all zeros or all ones would produce identical subkeys for each round. The same effect occurs if one half is all zeros, and the other all ones. There is another family of keys: *semi-weak keys*, for which pairs of keys cause the plaintext to be encrypted to the same ciphertext.

This aspect of weak (and semi-weak) keys has been seen as a failing of DES; in a reasonable algorithm, all possible keys should be equally secure, and any two different keys should produce different ciphertext.

One aspect of DES that took many years to decide was its *algebraic closure*. Suppose there are two keys k_1 and k_2, and a block of plaintext is first encrypted with key k_1 and then the resulting ciphertext further encrypted with k_2. Will there always be a third key k_3 for which this final encryption could have been reached by a single encryption of the plaintext using this key? That is, will

$$E_{k_2}(E_{k_1}(P)) = E_{k_3}(P)?$$

(This is trivially true for such ciphers as transposition ciphers, Vigenère ciphers, and matrix ciphers.) This question for DES has been answered in the negative; that is, the result is no [44].

8.12 Using DES

First note that DES can be used in any of the modes described in Section 8.4. It is good practice to use DES (or indeed, any block cipher) in a non-ECB mode. If a random initialization block is used to get the procedure underway, the resulting ciphertext is much harder to cryptanalyse than ciphertext obtained using ECB.

Given the security risks associated with the short key length, and the result above about the non-closure of DES as an algebraic operation, it seems reasonable to effectively double the key length by using DES twice; that is, choosing two keys k_1 and k_2, and encrypting a block of plaintext as

$$C = E_{k_2}(E_{k_1}(P)).$$

This method would appear to give an effective key length of 108 bits: far too long to attempt any sort of brute-force attack. Unfortunately, appearances can be deceiving! This method is vulnerable to an attack called emphmeet-in-the-middle, which basically means encrypting P with each possible key, decrypting E with each possible key, and seeing if there are any matches. This requires an enormous amount of memory, but cryptographers are of necessity highly paranoid people, and the theoretical existence alone of this attack greatly reduces confidence in the security of double encryption.

All is not lost; effective doubling of key length can be obtained by *triple encryption*, in which a block of plaintext is encrypted three times:

$$C = E_{k_3}(E_{k_2}(E_{k_1}(P))).$$

In fact, in order to fit in with certain cryptographic standards, and given that for DES, being a Feistel cipher, decryption is basically the same as encryption, triple encryption is more generally implemented as

$$C = E_{k_3}(D_{k_2}(E_{k_1}(P))).$$

That is, the *decryption algorithm* is used for the middle encryption. Then decryption of C is obtained by performing these steps in reverse order:

$$P = D_{k_1}(E_{k_2}(D_{k_3}(C))).$$

Variations on plain DES that involve triple encryption are called *triple-DES*. They are all much stronger and more secure than plain DES, at the cost of more required key bits, and of course more time required for encryption. There are variants that use triple encryption with only two keys. Although they are stronger than plain DES, they aren't as strong as triple DES with three independent keys. There is also *quintuple DES*. However, for most practical purposes, triple DES, with three independent keys, used in CBC mode, should provide sufficient security.

The main troubles here are the speed: triple DES is very slow, and the block size, 64 bits, is simply too small for serious use. For general purpose symmetric cryptography, it is better to use the AES, which will be discussed in Chapter 10.

Another approach, initially developed by Rivest, is called *key whitening*. Here, as well as the 56-bit key k, two 64-bit blocks k_1 and k_2 are used, one to XOR with the plaintext before encryption, and the other to XOR with the ciphertext. The complete encryption is then

$$c = k_2 \oplus E(p \oplus k_1, k).$$

Rivest's idea was that this method should provide computationally cheap protection against exhaustive brute-force attacks, and this idea has been proved to be sound [47].

8.13 Experimenting with DES

To experiment with DES in Sage, use Python's own DES implementation from the Crypto.Cipher standard library.

```
sage: from Crypto.Cipher import DES
```

For ease of discussion, strings of 16 hexadecimal characters will be used for plaintext, key and ciphertext. Suppose the plaintext is

`0xAAAABBBBCCCCDDDD`

and the key is

`0x0123456789ABCDEF`.

To perform an encryption, a new DES object must be instantiated with the key, but first the key will need to be converted into bytes:

```
sage: k = '0123456789ABCDEF'
sage: key = k.decode('hex')
```

To encrypt, the plaintext must also first be converted into bytes:

```
sage: plaintext = 'AAAABBBBCCCCDDDD'
sage: pl = plaintext.decode('hex')
```

Now create a new DES instance with the key, and use the encrypt method:

```
sage: D = DES.new(key);
sage: ct = D.encrypt(pl)
sage: ct.encode('hex')
  '6f32c06fb6ac37bf'
```

Decryption uses the `decrypt` method.

```
sage: p = D.decrypt(ct)
sage: p.encode('hex')
  'aaaabbbbccccdddd'
```

8.14 Lightweight ciphers

A recent development has been in the construction and analysis of *lightweight ciphers*, which are designed to be highly efficient in both computational requirements and memory, and are designed to run in low-powered embedded devices such as RFID chips. For such ciphers, long-term data storage (as is provided by a modern cipher such as the AES) is not as necessary as the requirement that they can be run with very little overhead, using a very small number of hardware processes such as logic gates.

Lightweight DES

There are in fact two versions: DESL and DESXL. DESL is simply DES with the 8 different S-boxes replaced by one single S-box which is used 8 times; DESXL is DESL used with key whitening. Details are given by Leander et al. [54]. The proposed S-box is

S															
14	5	7	2	11	8	1	15	0	10	9	4	6	13	12	3
5	0	8	15	14	3	2	12	11	7	6	9	13	4	1	10
4	9	2	14	8	7	13	0	10	12	15	1	5	11	3	6
9	6	15	5	3	8	4	11	7	1	12	2	0	14	10	3

To generate this S-box, the authors used Copperman's S-box criteria given above in Section 8.10, but replaced numbers 6 and 8 with the single criterion:

> If two inputs x and y differ in their first bit and are identical in their last two bits (so that $x \oplus y = 1 * * * 00$) then their outputs must differ.

They showed that DESL could encrypt a single block using 1868 gate equivalents (roughly the number of individual hardware processes), as compared to 2309 for DES and 3400 for AES.

Tiny Encryption Algorithm

The Tiny Encryption Algorithm, or TEA, was developed by Wheeler and Needham in 1994 [94]. The authors presented it as a very simple cipher, easy both to describe and implement. In fact it is so simple that it is easier to describe by a program than in words. Here is the original C program as given by the authors:

```
void code(long* v, long* k) {
    unsigned long y=v[0], z=v[1], sum=0,          /* set up */
     delta=0x9e3779b9, n=32;   /* a key schedule constant */
    while (n-->0) {                   /* basic cycle start */
        sum += delta;
        y += ((z<<4) + k[0]) ^ (z + sum) ^ ((z>>5) + k[1]);
        z += ((y<<4) + k[2]) ^ (y + sum) ^ ((y>>5) + k[3]);
    }                                         /* end cycle */
    v[0]=y ; v[1]=z ;
}
```

The hexadecimal constant `0x9e3779b9` is equal to $\lfloor(\sqrt{5}-1)2^{31}\rfloor$. The plaintext consists of two 32-bit words (hence the block size is 64 bits), and the key consists of four 32-bit words. Note that there is no key schedule; the words of the key are used in their original form at each round.

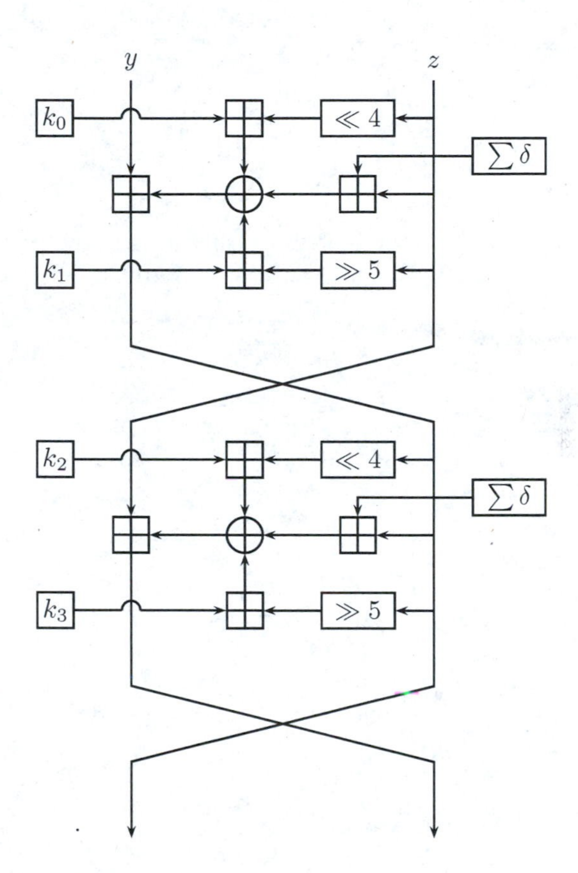

FIGURE 8.15: One cycle of TEA.

The cycle in the program is in fact two Feistel rounds, and a single cycle is shown in Figure 8.15.

The program above is easily translated into Sage:

```
def tea(plaintext,key):
  n = 32; N = 2^n
  delta = ZZ('0x9e3779b9'); sm = 0
  [z,y] = ZZ(plaintext).digits(N,padto=2)
  [k3,k2,k1,k0] = ZZ(key).digits(N,padto=4)
  for i in range(n):
      sm = (sm + delta)%N
      y = (y+ (((z<<4)+k0) ^^ (z+sm) ^^ ((z>>5)+k1)))%N
      z = (z+ (((y<<4)+k2) ^^ (y+sm) ^^ ((y>>5)+k3)))%N
  return hex(N*y+z)
```

In this implementation, all parameters are elements of the ring $\mathbb{Z}$ of integers, so that XOR and shifts can be immediately performed. Here is how it can be used:

```
sage: tea(0,0)
  '41ea3a0a94baa940'
sage: plaintext = '0x0123456789abcdef'
sage: key = '0x00112233445566778899aabbccddeeff'
sage: tea(ZZ(plaintext),ZZ(key))
  '126c6b92c0653a3e'
```

TEA has a very pleasing simplicity—in fact, too simple. It has been shown [46] that the full key of TEA can be recovered using 2^{23} plaintexts and roughly 2^{32} computations.

In response to the cryptanalysis of TEA, its authors first developed Block TEA, abbreviated as XTEA, and then a corrected version XXTEA. This last cipher can be described [99] as

```
void xxtea_full_cycle(uint32_t *v, int n, uint32_t*key, int cycle)
{
    uint32_t sum, z , y , e;
    int r;
    sum = cycle*0x9e3779b9;
    e = sum>>2;
    for (r = 0; r < n; r++)
    {
        z = v[(r+n1)%n]; // the left neighboring word
        y = v[(r+1)%n];   // the right neighboring word
        v[r] += ((z>>5^y<<2)+(y>>3^z<<4))^((sum^y)+(key[(r^e)%4]^z)) ;
    }
}
```

```
void xxtea (uint32_t *v, unsigned int n, uint32_t *k)
{
int i , cycles = 6+52/n;
for (i = 1; i < cycles; i++)
    xxtea_full_cycle(v , n , k , i);
}
```

This version is also breakable, although with more work required than with the original TEA. In spite of their deficiencies and vulnerabilities, the TEA family show very well how at least some security can be obtained with very simple means.

To finish this section, here is possibly the tiniest cipher yet developed: Treyfer, devised by Gideon Yuval in 1997 [100]. As with TEA, it uses a large number of rounds, and is to be considered, according to the author, for environments "for which even TEA and the SAFERs are gross overdesigns." It uses an S-box that is not prescribed; any 256×8 bytes will do, and these bytes can be obtained from anywhere, again minimizing the effective size of the cipher. The author charmingly claims that because of the use of a large number of rounds, "we hope to be able to scrounge an Sbox out of nowhere." Treyfer uses a 64-bit plaintext and key, considered as 8 bytes each.

Here is the code as given by its author:

```
for (r=0; r<NumRounds; r++) {
  text[8]=text[0];
  for (i=0; i<8; i++) {
    text[i+1]=(text[i+1]+Sbox[(key[i]+text[i])%256])<<<1;
  }
  text[0]=text[8];
}
```

8.15 Glossary

Block cipher. A bit-oriented cipher designed to encrypt a plaintext of fixed length with a fixed length key.

DES. The Data Encryption Standard, a Feistel-based substitution/permutation cipher.

Expansion permutation. A permutation that also involves repeating some of the input bits, so that the output is longer than the input.

Key schedule. The method by which a new round key is obtained for each round of the cipher.

Lightweight cryptosystem. A block cipher specifically designed to run in restricted memory or hardware, such as in a smart card or on an RFID chip.

S-Box. A list of substitutions, usually given as a list of length 2^n, so that a block b of n bits will be substituted by the b-th values of the list (b considered as a binary integer). Elements of the list are interpreted as binary blocks.

Substitution/permutation cipher. A cipher that obtains its strength from mixing two operations: substitution of a block of bits for another block, and permuting a block of bits.

Exercises

Review Questions

1. What are the differences between a block cipher and a stream cipher?
2. What is a "round"?
3. Give an informal definition of confusion and diffusion.
4. What is meant by a "mode of encryption"?
5. Why is ECB considered to be an insecure mode?
6. What is an initialization vector?
7. What is a "nonce value"?
8. Define a Feistel cipher.
9. What is an S-box?
10. What is triple DES?
11. Why is double DES not considered an improvement over (single) DES?
12. Why is DES no longer used?
13. What is an "avalanche effect"?
14. Why is it necessary to consider light-weight cryptosystems?

Beginning Exercises

15. Work through the sDES example in the notes, and check all the intermediate steps.

16. Work through the decryption of the example ciphertext.

17. For each of the following plaintext and key pairs, encrypt using sDES, and then decrypt the ciphertext.

    (a) $p = 00000000$, $k = 0000000000$,

    (b) $p = 00001111$, $k = 0000011111$,

    (c) $p = 11111111$, $k = 0011001100$,

    (d) $p = 11110000$, $k = 1111111111$.

18. Write down in full the final inverse permutation of DES.

19. (Assume the key is all zeros.) For each of the following 32-bit blocks, apply the expansion permutation and then the S-box permutations to the results:

    (a) $1111 \cdots 1111$,

    (b) $0000 \cdots 0000$,

    (c) $1010 \cdots 1010$,

    (d) $11001100 \cdots 11001100$,

    (e) $11110000 \cdots 11110000$,

    (f) $11 \cdots 1100 \cdots 00$ (16 1s followed by 16 0s).

20. (Again assume the key is all zeros.) For each 32-bit block s in the previous question, let $R_i = L_i = s$. Determine R_{i+1}.

21. Choose any 64-bit binary block, and any 56-bit key. Using your key as the input to a round of DES, put your block through one complete round.

22. Show that DES satisfies the *complementation property*: that $\overline{E(P, K)} = E(\overline{P}, \overline{K})$, where $\overline{X}$ is the bit complement of X.

Sage Exercises

23. Check the answers to question 17 using the implementation of sDES in Sage.

24. Test vectors for DES are available at `http://www.skepticfiles.org/faq/testdes.htm`. Here are several:

Key	Plaintext	Ciphertext
7CA110454A1A6E57	01A1D6D039776742	690F5B0D9A26939B
43297FAD38E373FE	762514B829BF486A	EA676B2CB7DB2B7A
025816164629B007	480D39006EE762F2	A1F9915541020B56
0101010101010101	0123456789ABCDEF	617B3A0CE8F07100

Check these four tests and, if you like, any others from the above site.

25. The TEA decryption routine as initially given is

```
void decode(long* v, long* k) {
unsigned long n=32, sum, y=v[0], z=v[1],
  delta=0x9e3779b9;
sum = delta << 5;
while (n-->0) {                            /* start cycle */
    z -= ((y<<4) + k[2]) ^ (y + sum) ^ ((y>>5) + k[3]);
    y -= ((z<<4) + k[0]) ^ (z + sum) ^ ((z>>5) + k[1]);
    sum -= delta;
    }                                      /* end cycle */
v[0]=y; v[1]=z;
}
```

Translate this into Sage and test your program on the output given above.

26. Calculate the Seberry robustness measure ϵ for S-boxes 2 to 8 of DES.

27. Calculate the Seberry robustness measure ϵ value for S-box S of DESL.

Chapter 9

Finite fields

Finite fields are mathematical structures that have enormous applications in cryptography. In fact, finite fields have been used already, but without investigating their algebraic underpinnings. This chapter provides the formal definitions of finite fields, and provides the basis for discussions of cryptosystems that require them, such as the Advanced Encryption Standard and elliptic curve cryptosystems. The theory of fields will be developed slowly, starting with a simpler structure: the group.

In particular, this chapter will explore:

- The definition of groups, rings and fields, with some of their most basic algebraic properties.
- Definition and creation of finite fields.
- Computation on finite fields.

9.1 Groups and rings

There are many structural similarities between the additive congruences modulo n, and multiplicative congruences modulo p, where p is prime:

Property	Addition $\pmod n$	Multiplication $\pmod p$
Closure	$a+b$ is a residue $\pmod n$	ab is a residue $\pmod p$
Commutativity	$a+b=b+a \pmod n$	$ab=ba \pmod p$
Associativity	$(a+b)+c=a+(b+c) \pmod n$	$(ab)c=a(bc) \pmod p$
Identity	The element 0 has the property that $a+0=a$ for all $a \pmod n$	The element 1 has the property that $a1=a$ for all $a \pmod p$
Inverses	For each a there is a "negative" b such that $a+b=0 \pmod n$	For each (non-zero) a there is an "inverse" b such that $ab=1 \pmod p$

Any mathematical structure endowed with a binary operation and identity element that satisfies all the above properties is called an *abelian group*; if it satisfies all properties except for commutativity it is a *group*. The group concept is one of the most profoundly unifying concepts in all of modern mathematics; groups are everywhere, and the study of their structures and properties is a very active branch of research.

Formally, a group is a set of elements S with a distinguishing element e, and a binary operation $\circ$ for which the following properties hold:

1. **Closure:** For every $x, y \in S$, $x \circ y \in S$.

2. **Associativity:** For every $x, y, z \in S$, $(x \circ y) \circ z = x \circ (y \circ z)$.

3. **Identity:** The element e is an identity element, in that $x \circ e = e \circ x = x$ for all $x \in S$.

4. **Inverses:** Every element x has an inverse y for which $x \circ y = y \circ x = e$.

If the group satisfies the extra property:

5. **Commutativity:** $x \circ y = y \circ x$ for all $x, y \in S$

then it is abelian. It is conventional to use addition to denote the operation in an abelian group, even though the operation may have no connection at all with arithmetic addition. The composition of an element with itself n times

$$\underbrace{x \circ x \circ x \cdots \circ x}_{n \text{ times}}$$

is denoted x^n, or nx if the group is abelian.

Historical note: Although aspects of group theory in particular settings goes back as far as Euler, group theory in its modern general form started early in the 19th century, with the work of Cauchy, Galois and Abel. Cauchy showed that groups of permutations constituted a theory to be studied in its own right, and Galois and Abel were able to use groups to demonstrate first the insolubility of the general fifth degree polynomial equation (the "quintic") using radicals alone, and then the conditions under which a polynomial equation could be solved. An introduction to the history of group theory is given by Kleiner [48], and a more in-depth discussion is given by Wussing [95].

An abelian group G is *cyclic* if there is an element $g \in G$ distinct from the identity for which the sequence

$$g, g^2, g^3, \ldots$$

generates all elements of G. If the group is finite then $x^m = e$ for some m. The integers modulo n with addition is a cyclic group, generated by 1. The non-zero integers modulo p (with p being prime) with multiplication is also a cyclic group, generated by any primitive root. In any finite group G, given an element x, the *order* of x is the smallest value of n for which $x^n = e$. In the group $\mathbb{Z}_{10}$ with identity $e = 0$, the orders of each element are

element	0	1	2	3	4	5	6	7	8	9
order	1	10	5	10	5	2	5	10	5	10

Given two groups $G_1 = (G, \circ)$, $G_2 = (H, \otimes)$, their *direct product* is defined to be the group defined on the set

$$G_1 \times G_2 = \{(g_1, g_2) \in G \times H\}$$

with the operation

$$(g_1, h_1) \bullet (g_2, h_2) = (g_1 \circ g_2, h_1 \otimes h_2).$$

If G_1 and G_2 are abelian, then it is a convention to refer instead to the *direct sum* and write $G_1 \oplus G_2$. The group $\mathbb{Z}_2 \oplus \mathbb{Z}_2$ is the smallest non-cyclic group.

Small groups can be studied by means of their *Cayley tables*, which list the value of $x \circ y$ for all x, y in the group. For the group $\mathbb{Z}_2 \oplus \mathbb{Z}_2$, the elements are $(0,0), (0,1), (1,0), (1,1)$ and the Cayley table is

$\circ$	$(0,0)$	$(0,1)$	$(1,0)$	$(1,1)$
$(0,0)$	$(0,0)$	$(0,1)$	$(1,0)$	$(1,1)$
$(0,1)$	$(0,1)$	$(0,0)$	$(1,1)$	$(1,0)$
$(1,0)$	$(1,0)$	$(1,1)$	$(0,0)$	$(0,1)$
$(1,1)$	$(1,1)$	$(1,0)$	$(0,1)$	$(0,0)$

If the elements $(0,0), (0,1), (1,0), (1,1)$ are denoted by $0, a, b, a+b$ respectively, and the operation as addition, the Cayley table shows the structure more clearly:

$+$	0	a	b	$a+b$
0	0	a	b	$a+b$
a	a	0	$a+b$	b
b	b	$a+b$	0	a
$a+b$	$a+b$	b	a	0

Note that the group cannot be cyclic as $x + x = 0$ for all x.

These tables can be generated in Sage:

```
sage: G = AbelianGroup(2, [2,2])
sage: G.cayley_table(names='elements')
   *       1     f1     f0 f0*f1
    +-----------------------
   1|      1     f1     f0 f0*f1
  f1|     f1      1 f0*f1     f0
```

```
    f0|     f0 f0*f1       1      f1
f0*f1| f0*f1      f0      f1       1

sage: G.is_cyclic()
  False
```

(Notice that Sage uses 1 and multiplication instead of 0 and addition.)

Two groups $(G, \circ)$ and $(H, \otimes)$ are *isomorphic* if the number of elements of G and H are equal and there is a bijective function

$$F : G \rightarrow H$$

that satisfies

$$f(x \circ y) = f(x) \otimes f(y)$$

for all $x, y \in G$. The function f is then an *isomorphism* between the groups. Two groups that are isomorphic are "essentially" the same.

The groups $G = \mathbb{Z}_2 \oplus \mathbb{Z}_2$ and $H = \mathbb{Z}_4$ are not isomorphic, as all elements of G have order 2, and two elements of H have order 4. There is thus no way of creating an isomorphism between them.

Rings

A group has only one operation of interest, but in fact for many groups there are other possible operations which can be defined on their elements. The set of integers $\mathbb{Z}$ forms a group under addition, but integers also have multiplication defined, although multiplicative inverses do not exist, except for 1. The *ring* concept extends the group concept to allow for another operation.

Formally, a ring is a set R with two operations, denoted $+$ and $\times$ (called *addition* and *multiplication*), and two distinguished elements 0 and 1 which satisfy

1. $(R, +)$ is an abelian group with 0 as the identity element.
2. The operation $x \times y$ is closed and associative over R.
3. The element 1 is an identity for $\times$: $x \times 1 = 1 \times x = x$ for all $x \in R$.
4. Multiplication is distributive over addition:

   $$x \times (y + z) = (x \times y) + (x \times z), \qquad (y + z) \times x = (y \times x) + (z \times x)$$

   for all $x, y, z \in R$.

Note that in a ring, multiplication is not required to be commutative (although it sometimes is). Examples of rings include

1. The integers $\mathbb{Z}$ with arithmetic addition and multiplication.

2. The rational numbers $\mathbb{Q}$ with arithmetic addition and multiplication.

3. The set $\mathbb{Z}_n$ of integers modulo n.

4. The set of square matrices of a given size, with the zero and identity matrices as the 0 and 1 elements. This is an example of a ring in which multiplication is not commutative.

5. The set of all polynomials

$$a_0 + a_1x + a_2x^2 + \cdots + a_kx^k$$

with each $a_0 \in \mathbb{Z}$.

6. Given any ring R, the set of polynomials whose coefficients are in R.

9.2 Introduction to fields

A *field* is simply a ring for which both $(R, +, 0)$ and $(R - \{0\}, \times, 1)$ are abelian groups, and in which multiplication is distributive over addition. That is, a field F must satisfy all of the following.

1. Addition and multiplication are *closed* over F; if x and y are elements of F, then so are $x+y$ and xy (note the use here of the standard convention that juxtaposition means multiplication: $xy = x \times y$).

2. Addition and multiplication are *commutative*: $x+y = y+x$ and $xy = yx$; and *associative*: $x + (y + z) = (x + y) + z$ and $x(yz) = (xy)z$.

3. There is an additive identity 0: $x + 0 = x$, and a multiplicative identity 1: $x1 = x$.

4. Each element x has an additive inverse $-x$ so that $x+(-x) = 0$, and each non-zero element x has a multiplicative inverse x^{-1} so that $xx^{-1} = 1$.

5. Addition is distributive over multiplication: $x(y + z) = xy + xz$.

Examples of fields are the rational numbers $\mathbb{Q}$, the real numbers $\mathbb{R}$, and the complex numbers $\mathbb{C}$, each with respect to standard arithmetic addition and multiplication.

There are some useful consequences of the above laws:

1. $x0 = 0$ for all x. To see this:

$$x.0 = x(x + (-x)) = x^2 - x^2 = 0.$$

2. If $xy = 0$, then at least one of $x = 0$ or $y = 0$. That is, it is not possible to have two non-zero elements whose product is zero. To see this, suppose $xy = 0$ with $x \neq 0$, so that x^{-1} exists (and is non-zero). Multiplying through by x^{-1} gives $y = 0x^{-1} = 0$. This result is often stated by saying that a field *has no zero divisors.*

Finite fields of prime order

All the above examples of fields are infinite. But for cryptographic purposes, it will be necessary to deal with fields with a finite number of elements; such fields are called *finite fields*, or *Galois fields*, after the French mathematician Évariste Galois, who was the first to investigate them. A Galois field with n elements is sometimes denoted $GF(n)$.

Historical note: Évariste Galois (1811–1832) was a remarkable mathematician by any standard. Almost entirely self taught, he mastered some of the great mathematical classics of his time, used the word "group" in the modern sense, as well as introducing Galois fields. He laid to rest an age-old problem about the solution of polynomial equations, by relating their roots to the actions of a permutation group. Although he published several papers in his short life, some of his greatest work exists in the form of notes and letters to friends and other mathematicians. He was killed in a duel at the age of only 20.

Consider the integers modulo 7 with the usual addition and multiplication. It is straightforward to set up addition and multiplication tables:

+	0	1	2	3	4	5	6
0	0	1	2	3	4	5	6
1	1	2	3	4	5	6	0
2	2	3	4	5	6	0	1
3	3	4	5	6	0	1	2
4	4	5	6	0	1	2	3
5	5	6	0	1	2	3	4
6	6	0	1	2	3	4	5

×	0	1	2	3	4	5	6
0	0	0	0	0	0	0	0
1	0	1	2	3	4	5	6
2	0	2	4	6	1	3	5
3	0	3	6	2	5	1	4
4	0	4	1	5	2	6	3
5	0	5	3	1	6	4	2
6	0	6	5	4	3	2	1

It may not be immediately obvious from these tables, but the set of integers {0,1,2,3,4,5,6}, with addition and multiplication mod 7, forms a field.

Now some of the above field laws can be checked: 0 and 1 are the additive and multiplicative identities, and the negatives and inverses can be read from the tables:

x	$-x$	x^{-1}
0	0	
1	6	1
2	5	4
3	4	5
4	3	2
5	2	3
6	1	6

Note that zero does not have a multiplicative inverse. The other laws, commutativity, associativity and distributivity, carry across from these laws for integers, and from the properties of congruences. This particular field is often denoted $\mathbb{Z}/7\mathbb{Z}$ or $\mathbb{Z}_7$, or $GF(7)$.

So, the integers mod 7 form a field. But what's so special about 7? Can we have fields of six elements, or of eight elements? Try 6. Here are the addition and multiplication tables:

$+$	0	1	2	3	4	5
0	0	1	2	3	4	5
1	1	2	3	4	5	0
2	2	3	4	5	0	1
3	3	4	5	0	1	2
4	4	5	0	1	2	3
5	5	0	1	2	3	4

$\times$	0	1	2	3	4	5
0	0	0	0	0	0	0
1	0	1	2	3	4	5
2	0	2	4	0	2	4
3	0	3	0	3	0	3
4	0	4	2	0	2	4
5	0	5	4	3	2	1

Clearly multiplication isn't as well behaved as it was for mod 7. In fact, most numbers do not have a multiplicative inverse. And this should come as no surprise, recalling that

$$x^{-1} \pmod{n}$$

exists only when x and n are relatively prime. Thus 2, 3 and 4 will not have inverses mod 6, since none of them are relatively prime to 6.

For the integers mod n to form a field, every integer $1 \leq x \leq n-1$ must have an inverse mod n; that is, it is relatively prime to n. But this will happen only when n is prime. This leads to a very important result:

The integers $0, 1, \ldots, n-1$ with the operations of addition and multiplication mod n form a field if and only if n is a prime number.

9.3 Fundamental algebra of finite fields

The characteristic of a field

The *characteristic* of a field is the smallest number n such that

$$\underbrace{1+1+\cdots+1}_{n} = 0$$

or 0 if no such positive integer exists. A fundamental result in the theory of fields is that if a field has only a finite number of elements, then its characteristic must be a prime number.

First, suppose that the characteristic is 0. This means that

$$\underbrace{1+1+\cdots+1}_{n} = n.1 \neq 0$$

for all n. A consequence of this is that all values $n.1$ must be different, for if

$$m.1 = n.1$$

for $m > n$, say, then

$$(m-n).1 = 0$$

from which it follows that $m = n$, as a field has no zero divisors. But if all values $n.1$ were distinct, then the field would be infinite in size.

Let n be the smallest number such that $n.1 = 0$. Suppose that n is composite, so that $n = ab$ with both $a, b > 1$. Then

$$0 = ab.1 = (a.1).(b.1)$$

where both $a.1$ and $b.1$ must be non-zero (because each of $a, b < n$ and by definition n is the smallest value for which $n.1 = 0$). But this means that the field must have zero divisors, which is a contradiction. Thus n must be prime.

The order of a field

Let F be a finite field. One example is the field consisting of all residues modulo a given prime p. This field has p elements. Thus there is a field of order p for all possible primes p. But what about other possible orders?

Suppose that F has characteristic p. Then for every $x \in F$, it follows that $px = 0$, (since $px = p.1.x = 0x = 0$). This means that x can be expressed as a sum

$$x = a_0v_0 + a_1v_1 + \cdots + a_{m-1}v_{m-1}$$

with each $a_i \in \mathbb{Z}/p\mathbb{Z}$. (Since F is finite, so must m be.) To see this, let

$$K = \{0, 1, 2, \ldots, p-1\} \subseteq F.$$

Set $v_0 = 1$. If $K \neq F$ then pick any element in $F - K$ and call it v_1. Now set

$$K^2 = \{a_0 v_0 + a_1 v_1\}$$

with each $a_0, a_1 \in \mathbb{Z}/p\mathbb{Z}$. It can be easily shown that $K^2 \subseteq F$ and has p^2 elements. If $K^2 \neq F$ then pick any element in $F - K^2$, call it v_2 and form the set

$$K^3 = \{a_0 v_0 + a_1 v_1 + a_2 v_2\}.$$

This process can be continued, producing K^4, K^5 and so on until eventually $K^m = F$ for some m. This must happen because

$$K \subset K^2 \subset K^3 \subset \cdots \subset F.$$

Suppose that $K^m \neq F$ for any m. Then since K is finite, there must be an integer q for which

$$K^q \subset F \subset K^{q+1}$$

and so there must be an element $x \in F - K^q$. This means that the set

$$K^{q+1} = \{z + a_q x\}$$

with $z \in K^q$ and $a_q \in \mathbb{Z}/p\mathbb{Z}$ must be a strict superset of F. But this is a contradiction.

This argument shows that any finite field F of characteristic p must have p^m elements. Alternatively, the language of linear algebra could have been used to say that F is a vector space of dimension m, with a basis

$$\{v_0, v_1, \ldots, v_{m-1}\}.$$

This means that, for example, there are no finite fields of orders 6 or 10. What remains to be shown is the existence of finite fields for all primes p and exponents m, as well as the relationships between fields of the same order. The following results, which are fundamental, will be stated but not proved here. Proofs can be found in many algebra texts, for example Hungerford [39] or Lang [53].

1. For every prime p and integer m there is a finite field of order p^m.

2. Any two finite fields of the same order are *isomorphic*; that is, there is a map that preserves addition and multiplication.

9.4 Polynomials mod 2

To create fields of non-prime order (that is, fields with a finite, but composite number of elements), elements will be polynomials rather then integers. For cryptographic purposes the greatest interest is in polynomials whose coefficients are just 0 or 1, and for which all arithmetic is done mod 2. This set of polynomials is denoted $\mathbb{Z}_2[x]$.

Polynomial arithmetic

Addition. Polynomials can be added very easily:

$$\begin{aligned}
&(x^4 + x^3 + x + 1) + (x^3 + x^2 + x) \\
&\quad = \quad x^4 + x^3 + x^3 + x^2 + x + x + 1 \qquad \text{removing brackets} \\
&\quad = \quad x^4 + (1 + 1)x^3 + x^2 + (1 + 1)x + 1 \qquad \text{collecting like terms} \\
&\quad = \quad x^4 + x^2 + 1 \qquad \text{simplifying coefficients mod 2.}
\end{aligned}$$

The last two lines can also be done by thinking of pairs of like powers cancelling out. Since addition and subtraction mod 2 are equivalent, the above result is also the answer to

$$(x^4 + x^3 + x + 1) - (x^3 + x^2 + x).$$

Multiplication. This can be done by the usual process of expanding brackets, and then simplifying the result:

$$\begin{aligned}
&(1 + x + x^3 + x^4)(1 + x^2 + x^3 + x^4) \\
&\quad = \quad 1(1 + x^2 + x^3 + x^4) + x(1 + x^2 + x^3 + x^4) \\
&\qquad\quad +x^3(1 + x^2 + x^3 + x^4) + x^4(1 + x^2 + x^3 + x^4) \\
&\quad = \quad 1 + x^2 + x^3 + x^4 + x + x^3 + x^4 + x^5 \\
&\qquad\quad +x^3 + x^5 + x^6 + x^7 + x^4 + x^6 + x^7 + x^8 \\
&\quad = \quad 1 + x + x^2 + x^3 + x^4 + x^8.
\end{aligned}$$

Division. Polynomial division, for example

$$(x^8 + x^7 + x^6 + x^2 + x + 1)/(x^4 + x^2 + x + 1)$$

and finding the remainder can be performed by the usual long division of polynomials, with all arithmetic mod 2:

$$
\begin{array}{r|l}
 & x^4 + x^3 \\
\hline
x^4 + x^2 + x + 1 & x^8 + x^7 + x^6 + 0x^5 + 0x^4 + 0x^3 + x^2 + x + 1 \\
 & x^8 \phantom{{}+x^7} + x^6 + x^5 + x^4 \\
\cline{2-2}
 & \phantom{x^8 +{}} x^7 \phantom{{}+x^6} + x^5 + x^4 \\
 & \phantom{x^8 +{}} x^7 \phantom{{}+x^6} + x^5 + x^4 + x^3 \\
\cline{2-2}
 & \phantom{x^8 + x^7 + x^6 + x^5 + x^4 +{}} x^3 + x^2 + x + 1
\end{array}
$$

and so the result of the division is $x^4 + x^3$, with remainder $x^3 + x^2 + x + 1$. Note that to do this it is easiest to write out *all* possible powers x^k, using 0 coefficients for powers not in the original polynomial. Such a polynomial long division as above can be more easily done by just writing down the coefficients, and not all the powers:

$$
\begin{array}{r|ccccccccc}
 & 1 & 1 & & & & & & & \\
\hline
1\,0\,1\,1\,1 & 1 & 1 & 1 & 0 & 0 & 0 & 1 & 1 & 1 \\
 & 1 & 0 & 1 & 1 & 1 & & & & \\
\cline{2-6}
 & & 1 & 0 & 1 & 1 & 0 & 1 & 1 & 1 \\
 & & 1 & 0 & 1 & 1 & 1 & & & \\
\cline{3-7}
 & & & & & & 1 & 1 & 1 & 1
\end{array}
$$

The *degree* of a polynomial is defined to be the highest power in it.

All these operations can be done in Sage, first defining x to be a variable in $\mathbb{Z}_2$:

```
sage: R.<x>=GF(2)[]
```

The variable R is now a polynomial ring defined over the finite field $\mathbb{Z}_2$. All the usual arithmetic on polynomials can now be performed in this ring:

```
sage: p = x^4+x^3+x+1
sage: q = x^3+x^2+x
sage: p+q
  x^4 + x^2 + 1
sage: p = x^8+x^7+x^6+x^2+x+1
sage: q = x^4+x^2+x+1
sage: p.quo_rem(q)
  (x^4 + x^3, x^3 + x^2 + x + 1)
```

9.5 A field of order 8

Since 8 is not a prime number, the integers modulo 8 with addition and multiplication do not form a field. However, polynomials of degree 2 or less can be used instead. In fact, there are simple equivalences between all numbers 0 to 7, their binary representations, and the polynomials for which the binary digits are the coefficients:

0	000	0
1	001	1
2	010	x
3	011	$x+1$
4	100	x^2
5	101	x^2+1
6	110	x^2+x
7	111	$x^2+x+1.$

Operations on these can be defined as follows: addition is just polynomial addition (mod 2); so that, for example:

$$\begin{aligned}(x+1)+(x^2+x) &= x^2+(1+1)x+1 \\ &= x^2+1.\end{aligned}$$

The binary representations can be used for a more compact display:

$$\begin{array}{rccc} & 0 & 1 & 1 \\ + & 1 & 1 & 0 \\ \hline & 1 & 0 & 1 \end{array}$$

from which it can be seen that addition is simply XOR of the bits.

Multiplication is performed mod x^3+x+1 (this polynomial will be discussed below). So that for example

$$\begin{aligned}(x^2+1)(x^2+x) &= x^2(x^2+x)+1(x^2+x) \\ &= x^4+x^3+x^2+x.\end{aligned}$$

Now this result must be divided by x^3+x+1 using polynomial long division (writing down the coefficients only):

$$\begin{array}{llllllll} & & & & 1 & 1 & & \\ \cline{5-8} 1 & 0 & 1 & 1\,\big) & 1 & 1 & 1 & 1 & 0 \\ & & & & 1 & 0 & 1 & 1 & \\ \cline{5-9} & & & & & 1 & 0 & 0 & 0 \\ & & & & & 1 & 0 & 1 & 1 \\ \cline{6-9} & & & & & & & 1 & 1 \end{array}$$

and so the remainder corresponds to the polynomial $x+1$. Thus

$$(x^2+1)\times(x^2+x) = x+1 \pmod{x^3+x+1}.$$

Multiplication can in fact be performed with a lot less work, and this will be shown later.

The equivalent of this last product using the corresponding numbers in the above table is

$$5\times 6 = 3$$

or in binary

$$101 \times 110 = 011.$$

Addition and multiplication tables of polynomials are easily shown, where to make the structure of these fields a little more apparent the polynomials are replaced with their decimal equivalents as given in the table on page 226:

+	0	1	2	3	4	5	6	7
0	0	1	2	3	4	5	6	7
1	1	0	3	2	5	4	7	6
2	2	3	0	1	6	7	4	5
3	3	2	1	0	7	6	5	4
4	4	5	6	7	0	1	2	3
5	5	4	7	6	1	0	3	2
6	6	7	4	5	2	3	0	1
7	7	6	5	4	3	2	1	0

×	0	1	2	3	4	5	6	7
0	0	0	0	0	0	0	0	0
1	0	1	2	3	4	5	6	7
2	0	2	4	6	3	1	7	5
3	0	3	6	5	7	4	1	2
4	0	4	3	7	6	2	5	1
5	0	5	1	4	2	7	3	6
6	0	6	7	1	5	3	2	4
7	0	7	5	2	1	6	4	3

Note that 0 and 1 are the additive and multiplicative identities, and that each non-zero object has a multiplicative inverse. From the last table, for example, see that

$$3\times 6 = 1$$

or in polynomial terms

$$(x+1)\times(x^2+x) = 1$$

and so in this field

$$(x+1)^{-1} = x^2+x.$$

This particular field is denoted $\mathbb{Z}_2[x]/(x^3+x+1)$, or sometimes just $GF(8)$.

This field can now be investigated and explored in Sage. To create it, first define x to be an element of a polynomial ring over $\mathbb{Z}_2$, and then define the field itself.

```
sage: R.<x> = GF(2)[]
sage: F.<x> = GF(8,name='x',modulus=x^3+x+1)
```

The field can now be listed:

```
sage: F.list()
  [0, x, x^2, x + 1, x^2 + x, x^2 + x + 1, x^2 + 1, 1]
```

Arithmetic is very easy:

```
sage: X = x^2+x
sage: Y = x^2+1
sage: X+Y
  x + 1
sage: X*Y
  x + 1
sage: X/Y
  x^2 + x + 1
sage: X^5
  x^2 + 1
sage: X^-1
  x + 1
sage: 1/X
  x + 1
```

Irreducible polynomials

To make multiplication work in the above field, multiplication of polynomials had to be performed mod x^3+x+1. The importance of this polynomial is that it is *irreducible*; that is, it cannot be written as a product of polynomials of lower degree. It is a sort of "prime" polynomial. The only other irreducible polynomial of degree three is x^3+x^2+1, and the field could have been just as well defined using that polynomial.

Other polynomials of degree three are not irreducible, for example:

$$\begin{aligned} x^3+1 &= (x+1)(x^2+x+1) \\ x^3+x^2+x+1 &= (x+1)(x^2+1). \end{aligned}$$

If a composite polynomial is used to define a field, then there will be elements that are not invertible. For example, suppose $p(x) = x^3+1$ is used. What is the inverse of $x+1$? To find this inverse, multiply $x+1$ by all non-zero polynomials in the hope of obtaining 1:

$q(x)$	$q(x)(x+1) \pmod{x^3+1}$
1	$x+1$
x	x^2+x
$x+1$	x^2+1
x^2	x^2+1
x^2+1	x^2+x
x^2+1	$x+1$
x^2+x+1	$0.$

Not only has this attempt to find an inverse failed, the table shows that there are two non-zero elements whose product is zero. So multiplication mod x^3+1 does not produce a field.

9.6 Other fields $GF(2^n)$

The previous section showed how to define and construct a field of order $8 = 2^3$. The same methods can be used to generate fields of order 2^n, for any n:

1. The elements of the field are all polynomials of degree $n-1$ or less. These polynomials correspond to binary strings of length n, these strings being the coefficients of the polynomials. These binary strings can also be considered as the binary representations of the decimal numbers between 0 and $2^n - 1$.

2. Addition is polynomial addition mod 2. Considered as binary strings, addition is bit-wise XOR.

3. Multiplication is performed modulo $p(x)$, where $p(x)$ is an irreducible polynomial of degree n. It doesn't matter which $p(x)$ we choose, as the fields they generate are all fundamentally the same.

Below is a short list of some irreducible polynomials, for different n

n	$p(x)$
3	x^3+x+1
4	x^4+x+1
5	x^5+x^2+1
6	$x^6+x^4+x^3+x+1$
7	x^7+x^6+1
8	$x^8+x^7+x^6+x+1$

9.7 Multiplication and inversion

Although the product of two polynomials $q(x)$ and $r(x)$ has been defined to be $q(x)r(x) \pmod{p(x)}$ where $p(x)$ is an irreducible polynomial, there is an easier way in practice. Start by setting up a table where every non-zero polynomial corresponds to a power of x.

Since by definition of modulus,

$$x^3 + x + 1 = 0 \pmod{x^3 + x + 1}$$

it follows immediately that

$$x^3 = x + 1 \pmod{x^3 + x + 1}.$$

So a table can be constructed consisting of increasing powers of x. To go from one power to the next, multiply its polynomial by x, and whenever there is an x^3 term, replace it with $x + 1$:

$$\begin{aligned}
x^0 &= 1 \\
x^1 &= x \\
x^2 &= x^2 \\
x^3 &= x + 1 \\
x^4 &= x(x + 1) \\
&= x^2 + x \\
x^5 &= x(x^2 + x) \\
&= x^3 + x^2 \\
&= x + 1 + x^2 \\
&= x^2 + x + 1 \\
x^6 &= x(x^2 + x + 1) \\
&= x^3 + x^2 + x \\
&= x + 1 + x^2 + x \\
&= x^2 + 1 \\
x^7 &= x(x^2 + 1) \\
&= x^3 + x \\
&= x + 1 + x \\
&= 1.
\end{aligned}$$

Having done all the hard work, the table can be written as:

power of x	polynomial
x^0	1
x^1	x
x^2	x^2
x^3	$x+1$
x^4	x^2+x
x^5	x^2+x+1
x^6	x^2+1
x^7	1

Since $x^7 = 1$, it follows that for any integer n, $x^n = x^{n \pmod 7}$. Note that in the table above, the second column contains *all* the non-zero polynomials of degree 2 or less.

Now multiplication, division, inversion are easy. For the polynomials, use their power representation from the above table. Perform the operation on the powers, turn the resulting power into a value modulo 7, and read back the polynomial from the table.

For example, a multiplication:

$$\begin{aligned}(x^2+1)(x^2+x) &= x^6\,x^4 \quad \text{from the table} \\ &= x^{10} \\ &= x^3 \quad \text{reducing the power modulo 7} \\ &= x+1 \quad \text{from the table again.}\end{aligned}$$

Inversion:

$$\begin{aligned}(x^2+x)^{-1} &= (x^4)^{-1} \quad \text{from the table} \\ &= x^{-4} \\ &= x^3 \quad \text{reducing the power modulo 7} \\ &= x+1 \quad \text{from the table.}\end{aligned}$$

Division:

$$\begin{aligned}(x+1)/(x^2+1) &= (x+1)(x^2+1)^{-1} \\ &= x^3(x^6)^{-1} \quad \text{from the table} \\ &= x^3\,x^{-6} \\ &= x^{-3} \\ &= x^4 \quad \text{reducing the power modulo 7} \\ &= x^2+x \quad \text{from the table.}\end{aligned}$$

Note that in all the above examples, the arithmetic is understood to be performed modulo x^3+x+1.

One more example, for the field $\mathbb{Z}_2[x]/(x^4+x^3+1)$.

power of x	polynomial	power of x	polynomial
x^0	1	x^8	x^3+x^2+x
x^1	x	x^9	x^2+1
x^2	x^2	x^{10}	x^3+x
x^3	x^3	x^{11}	x^3+x^2+1
x^4	x^3+1	x^{12}	$x+1$
x^5	x^3+x+1	x^{13}	x^2+x
x^6	x^3+x^2+x+1	x^{14}	x^3+x^2
x^7	x^2+x+1	x^{15}	1

Again note that all non-zero polynomials of degree three or less appear in the table, and that since $x^{15}=1$, all the powers will be reduced modulo 15.

Primitive elements

Recall from Chapter 6 that a primitive root a modulo a prime number p has the property that its powers $a, a^2, \ldots, a^{p-1} \bmod p$ generate all the non-zero residues modulo p in some order. Equivalently, a primitive root is a value a for which the smallest power d for which $a^d = 1 \pmod{p}$ is $p-1$.

Likewise, in a finite field, a *primitive element*, or *generator* is a field element a whose powers generate all the non-zero elements of the field. Or, in the finite field $GF(n)$ the smallest power d for which $a^d = 1$ is $n-1$. For example, recall from chapter 6 that 6 is a primitive root of 11. That means that 6 is a primitive element, or generator, of the field $GF(11) = \mathbb{Z}_{11}$.

The tables above showed that powers of x generate all the non-zero elements of the fields $\mathbb{Z}_2[x]/(x^3+x+1)$ and $\mathbb{Z}_2[x]/(x^4+x^3+1)$, so that x is a primitive element of both these fields. But is x *always* a primitive element? Try powers of x for the field $\mathbb{Z}_2[x]/(x^4+x^3+x^2+x+1)$:

power of x	polynomial	power of x	polynomial
x^0	1	x^8	x^3
x^1	x	x^9	x^3+x^2+x+1
x^2	x^2	x^{10}	1
x^3	x^3	x^{11}	x
x^4	x^3+x^2+x+1	x^{12}	x^2
x^5	1	x^{13}	x^3
x^6	x	x^{14}	x^3+x^2+x+1
x^7	x^2	x^{15}	1

Not all the polynomials are generated here: there is just a set of five distinct polynomials repeated. Also note that the lowest power d for which $x^d = 1$ is $d = 5$, rather than $d = 15$. This means that x is *not* a primitive element of $\mathbb{Z}_2[x]/(x^4+x^3+x^2+x+1)$.

In general, the *multiplicative order* of an element e is the smallest value d

for which $e^d = 1$. A primitive element will have a multiplicative order that is one less than the order of the field. Suppose the field has order p^n for some prime p and integer n. Then for any element e it follows that

$$e^{(p^n-1)} = 1$$

and so the order d of e must divide $p^n - 1$. In particular, if $p^n - 1$ is prime, all nonzero elements will be primitive.

With the above field then, instead of using powers of x, create a table of powers of $x + 1$ (mod $x^4 + x^3 + x^2 + x + 1$):

power of $x+1$	polynomial	power of $x+1$	polynomial
$(x+1)^0$	1	$(x+1)^8$	x^3+1
$(x+1)^1$	$x+1$	$(x+1)^9$	x^2
$(x+1)^2$	x^2+1	$(x+1)^{10}$	x^3+x^2
$(x+1)^3$	x^3+x^2+x+1	$(x+1)^{11}$	x^3+x+1
$(x+1)^4$	x^3+x^2+x	$(x+1)^{12}$	x
$(x+1)^5$	x^3+x^2+1	$(x+1)^{13}$	x^2+x
$(x+1)^6$	x^3	$(x+1)^{14}$	x^3+x
$(x+1)^7$	x^2+x+1	$(x+1)^{15}$	1

and since every non-zero polynomial is generated by powers of $x+1$, it follows that $x + 1$ is a primitive element of $\mathbb{Z}_2[x]/(x^4 + x^3 + x^2 + x + 1)$.

Having found a primitive element, and produced a table of powers, the table can be used to perform multiplications and divisions:

$$\begin{aligned}(x^3+1)(x^3+x+1) &= (x+1)^8(x+1)^{11} \quad \text{from the table}\\ &= (x+1)^{19}\\ &= (x+1)^4 \quad \text{reducing modulo 15}\\ &= x^3+x^2+x \quad \text{reading from the table.}\end{aligned}$$

Given any finite field, one of the easiest ways of multiplying and dividing elements is to set up a table of field elements and corresponding powers of a primitive element.

Here is how it could be done for the table above. Each element of the field is a polynomial of degree 4 or less, and so can be represented as a string of four bits, or as a single hexadecimal character. Likewise the corresponding power can be represented as a 4-bit string, or single hexadecimal character. Thus as strings

power of $x+1$	polynomial	power of $x+1$	polynomial
0000	0001	1000	1001
0001	0011	1001	0100
0010	0101	1010	1100
0011	1111	1011	1011
0100	1110	1100	0010
0101	1101	1101	0110
0110	1000	1110	1010
0111	0111	1111	0001

These lists can be turned into two tables: one that gives the power corresponding to a given polynomial, and one that gives the polynomial corresponding to a given power.

	00	01	10	11
00		0	c	1
01	9	2	d	7
10	6	8	e	b
11	a	5	4	3

polynomial → power
"logarithms"

	00	01	10	11
00	1	3	5	f
01	e	d	8	7
10	9	4	c	b
11	2	6	a	1

power → polynomial
"powers"

To save space, the powers or polynomials are given as hexadecimal characters. For example, consider polynomial $x^3 + x + 1$, which corresponds to the string 1011. The first two bits give the row, the second two bits give the column. The character in row 10, column 11 of the "logarithms" table is b, which has binary expansion 1011, or decimal value 11. And this can be verified from the original table of powers of $x + 1$.

To perform a multiplication then, find the logarithms from the first table, add them (modulo 15), and find the corresponding polynomial from the second "powers" table:

$$(x^2 + x)(x^3 + x + 1) = (0110)(1011) \quad \text{transforming into binary.}$$

The corresponding logarithms are d and b, or in binary 1101 and 1011. These can be added modulo 1111 to produce 1001. Reading from the powers table produces 4, or in binary 0100, which is the polynomial x^2.

9.8 Multiplication without power tables

For a very large field, it may not in fact be convenient first to compute and store all possible powers of a primitive element, and the corresponding lookup tables for going backwards. This section describes a more general method of computation, which can be implemented very efficiently in software or hardware.

Multiplication

Multiplication is fairly straightforward. To find the product of $a(x)$ by $(b(x)$ where these are both polynomials in a field for which the irreducible polynomial is $p(x)$ it will be convenient to work with the bit strings corresponding to their coefficients. Suppose the corresponding bit strings are a, b

and p. For example, to compute

$$(x^2+1)(x^2+x) \pmod{x^3+x+1}$$

the bit strings are

$$\begin{aligned} a &= 1\,0\,1 \\ b &= 1\,1\,0 \\ p &= 1\,0\,1\,1. \end{aligned}$$

Note that in the field $GF(2^n)$, both a and b will have length n (appending zeros on the left if necessary), and p will have length $n+1$. To perform the multiplication, first enumerate the strings corresponding to all products of the form

$$b_k = x^k b(x), \quad \text{for} \quad 0 \le k \le n-1$$

and then XOR all strings corresponding to the powers in $a(x)$.

Enumerating the strings is as follows:

1. $k=0$: $b_0 = x^k b(x)$ is just the string b.

2. Given the string b_k corresponding to $x^k b(x)$, the string b_{k+1} corresponding to $x^{k+1}b(x)$ is obtained as follows:

   Append a zero to the end of b_k to make a string of length $n+1$. If the first element is zero, then b_{k+1} is the last n elements of this string. If the first element is 1, then XOR this string with p, and remove the first element (which will be zero).

For example, suppose $b(x) = x+1$ in $GF(8)$ with $p(x) = x^3+x+1$. Then:

$$\begin{array}{lccccccc} b_0 & = & & & & 0 & 1 & 1 \\ b_1 & = & & & 0 & 1 & 1 & 0 \\ & & & & & 1 & 1 & 0 \\ b_2 & = & & & 1 & 1 & 0 & 0 \\ & & & \oplus & 1 & 0 & 1 & 1 \\ & = & & & 0 & 1 & 1 & 1 \\ & = & & & & 1 & 1 & 1. \end{array}$$

For the example above, the polynomials are $a(x) = x^2+1$ and $b(x) = x^2+x$.

The strings are

$$\begin{array}{rcrrrr}
b_0 & = & & 1 & 1 & 0 \\
b_1 & = & 1 & 1 & 0 & 0 \\
 & \oplus & 1 & 0 & 1 & 1 \\
 & = & 0 & 1 & 1 & 1 \\
 & = & & 1 & 1 & 1 \\
b_2 & = & 1 & 1 & 1 & 0 \\
 & \oplus & 1 & 0 & 1 & 1 \\
 & = & 0 & 1 & 0 & 1 \\
 & = & & 1 & 0 & 1.
\end{array}$$

Now to perform the multiplication, b_0 and b_2 will be XOR-ed, since these correspond to the powers in $a(x)$:

$$\begin{array}{rrrr}
 & 1 & 1 & 0 \\
\oplus & 1 & 0 & 1 \\
\hline
 & 0 & 1 & 1
\end{array}$$

which corresponds to the polynomial $x + 1$.

Although this method of multiplication is complicated to explain and to write, it is in fact very efficient to perform in software, and even more efficient in hardware. The only operations are bitwise shifts and XOR, and these can be programmed to be performed very efficiently.

Inversion

There are no simple operations on bit strings to compute inversions $b(x)^{-1} \pmod{p(x)}$, where $b(x) \neq 0$ and $p(x)$ is irreducible. However, inversion can also be performed using a more systematic method than simply computing all possible products $a(x)b(x)$ until one is found that is equal to 1.

A version of the extended Euclidean algorithm will be used. Start with the table:

i	q	r	u
-1		$p(x)$	0
0		$b(x)$	1

and for every line following, perform a division $r_{i-2}(x)/r_{i-1}(x)$. The quotient becomes q_i, and the remainder is r_i. And

$$u_i = u_{i-2} + u_{i-1}q_i.$$

When $r_k = 1$ is reached for some k, the required inverse is u_k.

For example, to find

$$(x^2)^{-1} \pmod{x^3 + x + 1}.$$

The initial table is

i	q	r	u
-1		x^3+x+1	0
0		x^2	1

Now when the division $(x^3 + x + 1)/x^2$ is performed, a quotient x and a remainder $x + 1$ are obtained:

i	q	r	u
-1		x^3+x+1	0
0		x^2	1
1	x	$x+1$	

Then u_1 is computed by

$$\begin{aligned} u_1 &= u_{-1} + u_0 q_1 \\ &= 0 + 1x \\ &= x \end{aligned}$$

and so the table with completed row 1 is

i	q	r	u
-1		x^3+x+1	0
0		x^2	1
1	x	$x+1$	x

For row 2, perform the division $x^2/(x + 1)$ to obtain a quotient of $x + 1$ and a remainder of 1:

i	q	r	u
-1		x^3+x+1	0
0		x^2	1
1	x	$x+1$	x
2	$x+1$	1	

Then u_2 is computed by:

$$\begin{aligned} u_2 &= u_0 + u_1 q_2 \\ &= 1 + x(x+1) \\ &= x^2 + x + 1 \end{aligned}$$

and so the table with completed row 2 is

i	q	r	u
-1		x^3+x+1	0
0		x^2	1
1	x	$x+1$	x
2	$x+1$	1	x^2+x+1

Since $r_2 = 1$, the algorithm stops here, with

$$(x^2)^{-1} = x^2 + x + 1 \pmod{x^3 + x + 1}.$$

As with multiplication, all operations can be computed very efficiently.

9.9 Glossary

Abelian group. A group for which the operation is also commutative.

Characteristic. The smallest integer p for which $px = 0$ for all $x \in F$, or 0 if not such p exists.

Field. A set S with two operations $+$ and $\cdot$, two distinguishing elements 0 and 1 for which $(S, +, 0)$ and $(S\backslash\{0\}, \cdot, 1)$ are abelian groups, and addition is distributive over multiplication.

Finite field. A field F with a finite number of elements.

Group. A mathematical object consisting of a set and a binary operation which is closed and associative on the set, has an identity element, and an inverse for every element.

Irreducible polynomial. A polynomial that can't be factored into polynomials of lower degrees.

Primitive element. An element $x \in F$ whose powers generate all non-zero elements of F.

Ring. A set S with two operations (addition $(+)$ and multiplication $(\cdot)$, for which $(S, +, 0)$ is an abelian group, multiplication is closed, associative and distributive over addition.

Exercises

Review Exercises

1. What is a "zero divisor"? Why does a field not have any zero divisors?

2. What is a "cyclic group"? Are all abelian groups cyclic? If not, give a counterexample.

3. Are all cyclic groups abelian?

4. Why is it not possible to have finite fields with 6 or 10 elements?

5. Give an example of a ring that is not a field.

6. Give an example of a ring for which multiplication is not commutative.

7. Let S be the set of all 2×2 matrices over $\mathbb{R}$ with determinant 1. With I being the identity matrix, show that $(S, \cdot, I)$ is a group. Is it abelian?

8. What is the "multiplicative order" of a field element $x \in F$, and how does it relate to the order of F?

Beginning Exercises

9. Consider the set with elements $\{1, a, b, c, d, e\}$ and an operation $\circ$ defined as

$\circ$	0	a	b	c	d	e
0	0	a	b	c	d	e
a	a	0	c	b	e	d
b	b	d	0	e	a	c
c	c	e	a	d	0	b
d	d	b	e	0	c	a
e	e	c	d	a	b	0

and show that this structure forms a non-abelian group.

10. Show that any cyclic group must be abelian.

11. The *dihedral group of order 4*, denoted D_4, is formed by the rotations and reflections of a square. Denoting a $90°$ rotation clockwise by R, and a vertical reflection by F, there are eight elements of this group:

$$1, R, R^2, R^3, F, FR, FR^2, FR^3.$$

They can be illustrated by:

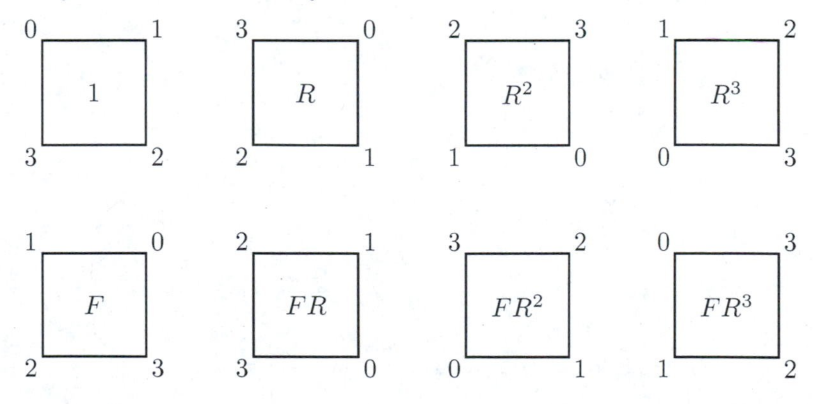

Show that these operations form a group, with juxtaposition as the operation, so that, for example $R^2 \circ FR$ consists of R^2 followed by FR, the result being FR^3.

12. Produce a Cayley table for D_4, and show that it is non-abelian.

13. Show that the group defined in question 9 is the dihedral group of order 3: the group formed by rotations and reflections of an equilateral triangle.

14. Let p and q be two distinct primes. Show that $\mathbb{Z}_p \oplus \mathbb{Z}_q$ is cyclic.

15. Show that if $f : G \to H$ is an isomorphism, then the identity element of G is mapped by f onto the identity element of H.

16. Using the field $\mathbb{Z}_7$ and the tables given on page 220, evaluate the following:

    (a) $(2+3)(4+5)^{-1}$, (b) $(3^4)/(4^5)$, (c) $1/2+3/4+4/5$,
    (d) $1/2+1/3+1/4+1/5+1/6$.

17. Using the field $\mathbb{Z}_2[x]/(x^3+x+1)$ and the tables given on page 227, evaluate the following:

    (a) $(2+3)(4+7)^{-1}$, (b) $(3^4)/(4^5)$, (c) $1/2+3/4+4/5+6/7$,
    (d) $1/2+1/3+1/4+1/5+1/6+1/7$.

18. Using the power table given on page 231, evaluate the following expressions $(\mathrm{mod}\, x^3+x+1)$:

    (a) $(x^2+1)(x+1)$, (b) $(x^2+x)(x^2+x+1)$,
    (c) $(x+1)(x^2+x)$, (d) $(x+1)/(x^2+1)$,
    (e) $(x^2+1)/(x^2+x+1)$, (f) $(x^2+x)/(x^2+1)$.

19. Verify the power table for the field $\mathbb{Z}_2/(x^4+x^3+1)$ given on page 232.

20. Using the power table for $\mathbb{Z}_2/(x^4+x^3+1)$, calculate the following expressions $(\mathrm{mod}\, x^4+x^3+1)$:

    (a) $x/(x^3+x+1)$, (b) $(x^3+x^2+1)(x^3+x+1)$,
    (c) $(x^3+x^2+1)/(x^3+x+1)$, (d) $(x^2+1)(x^3+x)$,
    (e) $(x+1)/(x^3+x+1)$, (f) $(x^2+x+1)/(x^3+x+1)$.

21. Attempt to write a power table mod x^3+x^2+x+1 for the powers $x^0,\ldots,x^7$. What does the table tell you about this polynomial?

22. For which of the following fields:

    (a) $\mathbb{Z}_2[x]/(x^4+x^3+x^2+x+1)$, (b) $\mathbb{Z}_2[x]/(x^4+x^3+1)$,
    (c) $\mathbb{Z}_2[x]/(x^4+x+1)$,

    is x^2 a primitive element?

Sage Exercises

23. Enter the finite field $\mathbb{Z}_2[x]/(x^5 + x^2 + 1)$.

    (a) List its elements.

    (b) Determine the values of $(x^2+1)^7$ and $(x^4+x^2+x+1)^3/(x^3+x+1)^5$.

    (c) Since $2^5 - 1$ is prime, what does that tell you about the multiplicative orders of each element?

24. Enter a finite field of 2^6 elements: $\mathbb{Z}_2[x]/(x^6 + x^4 + x^3 + x + 1)$.

    (a) What possible orders can elements of this field have?

    (b) Find an element for each possible order.

    (c) How many primitive elements does this field have?

25. Enter the field of 2^8 elements: $\mathbb{Z}_2[x]/(x^8 + x^4 + x^3 + x + 1)$.

    (a) Determine whether x is a primitive element in this field.

    (b) Is $x + 1$ a primitive element?

26. Sage can work with other prime powers:

    (a) Enter `F.<x> = GF(7^3)`

    (b) Find the irreducible polynomial by using `F.modulus()`

    (c) Find a primitive element for F.

27. Encrypt the message

    `TIME TO GO HOME`

    (including spaces) using a Hill cryptosystem, but over the field

    $$GF(3^3) = \mathbb{Z}_3/(x^3 + 2x + 1).$$

    The encryption matrix will be

    $$M = \begin{bmatrix} 2x^2+2 & x^2+x & 2x^2+2x+1 \\ x^2+2x & 2x+2 & 2 \\ x^2+x+2 & x^2+2x+1 & 1 \end{bmatrix}.$$

    (a) Enter the matrix M and find its inverse.

    (b) Find a primitive element p and use it to encode every letter of the alphabet to an element of the field. So, for example, `A` could be encoded as `p^(ord("A")-65`

    (c) What element does the space correspond to?

    (d) Encrypt the plaintext and decrypt the result.

28. The Chor-Rivest knapsack system [18] is based on finite fields. Here is a slightly simplified definition:

    (a) Choose a finite field of order $q = p^h$, with $f(x)$ being the irreducible polynomial. So the field is $\mathbb{Z}_p[x]/f(x)$. Let $g(x)$ be a primitive element in the field.
    (b) For each $i \in \mathbb{Z}_p$, determine the discrete logarithms $a_i = \log_{g(x)}(x+i)$.
    (c) Choose a random integer d so that $0 \leq d \leq p^h - 2$.
    (d) Compute $c_i = (a_i + d) \bmod (p^h - 1)$ for all $0 \leq i \leq p-1$.
    (e) Then your public key is $([c_0, c_1, \ldots, c_{p-1}], p, h)$ and your private key is $[f(x), g(x), d]$.

    Messages are binary strings m_i of length p with exactly h ones. Encryption is implemented by computing

$$C = \sum_{i=0}^{p-1} m_i c_i \bmod (p^h - 1).$$

    Decryption requires the following steps:

    (a) Compute $r = (C - hd) \bmod (p^h - 1)$.
    (b) Compute $u(x) = g(x)^r \pmod{f(x)}$.
    (c) Compute $s(x) = u(x) + f(x)$.
    (d) Factor $s(x)$ into linear factors

$$s(x) = \prod_{j=1}^{h} (x + t_j)$$

    where each $t_j \in \mathbb{Z}_p$. The values of the t_j in the factorization are the positions of the 1s in the message.

    Here is an example in Sage with small values:

```
sage: p = 7
sage: h = 4
sage: F.<x> = GF(p^h,name='x')
sage: f = F.modulus()
sage: g = F.multiplicative_generator()
sage: g.multiplicative_order()
2400
sage: a = [discrete_log(x+i,g) for i in range(p)]
sage: d = randint(0,p^h-2)
sage: c = [(i+d)%(p^h-1) for i in a]
```

Why does the multiplicative order of g show that g is a primitive element?

Here's an encryption:

```
sage: m = [0,1,1,1,0,0,1]
sage: C = sum(m[i]*c[i] for i in range(p))
```

and decryption:

```
sage: r = (C-h*d)%(p^h-1)
sage: u = g^r
sage: factor(u.polynomial()+f)
(x + 1) * (x + 2) * (x + 3) * (x + 6)
```

Try this out with a larger field, with $p = 23$ and $h = 10$. One other way of finding a random multiplicative generator is by entering

```
sage: g = F.random_element()
sage: g.multiplicative_order()==p^h-1
```

until you obtain "True." Also note that the computation of the discrete logarithms a may take a little while—just be patient!

Encrypt the plaintext

$$m = [0, 0, 0, 1, 1, 1, 1, 1, 0, 0, 0, 0, 0, 1, 1, 1, 1, 1, 0, 0, 0, 0, 0]$$

and decrypt the resulting ciphertext.

Further Exercises

29. Provide a formal proof that the Chor-Rivest cryptosystem "works," that the decryption routine does indeed recover the plaintext.

30. Show that if A is any square-free integer, then

$$F = \{a + b\sqrt{A} : a, b \in \mathbb{Q}\}$$

is a field.

31. Conway's "Nim-field" [20], or the field of "nimbers," is defined as follows:

    (i) The elements are the non-negative integers.

(ii) If $x = 2^m$ and $y = 2^n$ then

$$x + y = \begin{cases} 2^m + 2^n & \text{if } m \neq n, \\ 0 & \text{if } m = n. \end{cases}$$

This means, in effect, that addition is performed by XOR of the binary digits of x and y.

(iii) If $x = 2^{2^m}$ and $y = 2^{2^n}$ then

$$x \cdot y = \begin{cases} 2^{(2^m + 2^n)} & \text{if } m \neq n, \\ 3(2^{(2^n - 1)}) & \text{if } m = n. \end{cases}$$

(iv) All other additions and multiplications follow from the distributive law.

Using those definitions:

(a) List addition and multiplication tables for the subfield consisting of $\{0, 1, 2, 3\}$.

(b) Extend those tables to the subfield consisting of $\{0, 1, 2, \ldots, 13, 14, 15\}$.

(c) Write down a table of inverses for all non-zero elements of both subfields.

Chapter 10

The Advanced Encryption Standard

This chapter introduces the Advanced Encryption Standard, the current NIST-defined standard for secret key encryption. In particular:

- The history of choosing the AES cipher.
- The basic definition of Rijndael, the AES choice.
- Description of the components of Rijndael.
- A simplified version, suitable for hand computations.
- Some security discussion.

10.1 Introduction and some history

As a result of the problems inherent with DES, small keyspace, slow performance, fixed block size, in 1997 the National Institute of Standards and Technology (NIST) called for the creation of a new Advanced Encryption Standard. The basic requirements of this AES was that it should

- be publicly defined,
- be a symmetric block cipher,
- have a key length that may be increased as needed,
- be implementable in both hardware and software,
- be a) freely available or b) available under terms consistent with the American National Standards Institute (ANSI) patent policy.

By 1998, 15 cryptosystems were selected as possible candidates for this algorithm, and this list was whittled down to five. The five finalists were:

- MARS (submitted by IBM Corp.)
- RC6 (submitted by RSA Laboratories)

- Rijndael (submitted by Joan Daemen and Vincent Rijmen)
- Serpent (submitted by Ross Anderson, Eli Biham, Lars Knudsen)
- Twofish (submitted by Bruce Schneier, John Kelsey, Doug Whiting, David Wagner, Chris Hall, and Niels Ferguson).

After two years of intensive analysis and scrutiny, the Rijndael algorithm was selected to be the new AES, and its choice was announced in October 2000.

Before introducing the algorithm and its associated mathematics, a note on the pronunciation: *Rijndael* is pronounced similar to "Rhine-Dahl."

10.2 Basic structure

Rijndael shares with DES the property that encryption is done in a number of rounds (the number depends on the size of the block being used), and each round involves a number of steps. However, the similarity ends there. Rijndael is not a Feistel cipher; it is more of an algebraic cipher.

Bytes in Rijndael are added and multiplied according to the finite field

$$GF(2^8) = \mathbb{Z}_2[x]/(x^8 + x^4 + x^3 + x + 1).$$

This field consists of all polynomials of degree 7 or less; each polynomial can thus be identified with an 8-bit string, or a single byte. This particular field will be called the *Rijndael field.*

Although Rijndael is defined to operate on blocks of size 128, 192 and 256 bits, the AES standard is restricted to only 128 bit blocks, however with differing key sizes of 128, 192 or 256 bits. The description given here will consider only 128-bit keys; for definitions of the use of the other keys refer to the official standard [67].

Each round of Rijndael involves four steps called *layers*; they are:

ByteSub (BSB): This adds non-linearity to the algorithm, and adds resistance to both differential and linear cryptanalysis.

ShiftRow (SR): This mixes bits, and causes diffusion over repeated rounds.

MixColumn (MC): This is another mixing step.

AddRoundKey (ARK): The round key is added (using XOR) to the result of the MixColumn step.

An encryption of a single block starts with a single AddRoundKey step, using the 0th round key. Then there are nine rounds of all four of the above steps, finishing with all steps except for MixColumn. The reason for leaving out MixColumn from the last round will be discussed later. The scheme for Rijndael is shown in Figure 10.1.

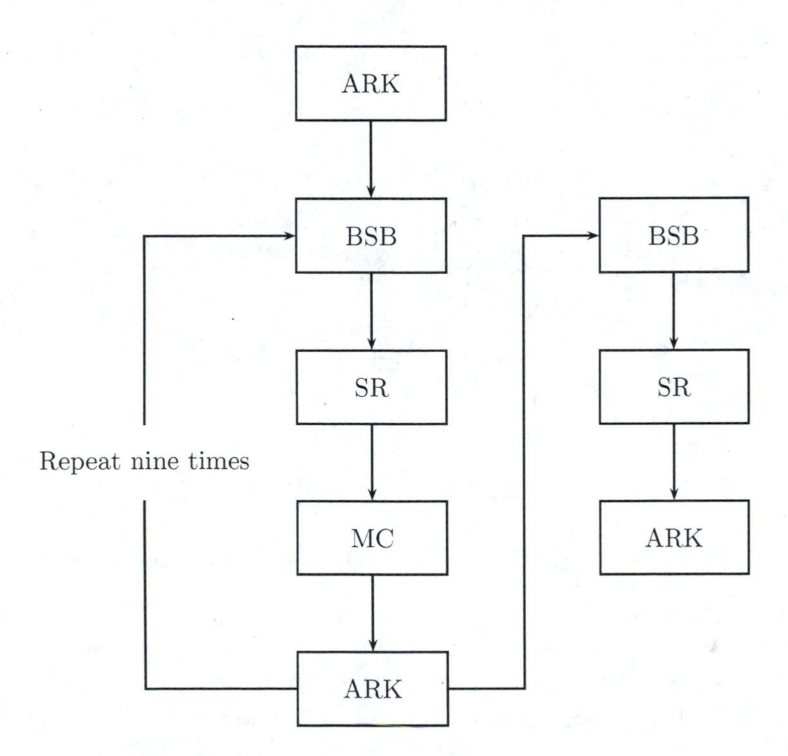

FIGURE 10.1: Rijndael encryption.

10.3 The layers in detail

For blocks of length 128 bits, a single block may be considered as consisting of 16 bytes: $a_0a_1a_2\dots a_{14}a_{15}$; these can be put into a 4×4 matrix, column by column,

$$A = \begin{bmatrix} a_0 & a_4 & a_8 & a_{12} \\ a_1 & a_5 & a_9 & a_{13} \\ a_2 & a_6 & a_{10} & a_{14} \\ a_3 & a_7 & a_{11} & a_{15} \end{bmatrix} = \begin{bmatrix} a_{0,0} & a_{0,1} & a_{0,2} & a_{0,3} \\ a_{1,0} & a_{1,1} & a_{1,2} & a_{1,3} \\ a_{2,0} & a_{2,1} & a_{2,2} & a_{2,3} \\ a_{3,0} & a_{3,1} & a_{3,2} & a_{3,3} \end{bmatrix}$$

where the second matrix shows the bytes indexed with row and column indices (the row and columns starting at zero rather than one).

Each layer then takes a matrix of bytes as input, and returns a matrix as output, as shown in Figure 10.2.

$$A \xrightarrow{\textbf{BSB}} B \xrightarrow{\textbf{SR}} C \xrightarrow{\textbf{MC}} D \xrightarrow{\textbf{ARK}} E$$

FIGURE 10.2: A single round of Rijndael.

ByteSub

Each byte $a_{i,j}$ is transformed to a byte $b_{i,j}$ as shown in Figure 10.3 by means of the S-box shown in Figure 10.4.

If the eight bits of $a_{i,j}$ correspond to the hexadecimal digits `XY`, then $b_{i,j}$ is the byte whose numerical value is in row `X` and column `Y` of the S-box, where rows and columns are numbered (in hexadecimal) from 0 to `F`.

For example, suppose $a_{i,j} = 10110101$. Then `X` = `0b1011` = `0xB`, or 11,

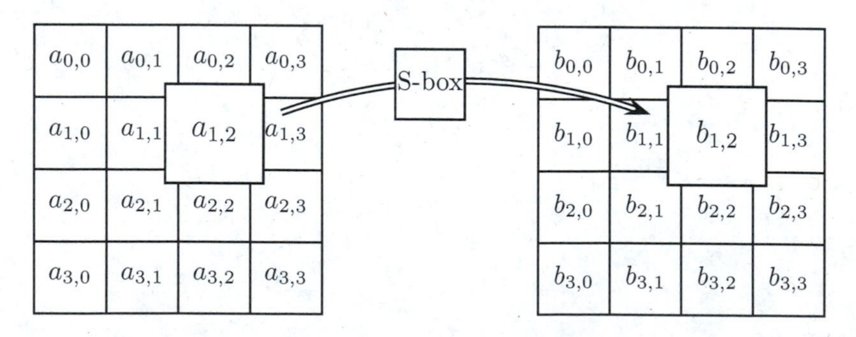

FIGURE 10.3: The ByteSub layer of Rijndael.

	0	1	2	3	4	5	6	7	8	9	A	B	C	D	E	F
0	63	7c	77	7b	f2	6b	6f	c5	30	01	67	2b	fe	d7	ab	76
1	ca	82	c9	7d	fa	59	47	f0	ad	d4	a2	af	9c	a4	72	c0
2	b7	fd	93	26	36	3f	f7	cc	34	a5	e5	f1	71	d8	31	15
3	04	c7	23	c3	18	96	05	9a	07	12	80	e2	eb	27	b2	75
4	09	83	2c	1a	1b	6e	5a	a0	52	3b	d6	b3	29	e3	2f	84
5	53	d1	00	ed	20	fc	b1	5b	6a	cb	be	39	4a	4c	58	cf
6	d0	ef	aa	fb	43	4d	33	85	45	f9	02	7f	50	3c	9f	a8
7	51	a3	40	8f	92	9d	38	f5	bc	b6	da	21	10	ff	f3	d2
8	cd	0c	13	ec	5f	97	44	17	c4	a7	7e	3d	64	5d	19	73
9	60	81	4f	dc	22	2a	90	88	46	ee	b8	14	de	5e	0b	db
A	e0	32	3a	0a	49	06	24	5c	c2	d3	ac	62	91	95	e4	79
B	e7	c8	37	6d	8d	d5	4e	a9	6c	56	f4	ea	65	7a	ae	08
C	ba	78	25	2e	1c	a6	b4	c6	e8	dd	74	1f	4b	bd	8b	8a
D	70	3e	b5	66	48	03	f6	0e	61	35	57	b9	86	c1	1d	9e
E	e1	f8	98	11	69	d9	8e	94	9b	1e	87	e9	ce	55	28	df
F	8c	a1	89	0d	bf	e6	42	68	41	99	2d	0f	b0	54	bb	16

FIGURE 10.4: Rijndael S-box.

and $Y =$ `0b0101` $=$ `0x5`, or 5. The value in row `B` and column 5 is `0xd5`, which has binary representation 11010101.

Although this S-box may seem like a random permutation of the bytes `0x0` to `0xFF`, it is in fact generated by a formal rule. Suppose the input is a byte b which may be considered as an element of the Rijndael field. Then define b' as

$$b' = \begin{cases} 0 & \text{if } b = 0 \\ b^{-1} & \text{otherwise.} \end{cases}$$

Define

$$a = y^4 + y^3 + y^2 + y + 1$$
$$c = y^6 + y^5 + y + 1$$

to be elements of the polynomial ring $R = \mathbb{Z}_2[y]/(y^8+1)$. Since $y^8 + 1$ is not irreducible, R is not a field. Define q to be

$$q = ab' + c.$$

The value q, treated as a byte, is the S-box value of b.

For example, take from above the byte $b = 10110101 = x^7+x^5+x^4+x^2+1$. Then its inverse is $x^6 + x^5 + x^4 + x^2 + 1$:

```
sage: K.<a> = GF(2)[]
sage: F.<x> = GF(256,name='x',modulus=a^8+a^4+a^3+a+1)
sage: b = x^7+x^5+x^4+x^2+1
```

```
sage: b1 = 1/b; b1
  x^6 + x^5 + x^4 + x^2 + 1
```

Now for the affine transformation:

```
sage: R.<y> = PolynomialRing(GF(2))
sage: a = y^4+y^3+y^2+y+1
sage: c = y^6+y^5+y+1
sage: q = (a*b1.polynomial().subs(x=y)+c).mod(y^8+1); q
  y^7 + y^6 + y^4 + y^2 + 1
```

This last polynomial corresponds to the byte 11010101 as above.

The S-box transformation can be equivalently described using matrices. Suppose b' is defined as above, and so

$$b' = b_7 b_6 b_5 b_4 b_3 b_2 b_1 b_0.$$

Then the S-box output is the byte $z = z_7 z_6 z_5 z_4 z_3 z_2 z_1 z_0$ for which

$$\begin{bmatrix} 1 & 0 & 0 & 0 & 1 & 1 & 1 & 1 \\ 1 & 1 & 0 & 0 & 0 & 1 & 1 & 1 \\ 1 & 1 & 1 & 0 & 0 & 0 & 1 & 1 \\ 1 & 1 & 1 & 1 & 0 & 0 & 0 & 1 \\ 1 & 1 & 1 & 1 & 1 & 0 & 0 & 0 \\ 0 & 1 & 1 & 1 & 1 & 1 & 0 & 0 \\ 0 & 0 & 1 & 1 & 1 & 1 & 1 & 0 \\ 0 & 0 & 0 & 1 & 1 & 1 & 1 & 1 \end{bmatrix} \begin{bmatrix} b_0 \\ b_1 \\ b_2 \\ b_3 \\ b_4 \\ b_5 \\ b_6 \\ b_7 \end{bmatrix} + \begin{bmatrix} 1 \\ 1 \\ 0 \\ 0 \\ 0 \\ 1 \\ 1 \\ 0 \end{bmatrix} = \begin{bmatrix} z_0 \\ z_1 \\ z_2 \\ z_3 \\ z_4 \\ z_5 \\ z_6 \\ z_7 \end{bmatrix}.$$

Note that the bytes y and z are listed in *reverse order.*

Again, take the byte 10110101 from above. In byte form its inverse is 01110101. Then

$$\begin{bmatrix} 1 & 0 & 0 & 0 & 1 & 1 & 1 & 1 \\ 1 & 1 & 0 & 0 & 0 & 1 & 1 & 1 \\ 1 & 1 & 1 & 0 & 0 & 0 & 1 & 1 \\ 1 & 1 & 1 & 1 & 0 & 0 & 0 & 1 \\ 1 & 1 & 1 & 1 & 1 & 0 & 0 & 0 \\ 0 & 1 & 1 & 1 & 1 & 1 & 0 & 0 \\ 0 & 0 & 1 & 1 & 1 & 1 & 1 & 0 \\ 0 & 0 & 0 & 1 & 1 & 1 & 1 & 1 \end{bmatrix} \begin{bmatrix} 1 \\ 0 \\ 1 \\ 0 \\ 1 \\ 1 \\ 1 \\ 0 \end{bmatrix} + \begin{bmatrix} 1 \\ 1 \\ 0 \\ 0 \\ 0 \\ 1 \\ 1 \\ 0 \end{bmatrix} = \begin{bmatrix} 1 \\ 0 \\ 1 \\ 0 \\ 1 \\ 0 \\ 1 \\ 1 \end{bmatrix}$$

which (read from the bottom up) is the correct S-box output.

Alternatively, given c as the byte 01100011, the transformation of b' to q can be defined bitwise as

$$q_i = b_i \oplus b_{(i+4) \bmod 8} \oplus b_{(i+5) \bmod 8} \oplus b_{(i+6) \bmod 8} \oplus b_{(i+7) \bmod 8} \oplus c_i.$$

The mapping $x \rightarrow x^{-1}$ was chosen to achieve non-linearity, and then combined with an affine transformation to resist certain attacks. Note also that the algebraic description of the S-box means that there is no possibility of any hidden structure, or of any backdoor, such as was an initial concern with DES.

ShiftRow

The rows in matrix B are shifted cyclically to the left by 0, 1, 2 and 3 places:

$$\begin{bmatrix} c_{0,0} & c_{0,1} & c_{0,2} & c_{0,3} \\ c_{1,0} & c_{1,1} & c_{1,2} & c_{1,3} \\ c_{2,0} & c_{2,1} & c_{2,2} & c_{2,3} \\ c_{3,0} & c_{3,1} & c_{3,2} & c_{3,3} \end{bmatrix} = \begin{bmatrix} b_{0,0} & b_{0,1} & b_{0,2} & b_{0,3} \\ b_{1,1} & b_{1,2} & b_{1,3} & b_{1,0} \\ b_{2,2} & b_{2,3} & b_{2,0} & b_{2,1} \\ b_{3,3} & b_{3,0} & b_{3,1} & b_{3,2} \end{bmatrix}.$$

This is the beginning of the diffusion of bits.

MixColumn

This can be described as a matrix product:

$$\begin{bmatrix} d_{0,0} & d_{0,1} & d_{0,2} & d_{0,3} \\ d_{1,0} & d_{1,1} & d_{1,2} & d_{1,3} \\ d_{2,0} & d_{2,1} & d_{2,2} & d_{2,3} \\ d_{3,0} & d_{3,1} & d_{3,2} & d_{3,3} \end{bmatrix} = \begin{bmatrix} 2 & 3 & 1 & 1 \\ 1 & 2 & 3 & 1 \\ 1 & 1 & 2 & 3 \\ 3 & 1 & 1 & 2 \end{bmatrix} \begin{bmatrix} c_{0,0} & c_{0,1} & c_{0,2} & c_{0,3} \\ c_{1,0} & c_{1,1} & c_{1,2} & c_{1,3} \\ c_{2,0} & c_{2,1} & c_{2,2} & c_{2,3} \\ c_{3,0} & c_{3,1} & c_{3,2} & c_{3,3} \end{bmatrix}.$$

In the matrix above, the numbers correspond to bytes; thus

$$\begin{aligned} 1 &= 00000001 \\ 2 &= 00000010 \\ 3 &= 00000011 \end{aligned}$$

and all multiplications and additions in the matrix product on the right are performed in the Rijndael field.

AddRoundKey

Each round has a *round key* of 128 bits, derived from the initial key. This will be described below. This round key can be put into a 4×4 matrix of bytes $k_{i,j}$, and then XOR-ed with the result of the previous step. The result is a matrix E for which

$$e_{i,j} = d_{i,j} \oplus k_{i,j}.$$

This completes one round of Rijndael.

To generate the round keys, start with the 4×4 matrix containing the bytes of the original 128-bit key. This matrix will be enlarged to a 4×44

matrix; in other words, 40 more columns will be added. Suppose the first four columns (that is, the columns for the original key) are W_1, W_2, W_3 and W_4. The other columns will be generated recursively.

If i is not a multiple of 4, then

$$W_i = W_{i-4} \oplus W_{i-1}.$$

If i is a multiple of 4, then

$$W_i = W_{i-4} \oplus T(W_{i-1})$$

where T is a transformation of the column. This transformation is performed as follows. Suppose that the elements of the column, the individual bytes, are a, b, c and d. Apply the S-box (from the ByteSub step) to each of these bytes to obtain e, f, g and h. Then compute the "round constant"

$$r_i = 2^{(i-4)/4}$$

where the power is to be calculated in the Rijndael field. Then

$$T(W_{i-1}) = (e \oplus r_i, f, g, h).$$

The round key for the i-th round of Rijndael consists of columns W_{4i}, W_{4i+1}, W_{4i+2}, W_{4i+3}.

10.4 Decryption

Rijndael can be decrypted simply by following all steps in reverse order. Each layer is easily invertible; AddRoundKey is its own inverse, ByteSub can be inverted using another lookup table, ShiftRow can be inverted by shifting the rows to the right instead of the left, and MixColumn can be inverted by multiplying by the inverse of the MixColumn matrix.

There are thus four operations for decryption: AddRoundKey (ARK), InverseByteSub (IBSB), InverseShiftRow (ISR) and InverseMixColumn (IMC). Decryption could be implemented by a schema similar to the encryption schema shown in Figure 10.1; this decryption schema is shown in Figure 10.5.

Equivalent inverse cipher

One of the many elegant attributes of Rijndael's construction is that the decryption can be implemented using a scheme almost exactly the same as for encryption.

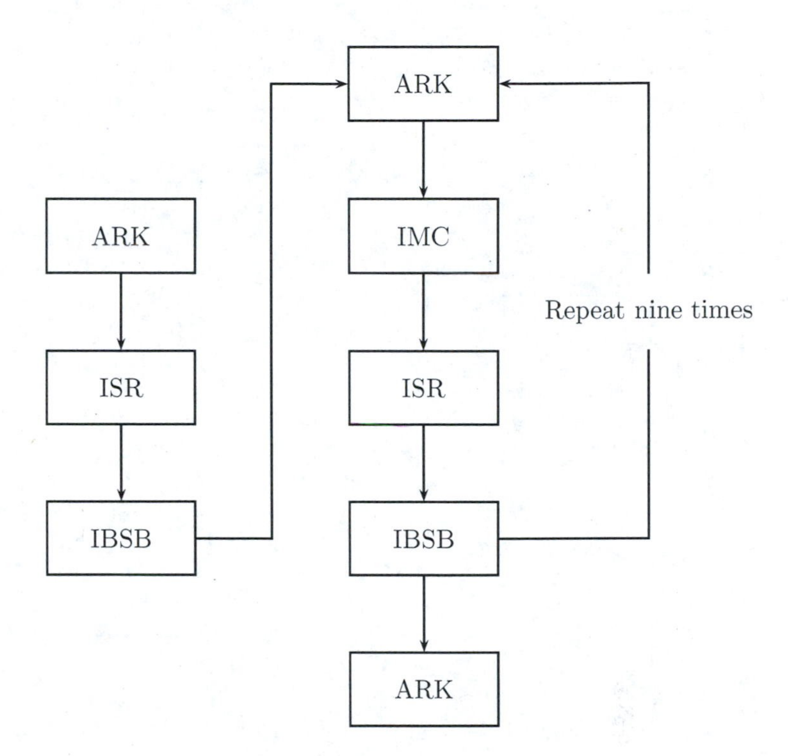

FIGURE 10.5: A schema for Rijndael decryption.

First note that the order of IBSB and ISR can be reversed, without affecting the result. And since both MC and IMC are linear operations on their input columns, it follows that

$$\text{IMC}(\text{state} \oplus \text{Round key}) = \text{IMC}(\text{state}) \oplus \text{IMC}(\text{Round key}).$$

This means that the order of ARK and IMC can also be reversed, provided that the columns of the round keys used in the decryption schedule are modified with IMC.

In other words, when ARK is followed by IMC, as it is for nine of the rounds as shown in Figure 10.5, these two operations can be replaced by IMC followed by IARK, where IARK is an XOR with $M^{-1}K$. Here M is the MixColumn matrix, and K is the current round key. Thus the schema shown in Figure 10.5 can be rewritten as shown in Figure 10.6.

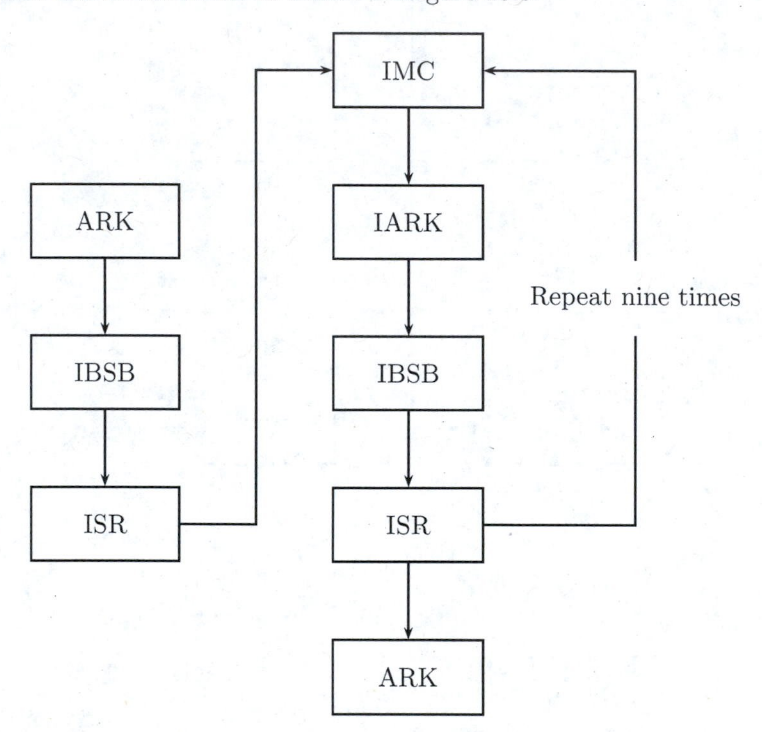

FIGURE 10.6: Equivalent inverse cipher for Rijndael decryption.

Now the operations in this new schema can be regrouped as shown in Figure 10.7.

If this schema is compared with the original encryption schema (Figure 10.1), it can be seen to be the same; the only difference is the operations. This also explains why the final round does not have a MixColumn step, for if one were included, then the decryption schema would not have the same form as for encryption.

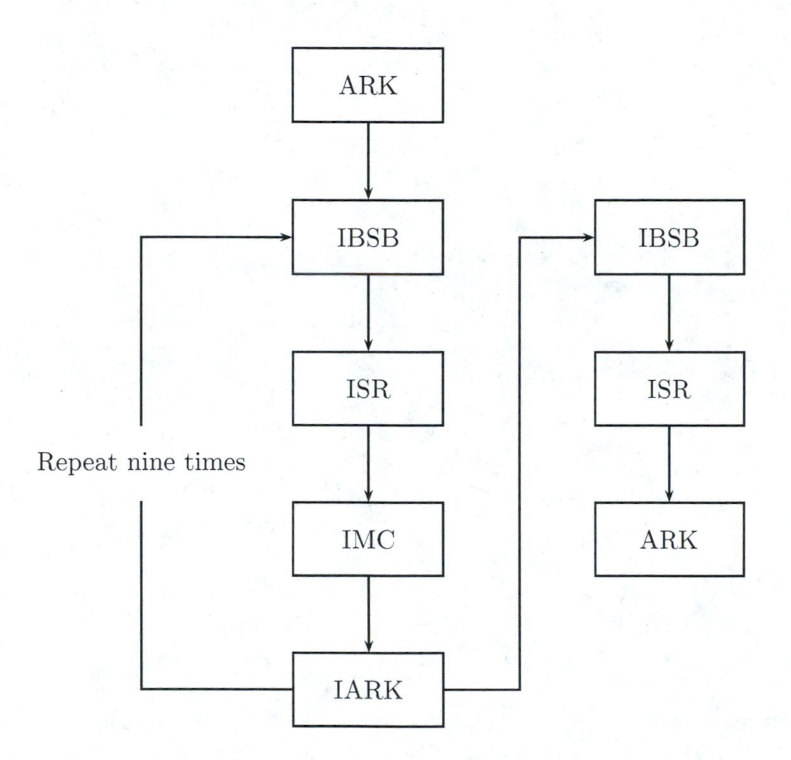

FIGURE 10.7: Final schema for Rijndael decryption.

10.5 Experimenting with AES

One way to experiment with AES in Sage is to use Python's own AES implementation from the Crypto.Cipher standard library. This can be done in the same way as was done for DES.

However, Sage includes a library for experimenting with *small-scale variants* of AES [19]. These are variants that allow for different numbers of rounds, of rows and columns in the rectangular state array, and of the size of the word. For the "full" AES, these values are 10, 4, 4, and 8 respectively.

Here's how this library can be used with the test vectors provided by NIST [67], with plaintext

`00112233445566778899aabbccddeeff`

and key

`000102030405060708090a0b0c0d0e0f`.

First the library must be imported into Sage:

```
sage: from sage.crypto import mq
sage: sr = mq.SR(10, 4, 4, 8, star=True,\
allow_zero_inversions=True,aes_mode=True)
sage: plain = '00112233445566778899aabbccddeeff'
sage: key = '000102030405060708090a0b0c0d0e0f'
sage: set_verbose(2)
sage: cipher = sr(plain, key)
R[01].start   00102030405060708090A0B0C0D0E0F0
R[01].s_box   63CAB7040953D051CD60E0E7BA70E18C
R[01].s_row   6353E08C0960E104CD70B751BACAD0E7
```

All the other state values are listed, but they won't be displayed here.

```
R[10].s_row   7AD5FDA789EF4E272BCA100B3D9FF59F
R[10].k_sch   13111D7FE3944A17F307A78B4D2B30C5
R[10].output  69C4E0D86A7B0430D8CDB78070B4C55A
sage: sr.hex_str_vector(cipher).lower()
'69c4e0d86a7b0430d8cdb78070b4c55a'
```

This library also makes it possible to explore the avalanche effect of AES, how a single bit change propagates through the cipher. To see this, create a new plaintext differing from the above by one bit, and encrypt it:

```
sage: plain2 = '00112233445566778899aabbccddeefe'
sage: cipher2 = sr(plain2, key)
```

Without rewriting the library, simply copy and paste the outputs into two separate text files, and remove all the initial material at the beginning of each line. So, for example, `cipher.txt` will look like:

```
00102030405060708090A0B0C0D0E0F0
63CAB7040953D051CD60E0E7BA70E18C
6353E08C0960E104CD70B751BACAD0E7
...
7AD5FDA789EF4E272BCA100B3D9FF59F
13111D7FE3944A17F307A78B4D2B30C5
69C4E0D86A7B0430D8CDB78070B4C55A
```

and `cipher2.txt` similarly. Now each file can be read into Sage:

```
sage: f = open("cipher.txt","r")
sage: f1 = f.read()
sage: f1 = f1.split('\n')
sage: f.close()
sage: f = open("cipher2.txt","r")
sage: f2 = f.read()
sage: f2 = f2.split('\n')
```

At this stage each of `f1` and `f2` will be lists of strings; each string being a state in the process of enciphering a plaintext. These lists can be stepped through simultaneously and their values compared using XOR:

```
sage: for i in range(50):
....:       x = ZZ('0x'+f1[i])
....:       y = ZZ('0x'+f2[i])
....:       xy = x^^y
....:       b = xy.binary()
....:       s = sum(ZZ(i) for i in b)
....:       if (i%5) != 4:
....:           print s,b.zfill(128)
....:
```

Since each fifth item is a round key value, they won't change and so there is no need to see them. The output (which has been slightly adjusted for readability) starts:

```
1 00000000000000000000000000000000000000000000000000000
  00000000000000000000000000000000000000000000000000000
  0000000000000000000001
4 00000000000000000000000000000000000000000000000000000
  00000000000000000000000000000000000000000000000000000
  0000000000000000101101
4 00000000000000000000000000101101000000000000000000000
```

```
   00000000000000000000000000000000000000000000000000000
   0000000000000000000000
18 00101101001011010111011101011010000000000000000000000
   00000000000000000000000000000000000000000000000000000
   0000000000000000000000
18 00101101001011010111011101011010000000000000000000000
   00000000000000000000000000000000000000000000000000000
   0000000000000000000000
19 11101110100001110100111110101100000000000000000000000
   00000000000000000000000000000000000000000000000000000
   0000000000000000000000
19 11101100000000000000000000000000000000000000000000000
   00010101100000000000000000001001111000000000000000010
   0001110000000000000000
71 11000111111011101110111000101001101011001010110011101
   11101000011010011111101000110011110010011111001001000
   0101011000011110000111
```

Even at this stage, which is only towards the end of the second round, over half the bits have changed. And from now to the end of the enciphering process the number of bits that differ between the two processes is about half the total number.

10.6 A simplified Rijndael

Just as DES can be better understood by working by hand through a simplified version, so can Rijndael. And in fact there are at least two such simplified versions, notably those by Phan [69] and by Musa et al. [64]. Of those two, Musa's version is the most "Rijndael-like" in such matters as the construction of its S-box, and in the key schedule. For that reason, it will be the version presented here. At the time of writing (late 2010), however Sage supports only the simplified version of Phan.

Basic outline

Simplified Rijndael operates on a 16-bit plaintext with a 16-bit key to produce a 1-bit ciphertext. States in the encryption are represented as 2×2 matrices of 4-bit blocks; such a block is called a "nybble"[1]. Nybbles are added using bit-wise XOR, and are multiplied using the field

$$F = \mathbb{Z}_2[x]/(x^4 + x + 1).$$

[1]In other contexts, writers refer to a 4-bit block as a "nibble."

Below is the addition table for this field

+	0	1	2	3	4	5	6	7	8	9	10	11	12	13	14	15
0	0	1	2	3	4	5	6	7	8	9	10	11	12	13	14	15
1	1	0	3	2	5	4	7	6	9	8	11	10	13	12	15	14
2	2	3	0	1	6	7	4	5	10	11	8	9	14	15	12	13
3	3	2	1	0	7	6	5	4	11	10	9	8	15	14	13	12
4	4	5	6	7	0	1	2	3	12	13	14	15	8	9	10	11
5	5	4	7	6	1	0	3	2	13	12	15	14	9	8	11	10
6	6	7	4	5	2	3	0	1	14	15	12	13	10	11	8	9
7	7	6	5	4	3	2	1	0	15	14	13	12	11	10	9	8
8	8	9	10	11	12	13	14	15	0	1	2	3	4	5	6	7
9	9	8	11	10	13	12	15	14	1	0	3	2	5	4	7	6
10	10	11	8	9	14	15	12	13	2	3	0	1	6	7	4	5
11	11	10	9	8	15	14	13	12	3	2	1	0	7	6	5	4
12	12	13	14	15	8	9	10	11	4	5	6	7	0	1	2	3
13	13	12	15	14	9	8	11	10	5	4	7	6	1	0	3	2
14	14	15	12	13	10	11	8	9	6	7	4	5	2	3	0	1
15	15	14	13	12	11	10	9	8	7	6	5	4	3	2	1	0

and the multiplication table

×	0	1	2	3	4	5	6	7	8	9	10	11	12	13	14	15
0	0	0	0	0	0	0	0	0	0	0	0	0	0	0	0	0
1	0	1	2	3	4	5	6	7	8	9	10	11	12	13	14	15
2	0	2	4	6	8	10	12	14	3	1	7	5	11	9	15	13
3	0	3	6	5	12	15	10	9	11	8	13	14	7	4	1	2
4	0	4	8	12	3	7	11	15	6	2	14	10	5	1	13	9
5	0	5	10	15	7	2	13	8	14	11	4	1	9	12	3	6
6	0	6	12	10	11	13	7	1	5	3	9	15	14	8	2	4
7	0	7	14	9	15	8	1	6	13	10	3	4	2	5	12	11
8	0	8	3	11	6	14	5	13	12	4	15	7	10	2	9	1
9	0	9	1	8	2	11	3	10	4	13	5	12	6	15	7	14
10	0	10	7	13	14	4	9	3	15	5	8	2	1	11	6	12
11	0	11	5	14	10	1	15	4	7	12	2	9	13	6	8	3
12	0	12	11	7	5	9	14	2	10	6	1	13	15	3	4	8
13	0	13	9	4	1	12	8	5	2	15	11	6	3	14	10	7
14	0	14	15	1	13	3	2	12	9	7	6	8	4	10	11	5
15	0	15	13	2	9	6	4	11	1	14	12	3	8	7	5	10

The S-box

Since the S-box is central to both encryption and the key-schedule, it needs to be discussed first. As with Rijndael, it can be presented as a permutation of all the integers 0 to 15:

$$s = [9, 4, 10, 11, 13, 1, 8, 5, 6, 2, 0, 3, 12, 14, 15, 7].$$

Also as with Rijndael, this S-box is generated by an affine transformation of inverses in the field. Suppose that n is a given nybble, which corresponds to an element of F. Define

$$n' = \begin{cases} 0 & \text{if } n = 0 \\ n^{-1} & \text{otherwise.} \end{cases}$$

Then

$$s_n = an' + b \bmod (y^4 + 1)$$

where

$$a = y^3 + y^2 + 1, \quad b = y^3 + 1$$

and both a and b are considered as elements of the ring $\mathbb{Z}_2[y]/(y^4 + 1)$. This is not a field as not all elements are invertible.

For example, consider the nybble $4 = 0100$, which corresponds to polynomial x^2. In F its inverse is $x^3 + x^2 + 1$:

```
sage: F.<x> = GF(16,name='x')
sage: p = 1/x^2; p
x^3 + x^2 +1
```

Then:

```
sage: R.<y> = PolynomialRing(GF(2))
sage: a = y^3+y^2+1
sage: b = y^3+1
sage: (a*R(p.polynomial())+b).mod(y^4+1)
y^3 + y^2 + 1
```

The coefficients of this polynomial correspond to the nybble 1101, or 13, which is indeed the value of s_4.

Key schedule

The key schedule is defined, similar as in Rijndael, in terms of the XOR of previously computed blocks. For simplified Rijndael, there are two extra operations: "RotNyb" which re-orders the position of two concatenated nybbles, so that

$$\text{RotNyb}(n_0 n_1) = n_1 n_0$$

and "SubNyb" which replaces each nybble by its S-box value, so that

$$\text{SubNyb}(n_0 n_1) = s(n_1)s(n_0).$$

The *round constant* RCON(i) for round i consists of a byte for which the first four bits are $x^{i+2} \in F$, and the second four bits are all zeros. So for $i = 1$,

$x^{i+2} = x^3 \equiv 1000$ so RCON(1) = 10000000. And for $i = 2$, $x^{i+2} = x^4 = x + 1 \equiv 0011$ so RCON(2) = 00110000. The initial key may be considered as consisting of two bytes w_0 and w_1. Two more round keys will be required, another 4 bytes in all, numbered w_2, w_3, w_4, w_5. They are created as follows:

$$\text{if } i = 0 \pmod 2 \text{ then } w_i = w_{i-2} \oplus \text{RCON}(i/2) \oplus \text{SubNyb}(\text{RotNyb}(W_{i-1}))$$
$$\text{if } i = 1 \pmod 2 \text{ then } w_i = w_{i-2} \oplus w_{i-1}.$$

This is in fact not as difficult as it looks. Suppose for example the the initial key is

00110011 00001111

so that $w_0 = 00110011$ and $w_1 = 00001111$. Then for $i = 2$:

$$\begin{aligned} w_2 &= w_0 \oplus \text{RCON}(2/2) \oplus \text{SubNyb}(\text{RotNyb}(w_1)) \\ &= 00110011 \oplus 10000000 \oplus \text{SubNyb}(11110000) \\ &= 00110011 \oplus 10000000 \oplus 01111001 \\ &= 11001010. \end{aligned}$$

For $i = 3$:

$$\begin{aligned} w_3 &= w_1 \oplus w_2 \\ &= 00001111 \oplus 11001010 \\ &= 11000101. \end{aligned}$$

The next two bytes w_4 and w_5 are created similarly. The three round keys are then

$$\begin{aligned} k_0 &= w_0 w_1, \\ k_1 &= w_2 w_3, \\ k_2 &= w_4 w_5. \end{aligned}$$

The layers

These are defined very similarly to Rijndael.

1. *AddRoundKey.* Simply adds all the bits of the state to the current round key, using xor.

2. *NybbleSubstitution.* Replaces each nybble in the current state with its value from the S-box.

3. *ShiftRow.* Treating the state as a 2×2 matrix, ShiftRow swaps the nybbles in the bottom row, so that

$$\begin{bmatrix} b_0 & b_2 \\ b_1 & b_3 \end{bmatrix} \xrightarrow{SR} \begin{bmatrix} b_0 & b_2 \\ b_3 & b_1 \end{bmatrix}.$$

4. *MixColumn.* Multiplies the current state by the fixed matrix

$$\begin{bmatrix} 1 & x^2 \\ x^2 & 1 \end{bmatrix}.$$

MixColumn may also be interpreted as an operation on bits. Suppose that one column in the current state is

$$\begin{bmatrix} b_0 + b_1x + b_2x^2 + b_3x^3 \\ b_4 + b_5x + b_6x^2 + b_7x^3 \end{bmatrix}.$$

Then the top value in the output will be

$$\begin{aligned} &b_0 + b_1x + b_2x^2 + b_3x^3 + (b_4 + b_5x + b_6x^2 + b_7x^3)x^2 \\ &\quad = (b_0 + b_6) + (b_1 + b_4 + b_7)x + (b_2 + b_4 + b_5)x^2 + (b_3 + b_5)x^3. \end{aligned}$$

The full encryption scheme is shown in Figure 10.8.

An example

Using the key 00110011 00001111 from above, encrypt the plaintext 01010110 11001
The key schedule can be shown to produce

$$\begin{aligned} k_0 &= 00110011\ 00001111, \\ k_1 &= 11001010\ 11000101, \\ k_2 &= 11100110\ 00100011. \end{aligned}$$

The first ARK produces

$$\begin{array}{rl} & 01010110\ \ 11001001 \\ + & 00110011\ \ 00001111 \\ \hline & 01100101\ \ 11000110 \end{array}$$

This can be placed into a 2×2 matrix column by column:

$$\begin{bmatrix} 0110 & 1100 \\ 0101 & 0110 \end{bmatrix}.$$

Now for round 1. First NybbleSub

$$\begin{bmatrix} 1000 & 1100 \\ 0001 & 1000 \end{bmatrix},$$

then ShiftRow

$$\begin{bmatrix} 1000 & 1100 \\ 1000 & 0001 \end{bmatrix},$$

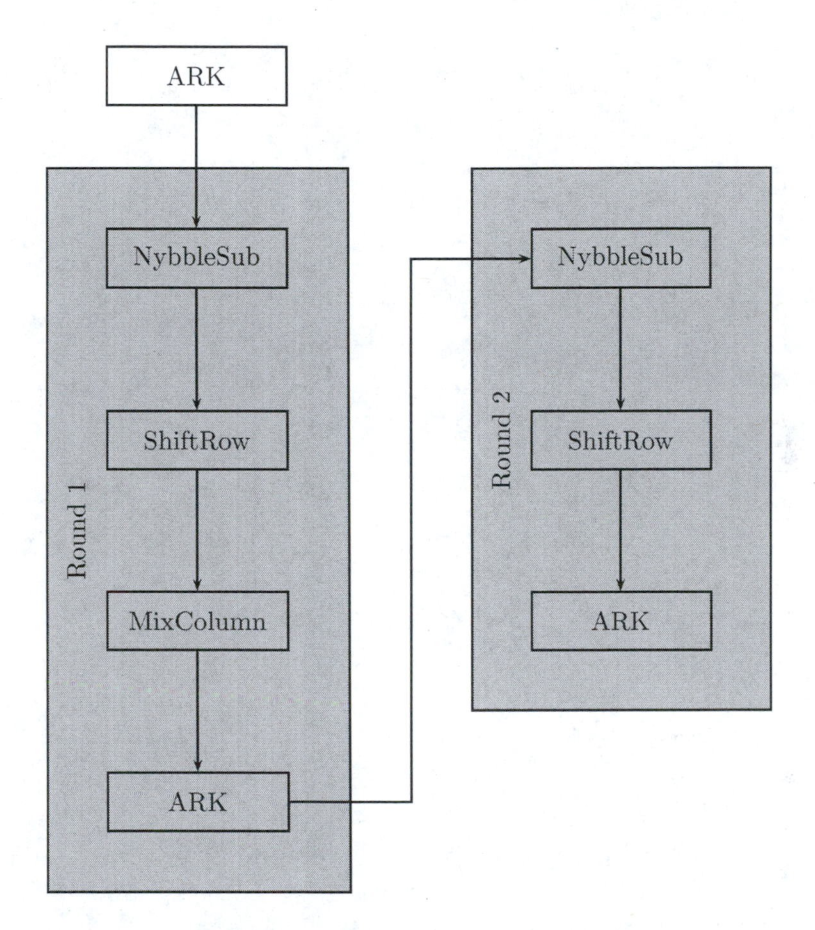

FIGURE 10.8: Simplified Rijndael.

then MixColumn

$$\begin{bmatrix} 1 & 4 \\ 4 & 1 \end{bmatrix} \begin{bmatrix} 8 & 12 \\ 8 & 1 \end{bmatrix} = \begin{bmatrix} 14 & 8 \\ 14 & 4 \end{bmatrix}.$$

Finally for this round, the next AddRoundKey

$$\begin{array}{rl} & 11101110\ 10000100 \\ + & 11001010\ 11000101 \\ \hline & 00100100\ 01000001 \end{array}$$

Now for round 2. NybbleSub:

$$\begin{bmatrix} 0010 & 0100 \\ 0100 & 0001 \end{bmatrix} \longrightarrow \begin{bmatrix} 1010 & 1101 \\ 1101 & 0100 \end{bmatrix}.$$

ShiftRow

$$\begin{bmatrix} 1010 & 1101 \\ 0100 & 1101 \end{bmatrix},$$

and the last AddRoundKey

$$\begin{array}{rl} & 10100100\ 11011101 \\ + & 11100110\ 00100011 \\ \hline & 01000010\ 11111110 \end{array}$$

This last row of bits is the required ciphertext.

10.7 Security of the AES

Since the selection of Rijndael to be the AES, it has undergone intensive scrutiny and analysis. So far, it has withstood every attack known, and gives the impression of being a very secure cryptosystem, especially if used in one of the non-ECB modes discussed in Chapter 8.

The main concern about AES is that Rijndael can be described in terms of operations on a finite field. The encryption can in fact be written entirely as a very large sequence of very large equations. If these equations can be solved, then Rijndael would be broken. Nobody knows if this can be made into a genuine attack, but the idea makes some cryptographers very worried and nervous.

However, with the Rijndael cryptosystem as an official standard, you should feel safe in using it.

10.8 Glossary

AES. The Advanced Encryption Standard, a NIST standard designed to take the place of DES.

Equivalent Inverse Cipher. An inverse (deciphering) scheme for Rijndael that has exactly the same scheme as encryption.

Layer. Any of the four processes: MixColumn, AddRoundKey, SubBytes and ShiftRow that form Rijndael.

Nybble. (Or "Nibble.") A four-bit block used in simplified Rijndael.

Rijndael. The Belgian cipher chosen by NIST to be the AES.

Exercises

Review questions

1. Why did NIST call for the creation of a new encryption standard?
2. What were the design criteria?
3. How many entries made it through the first round?
4. How many entries were in the final round?
5. How many rounds does Rijndael have?
6. How does the cipher Rijndael differ from the final standard definition of the AES?
7. Why is MixColumn left out of the last round?
8. What is a security concern with the AES?

Beginning Exercises

9. Apply Rijndael's ByteSub to each of the following bytes using each of (i) the S-box, (ii) the definition using polynomials, (iii) the transformation using matrices:

   (a) 11111111, (b) 10101010, (c) 00111100, (d) 01110010.

10. Show that the MixColumn operation of simplified Rijndael transforms a column of nybbles as

$$\begin{bmatrix} b_0b_1b_2b_3 \\ b_4b_5b_6b_7 \end{bmatrix} \longrightarrow \begin{bmatrix} b_0 \oplus b_6 & b_1 \oplus b_4 \oplus b_7 & b_2 \oplus b_4 \oplus b_5 & b_3 \oplus b_5 \\ b_2 \oplus b_4 & b_0 \oplus b_3 \oplus b_5 & b_0 \oplus b_1 \oplus b_6 & b_1 \oplus b_7 \end{bmatrix}.$$

11. Use simplified Rijndael to encrypt the following plaintext, key pairs:
    (a) 11111111 10101010, 10101010 11000011,
    (b) 00001111 00111100, 00000000 10000000,
    (c) 00000000 10000000, 11111111 00000001,
    (d) 11001100 00000010, 01000100 001000001.

12. Work back through the decryption of each of the ciphertexts you obtained in the previous question.

Sage Exercises

13. Use the polynomial definition to generate the S-box of simplified Rijndael.

14. Using the above method, generate the S-box of Rijndael.

15. Write Sage programs to implement simplified Rijndael, and test them on the inputs from question 11.

16. Show that the matrix used in the InverseMixColumn step has integer form

$$\begin{bmatrix} 14 & 11 & 13 & 9 \\ 9 & 14 & 11 & 13 \\ 13 & 9 & 14 & 11 \\ 11 & 13 & 9 & 14 \end{bmatrix}.$$

17. Investigate the avalanche effect using the same plaintexts, but keys that differ in only one bit.

Chapter 11

Hash functions

This chapter will investigate

- Definitions and requirements of cryptographic hash functions.
- Use of such functions as part of secure systems.
- Complete definition of one major bit-oriented hash function.
- Provably secure hash functions, with examples.
- Constructions of hash functions using block ciphers.
- Message Authentication Codes, and their constructions using hash functions.
- Some modern hash functions.

A cryptographic *hash function* is a function H that takes a message of arbitrary size as input, and returns a fixed length string of bits as output. Depending on the function used, this string can have 128, 160, 256 or 512 bits. The hash function output may be considered a sort of "fingerprint" of the message. To be useful, the function must also be fast to compute, both in hardware and software. There are also three more considerations:

- Given a hash output y, it should be computationally infeasible to find a message m for which $H(m) = y$. This is called the *one-way property* or *preimage resistance* of the hash function H: it is very easy to calculate y, but given y, it is very difficult to find the original m.
- For a given message m, it should be computationally infeasible to find another message m' for which $H(m) = H(m')$. This property is called *weak collision resistance.*
- It should be computationally infeasible to find *any* two messages m and m' for which $H(m) = H(m')$. This property is called *strong collision resistance.*

Note that as there are an infinite number of possible messages, and only a finite number of hash values, then theoretically there will be an infinite number of messages that hash to the same value. The point is that any collision should be hard to find. The hash value of a message is also called the *message digest.*

11.1 Uses of hash functions

Message authentication

One of the most powerful uses of hash functions is for *message authentication.* Suppose Alice sends a message to Bob, and Bob wants to make sure that the message he receives is really the message that Alice sent. Here's a protocol using a symmetric cryptosystem, with a key k known to both Alice and Bob:

1. Alice hashes her message m to produce $y = H(m)$, and encrypts y using the key k to obtain $e = E(y, k)$.

2. She appends e to the message and sends the lot to Bob.

3. Bob now separates the encrypted section from the received message m'. He decrypts this to obtain $y = D(e, k)$. He now hashes the message to obtain $y' = H(m')$.

4. If $y = y'$ then he knows that the message is indeed the correct message.

Note that this is quite a fast protocol; it doesn't require encrypting the entire message.

If key exchange is a problem, then the encryption and decryption above can be done using a public-key cryptosystem: Alice encrypts the hash value using Bob's public key, and Bob decrypts it with his private key.

There is another protocol that requires no encryption at all, but does require the use of some secret data S known only to Alice and Bob:

1. Alice places the secret data at the head of the message and hashes the lot to obtain $y = H(S\|m)$.

2. She then appends y to the original message and sends the lot to Bob.

3. Bob now separates the received message m' from the hash value y.

4. He now adds the secret value to the message m' and hashes the lot to obtain $y' = H(S\|m)$.

5. If $y = y'$ then he knows that the message is indeed the correct message.

Digital signatures

Recall that digital signatures require computation equivalent to performing a public key encryption. In general, If Alice is signing a message to Bob, she signs it with her private key, and Bob verifies the signature with her public key. Given the relative slowness of public-key encryption, it is much more

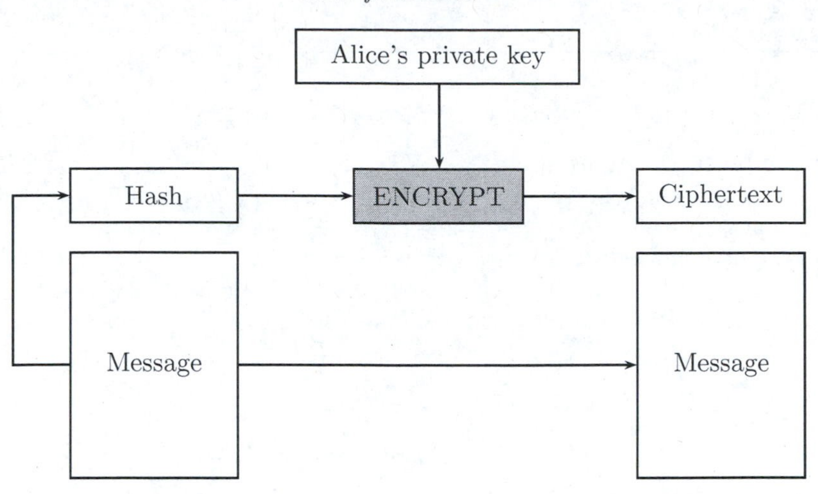

FIGURE 11.1: Signing a message using a hash function.

efficient to sign not the message itself, but the hash of the message. This protocol was discussed in Chapter 7, and is illustrated in Figure 11.1.

The process for Bob's verification of the signature is shown in Figure 11.2. There are other uses, some of which will be discussed later.

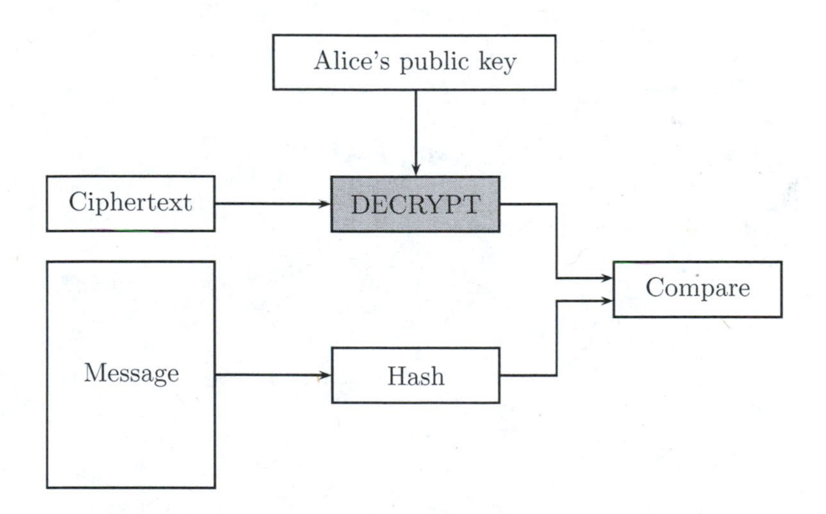

FIGURE 11.2: Authentication of a signature with a hash.

11.2 Security of hash functions

Suppose there is a hash function that hashes to values of 128 bits. How secure is this? The security is measured in the amount of work needed to find a collision, that is, two different messages that hash to the same value.

One way of considering this question is to start by looking at the *birthday paradox*: how many people are needed so that the probability that two of them share the same birthday will be greater than 1/2? If there are k people, the probability that all their birthdays are *different* is:

$$\begin{aligned}\prod_{i=1}^{k-1}\frac{365-i}{365} &= \frac{364}{365}\cdot\frac{363}{365}\cdots\frac{365-(k-1)}{365}\\ &= \left(1-\frac{1}{365}\right)\left(1-\frac{2}{365}\right)\cdots\left(1-\frac{k-1}{365}\right).\end{aligned}$$

So the probability that two people share a birthday will be 1 minus this value. This can easily be evaluated for different values of k:

```
sage: k=10
sage: float(1-prod((365-i)/365 for i in range(k)))
  0.11694817771107764
sage: k=40
sage: float(1-prod((365-i)/365 for i in range(k)))
  0.89123180981794892
```

By trial and error, or by writing a small loop, it can be found that only 23 people are needed to have a probability greater than 1/2. (With 57 people the probability is greater than 0.99).

Suppose there are N different hash values possible (if the hash value contains n bits, then $N = 2^n$). For k messages then, the probability of a collision is

$$P = \prod_{i=1}^{k-1}\frac{N-i}{N} = \prod_{i=1}^{k-1}\left(1-\frac{i}{N}\right).$$

Now assume that the fraction i/N is small enough so that the power series expansion of e^{-x} can be approximated with

$$e^{-x} \approx 1-x$$

if x is very small, and so

$$1-\frac{i}{N} \approx e^{-i/N}.$$

Thus the product above can be written

$$\prod_{i=1}^{k-1} e^{-i/N}$$

Since the sum of all the indices is

$$\sum_{i=1}^{k-1} \frac{i}{N} = \frac{k(k-1)}{2N}$$

the probability is approximately

$$P = 1 - \exp\left(\frac{-k(k-1)}{2N}\right) = 1 - \exp\left(\frac{k-k^2}{2N}\right).$$

Using the approximation $e^{-x} \approx 1 - x$ again, this reduces to

$$\frac{k^2 - k}{2N}.$$

Also assume that k is large enough so that it is negligible in comparison with k^2, and so

$$P \approx \frac{k^2}{2N}.$$

If $P = 0.5$ then this can be solved for k to obtain

$$k \approx \sqrt{N}.$$

Finally if this is applied this to a hash function that produces hash values of n bits, then with about

$$\sqrt{2^n} = 2^{n/2}$$

messages there will be a 50% probability of finding a collision. More messages would produce a higher probability, but it will still be of the order of the square root of N. Since a hash function has n-bit security if an attack requires 2^n steps, a 128-bit hash function in fact produces only 64-bit security. But 64 bits is far too low for modern security; 80 bits is considered the minimum. This means that for reasonable security a hash function should produce values at least 160 bits long.

11.3 Constructing a hash function

Many modern hash functions operate on an iterative principle developed by Ralph Merkle[1] and Ivan Damgård, called the *Merkle–Damgård construction*

[1] This is the same Ralph Merkle who developed the Merkle–Hellman knapsack cryptosystem. Another example of the same name cropping up in several different places.

and use an auxiliary function called a *compression function* f which takes as input two blocks, one of q bits and another of n bits, and produces as output an n-bit block. This construction also requires an n-bit initialization vector IV, and the message m will have to be split into k blocks of q bits each. This may require padding the last block (for example, with zeros and a binary representation of the size of the message) to bring it up to size. The hash value is the value of the last iteration. This construction is illustrated in Figure 11.3.

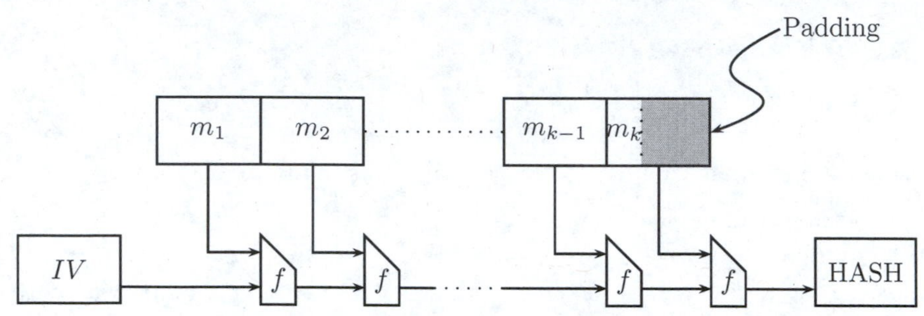

FIGURE 11.3: The Merkle–Damgård construction.

Using this construction, collision resistance of the hash is determined by the collision resistance of f: if the compression function can be shown to be collision resistant, then so is the hash. Conversely, if f has weaknesses, then the hash is not secure.

RIPEMD-160

This is one of the strongest of modern hash functions. It was developed by the German-Belgian team of cryptographers Hans Dobbertin, Antoon Bosselaers, and Bart Preneel. "RIPEMD" stands for "RIPE Message Digest," where "RIPE" stands for "RACE Integrity Primitives Evaluation" and where "RACE" stands for "Research and Development in Advanced Communications Technologies in Europe"—a nice example of a recursive abbreviation. As its name suggests, RIPEMD-160 produces 160 bit hashes. An earlier version, RIPEMD-128, has been found to be insecure; this newer version not only produces longer hashes, but is immune to the attacks to which RIPEMD-128 is vulnerable.

Like many modern hash functions, it is a descendent of MD4 (Message Digest 4) which was developed by Ronald Rivest in 1990. Although MD4 has since been shown to be very insecure, it spawned a number of offspring, of which RIPEMD-160 is one.

Many hash functions use *chaining variables*. These have initial values specified by the algorithm, and are updated each round according to the algorithm and to the values of the current message block. The final values of the chaining variables form the hash. Most hash functions based on MD4 have just one strand of operations through which the chaining variables pass, but RIPEMD-160 has two strands, a left and right, through which the variables pass independently, to be joined at the end. Figure 11.4 shows a single round of RIPEMD-160 at the most basic level.

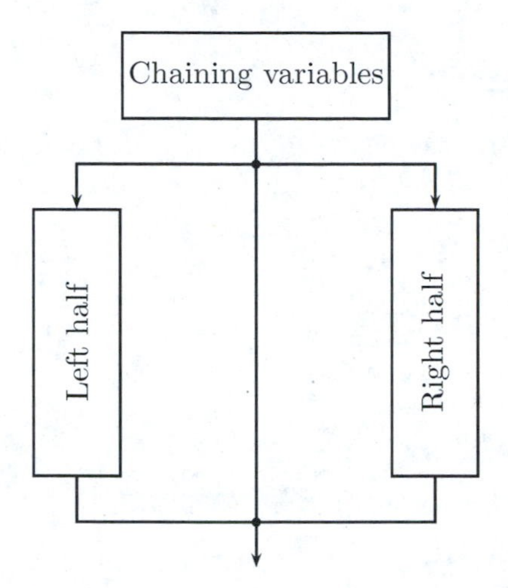

FIGURE 11.4: The RIPEMD-160 hash function, all details removed.

RIPEMD-160 uses message blocks of size 512 bits. In order to pad the last block, a single 1 is added followed by enough 0s to produce 448 bits. The last 64 bits contain the binary representation of the length of the entire message. If the message is longer than 2^{64} bits, then only the least significant 64 bits are used. (Note that the entire US Library of Congress contains about 2^{44} bits of information, so a message larger than 2^{64} bits would be very rare indeed!)

There are five chaining variables, A, B, C, D and E initialized as follows (in hexadecimal):

$$A = \texttt{0x67452301}$$
$$B = \texttt{0xEFCDAB89}$$
$$C = \texttt{0x98BADCFE}$$
$$D = \texttt{0x10325476}$$
$$E = \texttt{0xC3D2E1F0}$$

Each value is just a permutation of a subset of all of the 16 hexadecimal digits.

Each of the right and left halves consists of five steps, called *rounds*, and each round involves a boolean function. They are

$$
\begin{aligned}
f_1(x, y, z) &= x \oplus y \oplus z \\
f_2(x, y, z) &= (x \wedge y) \vee (\neg y \wedge z) \\
f_3(x, y, z) &= (x \vee \neg y) \oplus z \\
f_4(x, y, z) &= (x \wedge z) \vee (y \wedge \neg z) \\
f_5(x, y, z) &= x \oplus (y \wedge \neg z).
\end{aligned}
$$

Consider a single message block, of 512 bits. It is divided into 16 32-bit words $(m_0, m_1, m_2, \ldots, m_{15})$. (Note that each of the chaining variables is 32 bits long.) For each round, the message words are permuted according to two permutations based on

$$
\begin{aligned}
\rho &= [7, 4, 13, 1, 10, 6, 15, 3, 12, 0, 9, 5, 2, 14, 11, 8] \\
\pi &= [5, 14, 7, 0, 9, 2, 11, 4, 13, 6, 15, 8, 1, 10, 3, 12].
\end{aligned}
$$

The second permutation is defined by $\pi_i = 9i + 5 \pmod{16}$. From these two permutations, eight others are produced, so that the left and right halves, in more detail, can be described as in Figure 11.5, where *id* is the identity permutation.

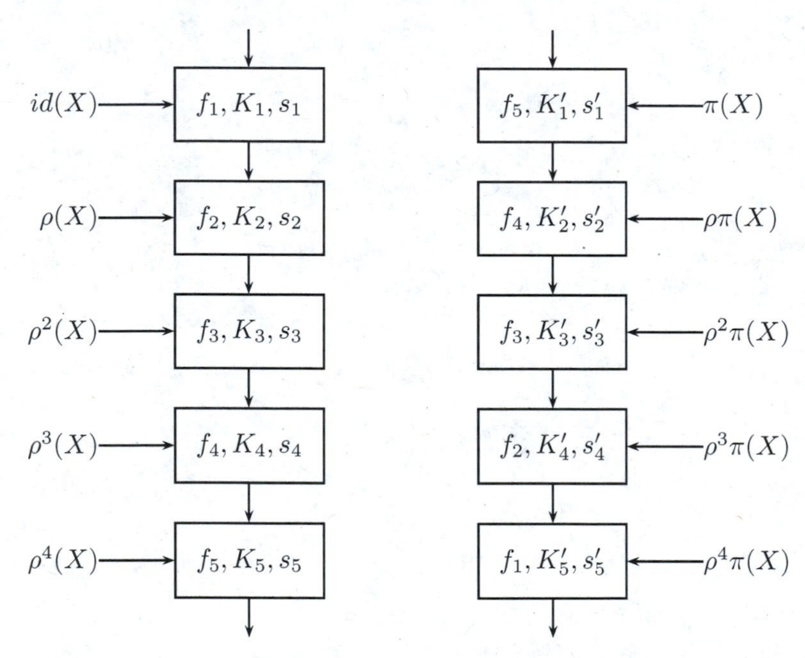

FIGURE 11.5: Left and right halves of RIPEMD-160 in more detail.

The extra constants K_i and K'_i, whose use will be explained below, are defined as

$$\begin{aligned}
K_1 &= \texttt{0x00000000} \\
K_2 &= \texttt{0x5A827999} = \lfloor 2^{30}\sqrt{2} \rfloor \\
K_3 &= \texttt{0x6ED9EBA1} = \lfloor 2^{30}\sqrt{3} \rfloor \\
K_4 &= \texttt{0x8F1BBCDC} = \lfloor 2^{30}\sqrt{5} \rfloor \\
K_5 &= \texttt{0xA953FD4E} = \lfloor 2^{30}\sqrt{7} \rfloor \\
K'_1 &= \texttt{0x50A28BE6} = \lfloor 2^{30}\sqrt[3]{2} \rfloor \\
K'_2 &= \texttt{0x5C4DD124} = \lfloor 2^{30}\sqrt[3]{3} \rfloor \\
K'_3 &= \texttt{0x6D703EF3} = \lfloor 2^{30}\sqrt[3]{5} \rfloor \\
K'_4 &= \texttt{0x7A6D76E9} = \lfloor 2^{30}\sqrt[3]{7} \rfloor \\
K'_5 &= \texttt{0x00000000}.
\end{aligned}$$

To check this, look at K'_3:

```
sage: hex(floor(2^30*5.0^(1/3))).upper()
  '6D703EF3'
```

The values s_i and s'_i are shifts that are applied to each of the words from the message block; they are given below.

Now to peer inside each of the boxes. The inputs to each box are the current values of the chaining variables A, B, C, D, E, and the sixteen 32-bit words of the message block, permuted according to the definitions above. Suppose that

$$[X_0, X_1, X_2, \ldots, X_{15}]$$

are the permuted words. Then in each box the following pseudo-code is implemented, where $\boxplus$ is addition modulo 2^{32}, and $\lll s$ means left rotation by s bits:

for $j = 0$ to 15 {
  $T \leftarrow ((A \boxplus f(B, C, D) \boxplus X_j \boxplus K) \lll s) \boxplus E$
  $A \leftarrow E$
  $E \leftarrow D$
  $D \leftarrow C \lll 10$
  $C \leftarrow B$
  $B \leftarrow T$
}

This flow of information is shown in Figure 11.6.

The value of K, the function f and the shift s are chosen according to the scheme shown in Figure 11.5. The values of the shift s for the right side are

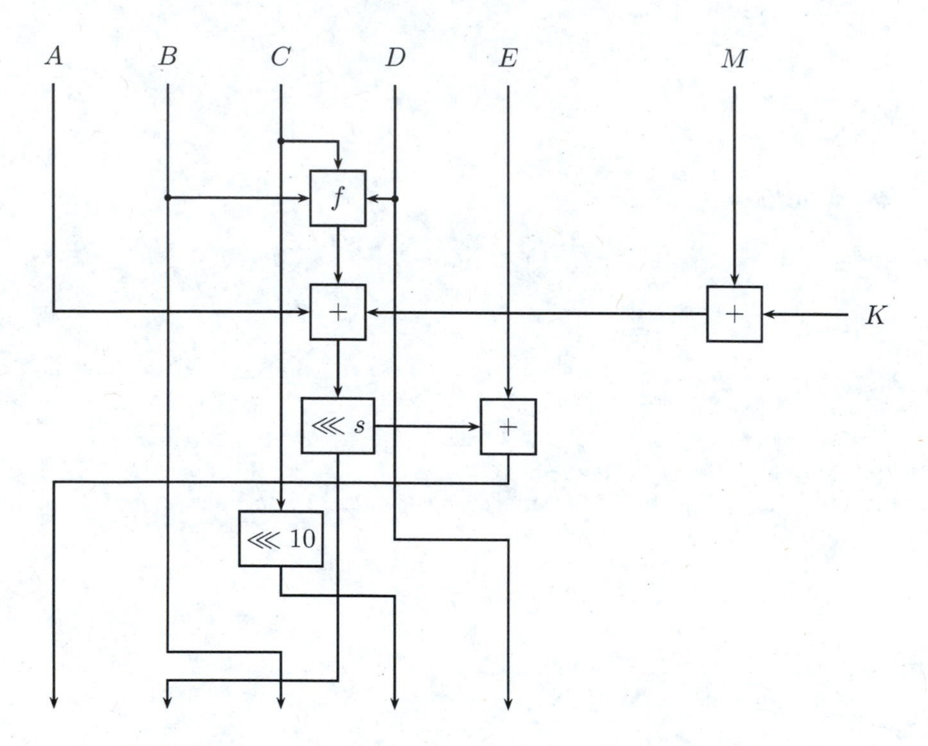

FIGURE 11.6: The mixing function used in RIPEMD-160.

given by

$j =$	0	1	2	3	4	5	6	7	8	9	10	11	12	13	14	15
s'_1	11	14	15	12	5	8	7	9	11	13	14	15	6	7	9	8
s'_2	7	6	8	13	11	9	7	15	7	12	15	9	11	7	13	12
s'_3	11	13	6	7	14	9	13	15	14	8	13	6	5	12	7	5
s'_4	11	12	14	15	14	15	9	8	9	14	5	6	8	6	5	12
s'_5	9	15	5	11	6	8	13	12	5	12	13	14	11	8	5	6

and for the left side by

$j =$	0	1	2	3	4	5	6	7	8	9	10	11	12	13	14	15
s_1	8	9	9	11	13	15	15	5	7	7	8	11	14	14	12	6
s_2	9	13	15	7	12	8	9	11	7	7	12	7	6	15	13	11
s_3	9	7	15	11	8	6	6	14	12	13	5	14	13	13	7	5
s_4	15	5	8	11	14	14	6	14	6	9	12	9	12	5	15	8
s_5	8	5	12	9	12	5	14	6	8	13	6	5	15	13	11	11

The final requirement is putting together the two strands from the left and right sides. Suppose that h_0, h_1, h_2, h_3, h_4 are the original chaining variables, and A, B, C, D, E are the right side outputs, and A', B', C', D', E' are the left side outputs. This is done according to

$$\begin{aligned}
T &= h_1 \boxplus C \boxplus D' \\
h_1 &= h_2 \boxplus D \boxplus E' \\
h_2 &= h_3 \boxplus E \boxplus A' \\
h_3 &= h_4 \boxplus A \boxplus B' \\
h_4 &= h_0 \boxplus B \boxplus C' \\
h_0 &= T.
\end{aligned}$$

Although RIPEMD-160 is complicated to explain, it is important to realize that all the individual steps are very fast, consisting of rotations and additions modulo 2^{32} (which is just an addition and taking only the right-most 32 bits).

The use of RIPEMD-160 can be checked in Sage by invoking the Python "hashlib" library, which implements several standard hash functions.

```
sage: import hashlib
sage: r = hashlib.new('ripemd160')
sage: r.update("a")
sage: r.hexdigest()
  '0bdc9d2d256b3ee9daae347be6f4dc835a467ffe'
```

A gallery of hash functions

There are many hash functions available. Since all are quite complicated to explain (although efficient to compute), none of them will be described in detail. Full information is available on the Internet.

The MD family. These hash functions are MD2, MD4 and MD5; all developed by Ron Rivest (the same Rivest who is the "R" in RSA). There are also MD1 and MD3, which have never been published. MD5 has been very popular, and was designed specifically to overcome flaws in its earlier version MD4. However, its internal compression function has been found to have flaws: thus it is not a secure function. Also, MD5 hashes to 128-bit values, which as discussed above is not sufficient for modern security.

MD2 is particularly easy to describe; it is based around a permutation S of the values 0 to 255, and consists of three steps:

1. Padding. The message is increased to be a multiple of 16 bytes: i copies of the byte with value i are appended to the message, with $1 \leq i \leq 16$.

2. Checksum. The padded message is increased with another 16 bytes C_i called the checksum. With N the length of the padded message (in bytes), set each $C_i = 0$, and also $L = 0$. Then

```
for i in range(N/16):
    for j in range(16):
        c = M[16i+j]
        Cj = Cj ⊕ S[c ⊕ L]
        L = Cj
```

3. The hash. Start by initializing 48 bytes X_i to 0. Then with N' being the length of the message M with checksum:

```
for i in range(N′/16):
    for j in range(16):
        X[j+16] = M[16*i+j]
        X[j+32] = X[j+16] ⊕ X[j]
    t = 0
    for j in range(18):
        for k in range(48):
            t = X[k] ⊕ S[t]
            X[k] = t
        t = (t+j)%256
```

The final hash is the first 16 bytes of X.

This hash is not secure: it is too small, and collisions have been found. However the checksum step adds a measure of security that makes collisions harder to find. Its simplicity and ease of description is unique among bit-oriented hashes. Full details about its definition are given by Kaliski [45] with errata available at [5].

SHA-1. The *Secure Hashing Algorithm* was developed by the US National Institute of Standards and Technology (NIST) in conjunction with the US National Security Agency (NSA). It produces 160-bit values, and was considered to be very secure. It works by breaking the message blocks into 16 32-bit words (so $k = 512$), and by breaking the most recently computed hash value into five 32-bit words. The compression function takes 80 steps, each involving an XOR with some constant values, combined with shifts. In spite of the large number of steps, each can be performed very efficiently, so this algorithm is very fast.

Values of SHA-1 hashes can be obtained again with the hashlib library.

```
sage: import hashlib
sage: sage: hashlib.sha1('').hexdigest()
  'da39a3ee5e6b4b0d3255bfef95601890afd80709'
sage: hashlib.sha1('a').hexdigest()
  '86f7e437faa5a7fce15d1ddcb9eaeaea377667b8'
sage: hashlib.sha1('abc').hexdigest()
  'a9993e364706816aba3e25717850c26c9cd0d89d'
sage: hashlib.sha1('abd').hexdigest()
  'cb4cc28df0fdbe0ecf9d9662e294b118092a5735'
sage: hashlib.sha1('Now is the winter of our discontent,\
made glorious summer by this sun of York;').hexdigest()
  '80ad11904ac2b72ac046673dec7874fe3b3e8916'
```

Note that when two inputs differ by only a very small amount, such as with "abc" and "abd," their hash values are completely different. There are also companion hash functions SHA-256 and SHA-512 with correspondingly larger hash values.

As of February 2005, a research team showed that they had broken SHA-1 [91]. That is, they showed that collisions could be found in only 2^{69} operations, rather than the 2^{80} operations required for a brute-force attack. Although this may seem like an insignificant difference, the ramifications are large. It means that SHA-1 is not as secure a hash algorithm as its length indicates. However, the longer versions of SHA are not known to be in any way insecure.

Hash functions based on block ciphers. Any block cipher can be used to create a hash function. If the block size for the cipher is the same as the hash size, then there are a number of different schemes that can be used. To create the hash function H, start by choosing an initial random block h_0. Suppose the message to be hashed is broken up into blocks $m_1, m_2, m_3, \ldots, m_t$, the last block m_t being padded if necessary to bring it up to size. The compression function takes as input the current hash value h_{i-1} and the current message block m_i to create the next hash value h_i, so that

$$h_i = f(m_i, h_{i-1}).$$

The final hash value h_t is the hash $H(m)$ of the message. All such schemes have the form

$$h_i = E(x, y) \oplus z$$

where E is the block cipher encryption, with x being the plaintext and y the key. Each of x, y, z can take on the values $0, m_i, h_{i-1}, m_i \oplus h_{i-1}$. There are thus $4^3 = 64$ different possible schemes, most of which are insecure. All the schemes have been extensively analyzed by Preneel [72].

Of the 12 schemes known to be secure, the most popular are

- $h_i = E(m_i, h_{i-1}) \oplus m_i$: the Matyas–Meyer–Oseas hash,
- $h_i = E(h_{i-1}, m_i) \oplus h_{i-1}$: the Davies–Meyer hash,
- $h_i = E(m_i, h_{i-1}) \oplus m_i \oplus h_{i-1}$: the Miyaguchi–Preneel hash.

These are illustrated in Figure 11.7. These functions can also be applied for ciphers for which the block and key sizes are different. For such ciphers, the input to the key can be padded or compressed to obtain the correct size.

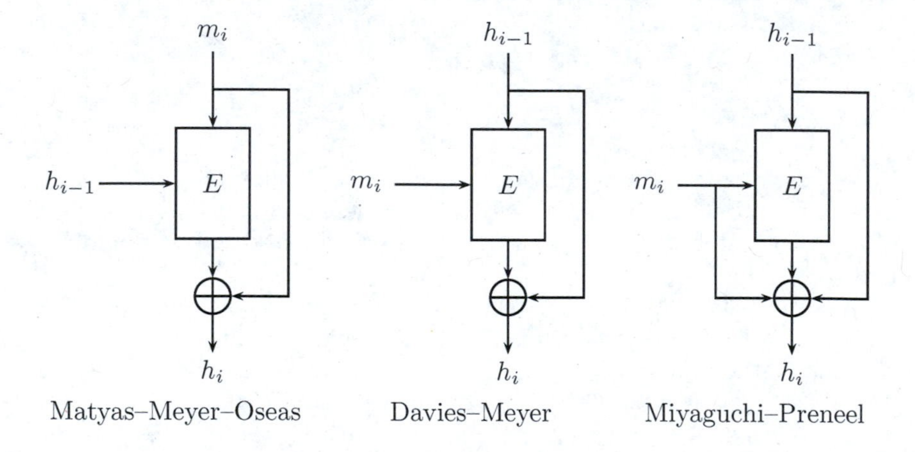

FIGURE 11.7: Hash functions from block ciphers.

Here's an example of the use of the Matyas–Meyer–Oseas (MMO) hash, using the TEA algorithm from Chapter 8. Since TEA has a block size of 16 bytes and a key size of 32 bytes, to use it in the MMO hash with block sizes of 16 bytes requires repeating the current value of h_i to use as the key. Start by taking some text and turning it into a hexadecimal string which is a multiple of 16 bytes long:

```
sage: m = 'The quick brown fox jumps over the lazy dog'
sage: mh = m.encode('hex')
sage: M = mh+((-len(mh))%16)*'0'
```

The message is now represented as a string `M` of length 96. Before computing the hash, here's a little program that computes the XOR of two strings of hexadecimal characters:

```
sage: def xor(a,b):
....:     return hex(ZZ('0x'+a)^^ZZ('0x'+b))
```

Now the hash can be computed, starting with a string of zeros:

```
sage: h=16*'0'
sage: for i in range(len(M)//16):
....:     m = M[16*i:16*i+16]
....:     E = tea(ZZ('0x'+m),ZZ('0x'+2*h))
....:     h = xor(E,m)
....:
sage: h
  '3eb229c901671c74'
```

If the above steps are repeated, but for a slightly different message

```
sage: n = 'The quick brown fox jumps over the lazy doh'
```

the resulting hash will be `593caf2eb3909293`, which is very different from the first hash.

11.4 Provably secure hash functions

Although the fastest and most used hash functions are those that have been specifically designed for the purpose, such as RIPEMD-160, there are other hash functions that are based on "hard" problems, such as factorization and discrete logarithms, which can be shown to be provably secure.

Shamir's hash function

This hash has been attributed to Shamir, but see also Gibson [32]. Choose two large primes p and q, form their product $n = pq$, and let a be an integer of maximum order modulo $n = pq$. This maximum order can be shown to be $\operatorname{lcm}(p-1, q-1)$. Let the message to be hashed be represented as an integer $x < n$. Then the hash is defined as

$$h = a^x \pmod{n}$$

where p and q are not made publicly available. This hash function is provably collision resistant. Suppose that x and y are to have the same hash. This means that

$$a^x = a^y \pmod{n}$$

or that

$$a^{x-y} = 1 \pmod{n}$$

and so

$$x - y = 0 \pmod{\operatorname{lcm}(p-1, q-1)}.$$

Therefore x and y can't be found unless p and q can be determined, which requires factoring n.

For example:

```
sage: p=next_prime(2^20);p
  1048583
sage: q=next_prime(3^13);q
  1594331
```

By some trial and error, it can be found that $a = 2$ has the required order:

```
sage: lcm(p-1,q-1)
  835892870030
sage: mod(2,n).multiplicative_order()
  835892870030
```

Now for some experiments with hashing. One expected result is that two similar inputs should have very dissimilar outputs:

```
sage: a=Mod(2,n)
sage: a^str2num("A dog")
  47971055483
sage: a^str2num("A cog")
  211138465074
```

Chaum, van Heijst, Pfitzmann hash

Let q be a large prime such that $p = 2q + 1$ is also prime, and let α, β be two primitive roots of p for which the discrete log $log_\alpha \beta \pmod{p}$ is computationally difficult. Inputs to this function are pairs (x, y) where $x < q$ and $y < q$. Then

$$H(x) = \alpha^x \beta^b \pmod{p} \text{ [17]}.$$

Following Buchmann [14], it can be shown that the problem of finding a collision is equivalent to finding a discrete logarithm. Suppose there is a collision, that is two pairs (x, y) and (w, z) with the same hash:

$$\alpha^x \beta^y = \alpha^w \beta^z \pmod{p}.$$

This can be rewritten as

$$\alpha^{x-w} = \beta^{z-y} \pmod{p}.$$

Suppose that $\lambda = log_\alpha \beta \pmod{p}$, so that

$$\alpha^{x-w} = \alpha^{\lambda(z-y)} \pmod{p}.$$

Since α is a primitive root of p, it follows that

$$x - w = \lambda(z - y) \pmod{p-1}. \tag{11.1}$$

This means that $d = \gcd(z - y, p - 1)$ must divide $x - w$. Since each of z and y is less than q, then $|z - y| < q$, and since $p - 1 = 2q$, the only two possible values for d are 1 or 2. From Buchmann (with slightly changed notation):

> "If $d = 1$, the equation 11.1 has a unique solution modulo $p - 1$. The discrete logarithm λ can be determined as the smallest nonnegative solution of this congruence. If $d = 2$ the congruence has two different solutions mod $p-1$ and the discrete logarithm can be found by trying both."

More details of this hash function can be found in the original paper [17].

The Zémor–Tillich hash function

The Zémor–Tillich hash function [87] is one of the newest provably secure hash functions, and has been the subject of some intense research and investigation. It remains, with some reservations, a very strong hash function. It has the virtue of being very easy to describe.

Let p be a large prime, and define the two matrices

$$A_0 = \begin{bmatrix} 1 & 2 \\ 0 & 1 \end{bmatrix} \pmod{p}, \qquad A_1 = \begin{bmatrix} 1 & 0 \\ 2 & 1 \end{bmatrix} \pmod{p}.$$

Let the input be a binary string $m_1 m_2 m_3 \ldots m_k$ of arbitrary length. Then the hash is defined as

$$\prod_{i=1}^{k} A_{m_i}.$$

The collision resistance can be shown to be based on the property that matrix multiplication is associative but not commutative (which has led to some generalizations over other non-abelian semi-groups).

This can be easily implemented; first enter the prime and the matrices:

```
sage: p = next_prime(2^60)
sage: A = [matrix(GF(p),[[1,2],[0,1]]),\
....: matrix(GF(p),[[1,0],[2,1]])]
```

To obtain a binary version of the message, use the `str2num` command from earlier chapters, and output its base 2 digits:

```
sage: m = "message digest"
sage: s = str2num(m).digits(2)
sage: prod(A[i] for i in s)
  [657700567577292751 176428445714224241]
  [703921236651774646 784092080455816438]
sage: m = "nessage digest"
sage: s = str2num(m).digits(2)
sage: prod(A[i] for i in s)
  [109891493661021540 891025230172718584]
  [843997821449324052 212072828953191623]
```

As with previous hashes, a small change in the input yields a completely different hash value.

MASH

The MASH (Modular Arithmetic Secure Hash) is in fact two algorithms; they are the only hash functions based on modular arithmetic that are an ISO standard.

These hash functions start with two large primes p and q and their product $M = pq$ of bit-length m. As with other hash functions, the input is broken into small chunks, and each chunk expanded to a length n based on m (in the standard, n is the largest multiple of 16 which is less than m). Given an initial value $H_0 = 0$, and blocks $y_0, y_1, \ldots y_t$, the standard for MASH-1 defines

$$H_i = (((H_{i-1} \oplus y_i) \vee A))^2 \bmod M \dashv n) \oplus H_{i-1}.$$

Here A is a constant whose hexadecimal value is `0xF00...0`, and $\dashv n$ means keep the rightmost n bits of the m-bit result. (This can be implemented by taking the residue modulo 2^n.) Notice that this formulation requires arithmetic on bits. MASH-2 is defined similarly, except that the exponent becomes $2^8 + 1 = 257$ instead of 2.

In Sage, this can be done easily using standard operators.

```
sage: p = next_prime(2^20);
sage: q = next_prime(3^13);
sage: M = p*q
sage: m = len(M.bits());m
  41
```

So a good choice would be $n = 32$. To create the y values, start with a message:

```
sage: n=16*(m//16);n
  32
sage: x=str2num("message digest")
sage: y=x.digits(2^n);y
  [1936942445, 543516513, 1701276004, 29811]
```

Now for the hash computation:

```
sage: H=0
sage: A=15*16^(n/4)
sage: for i in y:
....: H=((((H^^i)|A)^2)%M)%2^n^^H
sage: H
  1145088928
```

If the initial message was "nessage digest" then the final hash is 2980735530, very different from the previous value.

Note that the full ISO definition of MASH requires that each y_i is in fact made from an $n/2$ bit block that is padded; for ease of implementation this has been dropped from the above discussion. See Menezes et al. [61] for full details.

11.5 New hash functions

At the time of writing (late 2010), there is a search for a hash function to be the new standard, and to take the place of SHA-1, which is now considered insecure. Several different functions have been proposed, and five have been chosen as final round candidates, to be analyzed before the one is chosen to be the new hash standard SHA-3. The five functions are:

- BLAKE (submitted by Jean-Philippe Aumasson, Luca Henzen, Willi Meier and Raphael C.-W. Phan)
- Grøstl (submitted by Praveen Gauvaram, Lars Knudsen, Krystian Matusiewicz, Florian Mendel, Christan Rechberger, Martin Schläffer and Søren Thomsen)
- JH (submitted by Hongjun Wu)
- Keccak (submitted by Guido Bertoni, Joan Daemen, Michaël Peeters and Giles van Assche)

- Skein (submitted by Niels Ferguson, Stefan Lucks, Bruce Schneier, Doug Whiting, Mihir Bellare, Tadayoshi Kohno, Jon Callas and Jesse Walker).

Of the five, Grøstl is particularly easy to describe as it is based in part on Rijndael, and borrows much of its internal machinery from it. Grøstl is based on a Merkle–Damgård construction as shown in Figure 11.8.

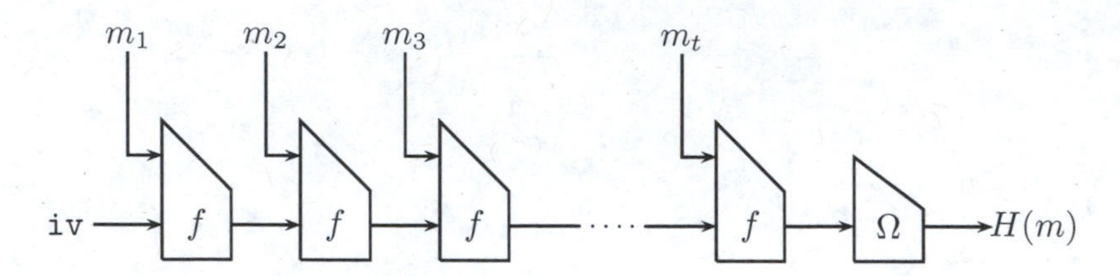

FIGURE 11.8: The Grøstl hash function.

A pronunciation note: the ø in "Grøstl" is pronounced similar to the vowel in "bird."

The compression function f is based on two "permutation functions" P and Q, and is defined as

$$f(h, m) = P(h \oplus m) \oplus Q(m) \oplus h.$$

The final output transformation $\Omega(x)$ takes the trailing n bits of $P(x) \oplus x$, where n is the length of the final hash and can be any number of bytes from 1 to 64.

Each of P and Q is defined using very similar operations to Rijndael. In fact each is based on a *round function* R which is defined in terms of four *round transformations* as

$$R = \text{MixBytes} \circ \text{ShiftBytes} \circ \text{SubBytes} \circ \text{AddRoundConstant}.$$

This round function R is repeated for r rounds, where $r = 10$ for an output of 256 bits or less, and $r = 14$ for an output greater than 256 bits.

As explained below, P and Q only differ in the definition of AddRoundConstant. For ease of explanation, consider only Grøstl-256, which produces a 256 bit hash, or 32 bytes. For this function the internal state may be considered as an 8×8 matrix A of bytes, where the bytes are placed into the matrix column by column. Then the round transformations are defined as

AddRoundConstant. For round number i, A is replaced with $A + C_i$, where C is an 8×8 matrix. Each C_i consists entirely of zeros, except for one byte: for round i of P, $C_i[0, 0] = i$. That is, the top left entry of C_i is i. For round i of Q, the bottom left entry is set equal to $i \oplus \texttt{0xff}$; that is, $C_i[7, 0] = i \oplus \texttt{0xff}$.

SubBytes. This simply applies Rijndael's S-box to each byte in A.

ShiftBytes. Each row of the current state A is rotated left by a fixed amount; of the eight rows (indexed 0 to 7), row i is rotated left by i places.

MixBytes. This is a matrix product using the Rijndael field

$$\mathbb{Z}_2[x]/(x^8 + x^4 + x^3 + x + 1)$$

and where the state A is replaced by BA, with B being defined as

$$B = \begin{bmatrix} 02 & 02 & 03 & 04 & 05 & 03 & 05 & 07 \\ 07 & 02 & 02 & 03 & 04 & 05 & 03 & 05 \\ 05 & 07 & 02 & 02 & 03 & 04 & 05 & 03 \\ 03 & 05 & 07 & 02 & 02 & 03 & 04 & 05 \\ 05 & 03 & 05 & 07 & 02 & 02 & 03 & 04 \\ 04 & 05 & 03 & 05 & 07 & 02 & 02 & 03 \\ 03 & 04 & 05 & 03 & 05 & 07 & 02 & 02 \\ 02 & 03 & 04 & 05 & 03 & 05 & 07 & 02 \end{bmatrix}.$$

The initial value `IV` is fixed, and for Grøstl-256 it consists of 32 bytes, where all bytes are zero except for the second right-most byte, which is equal to `0x01`.

For further details, including how padding is performed, see the original description [30] and the newer addendum [31], which describes some aspects of its attack resistance, as well as other ways of describing the function.

11.6 Message authentication codes

A *Message authentication codes* or MAC, is a kind of hash function that uses a key as well as the message data. Thus a MAC takes a message of arbitrary length and a key of fixed length as input, and returns a fixed length string as output.

MACs are vital for message authentication, but they can be used in other ways. For example, suppose you wish to protect your computer against a malicious virus that can change your files. One way would be to hash each file, and store the hash values in another file. The trouble is that the virus, when it changes the files, can easily hash the new files and place the new hash over the old one. There is no way you could determine if the file had been corrupted, because its stored hash value would be correct. Far better is to store a MAC of each file. A virus can change any file it likes, but it can't compute a MAC because it doesn't know the key. Thus you can be sure when a file has been corrupted: its new MAC won't match the MAC in the table.

Creating a MAC

A MAC can be created from a hash function; the message and the key are concatenated in various ways. A very simple way is

$$M = H(k \| H(k \| m)).$$

In other words, given the message and the key, the MAC can be created by these steps:

1. Concatenate the key with the message.
2. Hash the result.
3. Place the key in front of the hash value.
4. Hash again.

A more powerful scheme is called HMAC. It can be used with any hash function and the steps to perform it are

1. Add zeros to the end of the key k to create a string of length 64 bytes (or 512 bits).
2. XOR the string from step 1 with the string 00110110 repeated 64 times.
3. Append the message m to the result of step 2.
4. Hash the result of step 3.
5. XOR the string from step 1 with the string 01011100 repeated 64 times.
6. Append the hash function from step 4 to the 64-byte string obtained from step 5.
7. Hash the result of step 6.

If the 64 copies of 00110110 are abbreviated to "36," and the 64 copies of 01011100 to "5C" (which are the hexadecimal equivalents of those bytes), then HMAC can be written as

$$\text{HMAC}(m, k) = H(k \oplus \text{“5C”} \| H(k \oplus \text{“36”} \| m)).$$

11.7 Using a MAC

Recall that a hash function can be used for digital signatures. Indeed, seeing that it is more efficient to sign a hash than the message itself, most digital signature algorithms work with $H(m)$ rather than with m itself.

The principal use of a MAC is used to detect message tampering and forgery. Suppose Alice is sending a message m to Bob. Using a key k known to both of them, she forms the MAC using m and k, and sends both the message and its MAC to Bob.

When Bob gets them, he calculates the MAC of his received message m', using k, and compares it with the MAC that Alice sent. If the two MACs are equal, then Bob may be sure that the message has arrived safely, and with no tampering.

11.8 Glossary

Chaining variable. An internal variable in a hash function that is updated over the course of the function's execution.

Compression function. The internal function used by a hash algorithm and that operates on a fixed length block.

Cryptographic hash function. A function that produces a fixed length string of bits for arbitrary input.

Message Authentication Code. A hash function that requires a key as input as well as the data.

Provably secure hash functions. Hash functions for which there are mathematical proofs of their security.

SHA, MD, RIPEMD. Families of hash functions.

Exercises

Review Exercises

1. What are some of the uses of a hash function?

2. What is preimage resistance?

3. What are weak and strong collision resistance, and how do they differ?

4. How are hash functions used in a digital signature protocol?

5. How are hash functions used for message authentication?

6. What is the birthday paradox, and how is it used to measure the strength of a hash function?

7. What is the least number of bits for a hash below which the hash function could not be considered secure?

8. What does it mean to "break" a hash function?

9. List all the arithmetic operations used in each of RIPEMD-160 and MD2.

10. List all the bit operations used in each of RIPEMD-160 and MD2.

11. What is the Merkle–Damgård construction?

12. What is Shamir's hash function?

13. Why are provably secure hash functions little used in practice?

14. How can a message authentication code be constructed from a hash function?

15. How can a message authentication code be used to detect tampering with files?

Beginning Exercises

16. Define a simple hash function as follows: write the plaintext as rows of 8 bits—one row for the ASCII value of each character in binary, and add the rows modulo 2 (this is the same as "XOR-ing" all the columns). For example, the hash of `Tuesday` is

T	84	0	1	0	1	0	1	0	0
u	117	0	1	1	1	0	1	0	1
e	101	0	1	1	0	0	1	0	1
s	115	0	1	1	1	0	0	1	1
d	100	0	1	1	0	0	1	0	0
a	97	0	1	1	0	0	0	0	1
y	121	0	1	1	1	1	0	0	1
		0	1	0	0	1	0	1	1

   (a) Check the working in the above example.

   (b) Suppose the text is changed to `Tuesdax`. How is the hash changed?

   (c) Show that this function satisfies neither the one-way nor the collision resistance properties.

17. Can you think of a way of enhancing the function described in the previous question so that a single change in the input will affect approximately half the characters in the output?

18. Consider the Matyas–Meyer–Oseas scheme $h_i = E(m_i, h_{i-1}) \oplus m_i$ for using a block cipher to construct a hash. This can be used with a matrix cipher by breaking the plaintext into $n \times n$ blocks M_i, and letting H_0 be a previously determined $n \times n$ matrix. Then $H_i = M_i H_{i-1} + M_i \bmod 2$.

    Rewrite the Davies–Meyer and Miyaguchi–Preneel schemes in matrix form.

19. Show that the Davies–Meyer hash, applied to the Hill cipher, satisfies

    $$h_k = h_0(M_1 + I)(M_2 + I) \cdots (M_k + I$$

    where M_i, h_i are square matrices.

20. Find similar expressions for the Matyas–Meyer–Oseas and the Miyaguchi–Preneel hashing schemes.

21. Show that each scheme is in fact not secure when used with the Hill cipher; that is, show how to obtain collisions.

    For more discussion about the Hill cipher used for block cipher examples see [58].

22. One possible hash based on a block cipher has the compression function

    $$h_i = E(M_i, h_{i-1}).$$

    Show why this is not secure by using the decryption function to obtain a collision.

23. Create a simple MASH-type hash as follows: choose two primes p and qm, and compute their product $n = pq$. Break up the data into blocks $y_0, y_1, \ldots, y_t$, each less than n.

    Define $H_0 = 0$ and for each $i = 1, 2, \ldots t$ define

    $$H_i = (H_{i-1} + y_i)^2 + H_{i-1} \pmod{n}.$$

    Let $p = 31$ and $q = 47$. Divide the data

    123456789

    into chunks of three digits each, and apply this hash function, starting from the left. What is the hash?

24. Suppose a simple hash is defined using a prime p and primitive root a as

    $$h = a^m \pmod{p}.$$

    Why is this not a good choice for a hash function?

Sage Exercises

25. Mars orbits the sun and rotates in such a way that there are 668 Martian days in a Martian year. How many Martians are required so the probability that two of them share a birthday is (a) 0.5? (b) 0.9?

26. Use the simplified MASH (SMASH) of question 23 as follows: for a message m, considered as a large integer, define y_i to be the "digits" of m in base n. This can be obtained in Sage with

```
sage: y = m.digits(n)
```

Define

```
sage: p = next_prime(2^40)
sage: q = next_prime(3^26)
```

Use these values to hash the phrase:

`Now is the winter of our discontent made glorious summer by this sun of York.`

and use the `str2num` command used in earlier chapters.

27. Apply the Zémor–Tillich hash to the phrase in the previous question, using $p = 2^{61} - 1$.

28. Using the definition given on page 278 and the S permutation defined in [45], write an implementation in Sage of MD2.

29. Use the Shamir-Gibson hash function with primes

```
sage: p = next_prime(87654321)
sage: q = next_prime(98765432)
```

and $g = 17$ to hash the following strings:

`My name is Sam`
`My name is San.`

Even though the strings are very similar, how similar are the hash values?

30. Try hashing some similar strings with SHA-1:

```
The quick brown fox jumps over the lazy dog
The quick brown fox jumps over the lazy doh
```

By how much do their hashes differ?

31. Repeat the above question but using RIPEMD-160 in place of SHA-1.

32. Below is one way to find the exact number of bits at which the two hashes differ:

```
sage: s1 = "The quick brown fox jumps over the lazy dog"
sage: s2 = "The quick brown fox jumps over the lazy doh"
sage: h1 = sha1(s1).hexdigest()
sage: h2 = sha1(s2).hexdigest()
sage: b = ZZ('0x'+h1)^^ZZ('0x'+h2)
```

(Note: in `'0x'` the first character is the number zero.)

To explain the last command: since both `h1` and `h2` are strings of hexadecimal characters, the addition of the characters `0x` to both will allow them to be interpreted as hexadecimal *numbers*.

The function `ZZ` produces an integer value, and the operator `^^` is XOR in Sage.

The idea is that the numbers of 1s in the XOR will be the number of places at which the two values differ:

```
sage: b.bits().count(1)
```

33. Repeat the above question but using RIPEMD-160 in place of SHA-1.

34. Determine the number of places at which the two Matyas–Meyer–Oseas hashes computed on page 280 differ.

35. Using the TEA algorithm, compute the Davies–Meyer and the Miyaguchi–Preneel hashes of those same two strings.

    For each pair of hashes, determine the number of bits at which they differ.

36. Repeat the last question, but use DES in place of TEA.

Chapter 12

Elliptic curves and cryptosystems

This chapter will introduce:

- Elliptic curves over the reals: points arithmetic and the associated group.
- Different coordinate representations of elliptic curves and their points.
- Elliptic curves over finite fields, including point counting.
- Elliptic curve cryptosystems, including the El Gamal system and the Menezes–Vanstone variation.
- Signature schemes using elliptic curves.
- An introduction to pairing-based protocols using elliptic curves.

Although elliptic curves and their associated mathematics have been studied intensively for a long time—for example they are central to Andrew Wiles' proof of Fermat's last theorem—only more recently have they become used for cryptographic purposes. Elliptic curve cryptosystems are analogous to systems based on discrete logarithms, such as the El Gamal system, or Diffie–Hellman key exchange, but use far smaller keys to achieve the same levels of security. Arithmetic on elliptic curves can be made extremely efficient, both in terms of memory and time, and so have shown great promise for embedded systems on low power components such as smart cards. This chapter will develop the basic theory of elliptic curves, leaving out most proofs (instead giving references to where proofs can be found), and show how elliptic curve cryptosystems are defined.

12.1 Basic definitions

Suppose K is a field not of characteristic 2 or 3. Let $a, b \in K$ satisfy $4a^3+27b^2 \neq 0$. Then an *elliptic curve* over K is the set of points $(x, y) \in K \times K$ satisfying

$$y^2 = x^3 + ax + b.$$

The condition $4a^3 + 27b^2 \neq 0$ is vital. The cubic polynomial

$$x^3 + ax + b$$

has discriminant $-16(4a^3 + 27b^2)$, which is equal to zero if and only if the polynomial has equal roots. Requiring the discriminant to be non-zero means that all roots of the polynomial are distinct, and this guards against "singular points" of the curve. This particular form of an equation for an elliptic curve is called the *Weierstrass form* or sometimes the *simplified Weierstrass form* of the curve. There are many other equations that describe elliptic curves.

Before investigating curves defined over finite fields, some basic insight and understanding can be obtained by looking at elliptic curves over $\mathbb{Q}$. Such curves are not used for cryptography, but can and are used to build up some of the basic theory.

Here are two examples: the curve E defined by

$$y^2 = x^3 - 5x + 2$$

and F defined by

$$y^2 = x^3 - 2x + 4.$$

These can easily be plotted in Sage:

```
sage: E = EllipticCurve([-5,2])
sage: E.plot(aspect_ratio=1)
sage: F = EllipticCurve([-2,4])
sage: F.plot(aspect_ratio=1)
```

and the plots are shown in Figure 12.1.

An elliptic curve can be defined over any field; for example:

```
sage: G = EllipticCurve(GF(11),[-5,2])
sage: print G.list()
  [(0 : 1 : 0), (1 : 3 : 1), (1 : 8 : 1), (2 : 0 : 1),
  (3 : 5 : 1), (3 : 6 : 1), (5 : 5 : 1), (5 : 6 : 1),
  (6 : 1 : 1), (6 : 10 : 1), (8 : 1 : 1), (8 : 10 : 1),
  (9 : 2 : 1), (9 : 9 : 1)]
sage: G.plot(pointsize=40)
```

and the plot is shown in Figure 12.2. The reason that the output of points of an elliptic curve have three elements each will be discussed later.

Having given the basic definition of an elliptic curve, the next step is to define addition of points on the curve. This can be done geometrically by a method known as the "chord and tangent method." Supposing P and Q are distinct points on the curve, as shown on the left hand diagram in Figure 12.3, their sum $P + Q$ is obtained by the following steps:

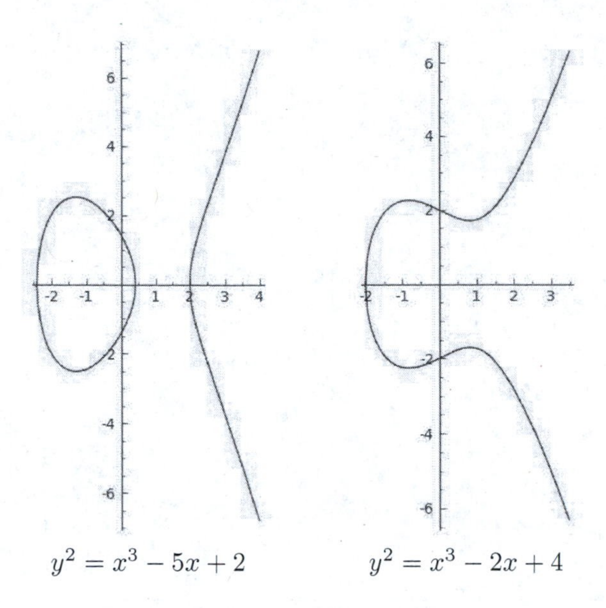

$y^2 = x^3 - 5x + 2$ $\quad$ $y^2 = x^3 - 2x + 4$

FIGURE 12.1: Elliptic curves.

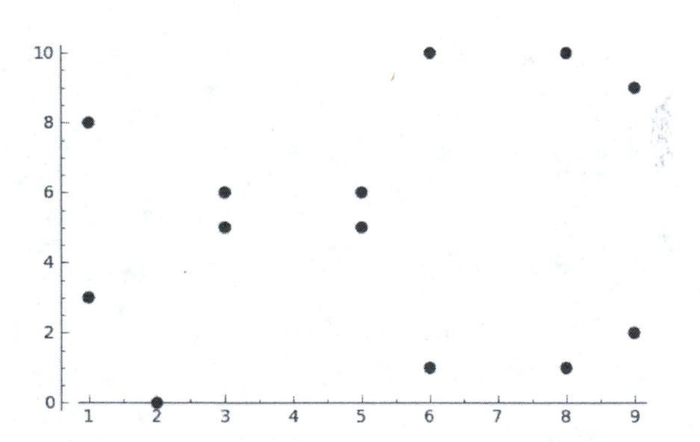

FIGURE 12.2: An elliptic curves over the field $\mathbb{Z}/11\mathbb{Z}$.

1. Find the line that passes through P and Q,
2. Find the point R at which the line intersects the curve in a third point,
3. The point $P+Q$ is the reflection of R in the x-axis.

Doubling of a point is obtained similarly, except the line is obtained by taking the tangent to the curve at P. Both these operations are demonstrated in Figure 12.3.

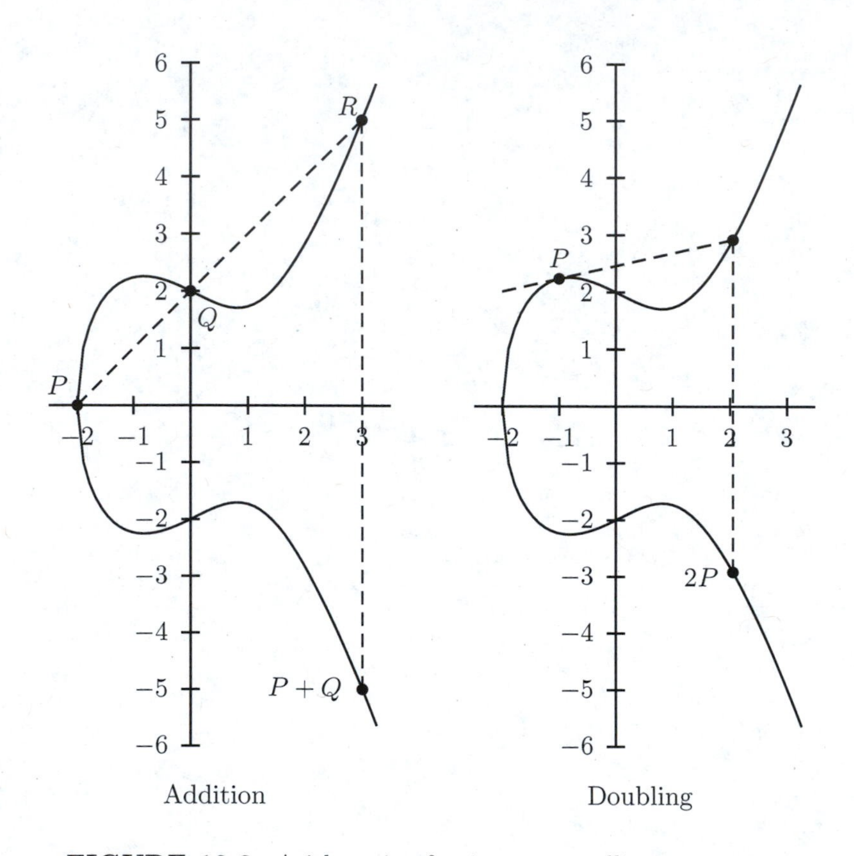

FIGURE 12.3: Arithmetic of points on an elliptic curve.

To develop formulas for these operations, suppose $P = (x_1, y_1)$ and $Q = (x_2, y_2)$. The line between them is given by

$$y - y_1 = \frac{y_2 - y_1}{x_2 - x_1}(x - x_1).$$

Let the gradient be denoted by g, so that the line can be expressed as $y = gx + h$. Finding the intersection with the curve requires solving

$$(gx + h)^2 = x^3 + ax + b$$

or

$$x^3 - g^2x^2 + (a - 2gh)x + b - h^2 = 0.$$

However, there are already two known solutions to this cubic, x_1 and x_2, and from the formula relating the roots of a cubic to its coefficients

$$(x - \alpha)(x - \beta)(x - \gamma) = x^3 - (\alpha + \beta + \gamma)x^2 + (\alpha\beta + \alpha\gamma + \beta\gamma)x - \alpha\beta\gamma$$

it follows that the sum of the roots is equal to the negative of the coefficient of x^2. Let (x_r, y_r) be the coordinates of the point R. Then

$$x_1 + x_2 + x_r = g^2$$

or

$$x_r = g^2 - x_1 - x_2.$$

The y value can be found by substituting into the equation of the line:

$$y_r = g(x_r - x_1) + y_1.$$

So if the coordinates of $P + Q$ are $(x_3, y_3) = (x_r, -y_r)$, then

$$x_3 = g^2 - x_1 - x_2$$
$$y_3 = g(x_1 - x_3) - y_1.$$

An example

For the elliptic curve F above, let $P = (-2, 0)$ and $Q = (0, 2)$. Then

$$g = \frac{0 - 2}{-2 - 0} = 1.$$

Then $x_3 = 1^2 - (-2) - 0 = 3$ and $y_3 = 1(-2 - 3) - 0 = -5$. In Sage:

```
sage: F([-2,0])+F([0,2])
  (3 : -5 : 1)
```

The formula for doubling is obtained similarly, except that the gradient of the line is found by implicit differentiation. If $P = (x_1, y_1)$, then

$$g = \frac{3x_1^2 - a}{2y_1}$$

and the values (x_3, y_3) are computed using the same formulas as for addition.

12.2 The group on an elliptic curve

Recall from section 9.1 that a group is a set with a binary operation that satisfies closure, associativity, an identity element, inverses, and (for an abelian group) commutativity. In order for the addition operation previously defined to form a group on the elliptic curve, one more point is needed: the "point at infinity." This is denoted $\mathcal{O}$[1], and may be considered to be at the end of every vertical line. (This makes more sense in terms of projective geometry than Cartesian geometry; however it is a perfectly reasonable point.) One way of interpreting this point at infinity is with a projection of the real line onto a circle as shown in Figure 12.4.

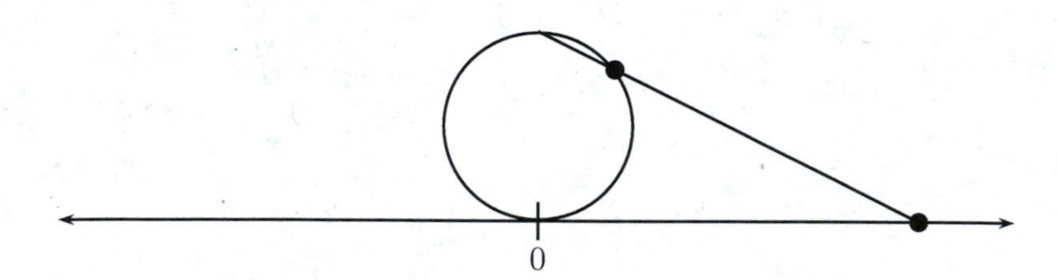

FIGURE 12.4: Projecting the real line onto a circle.

The point at infinity on the real line corresponds to the uppermost point on

[1]Some authors denote this point with ∞, but the most important thing about this point for cryptographic purposes is not that it is "at infinity," but that it is the identity of an abelian group.

the circle. With the elliptic curve and this new point, addition on the curve can be formally defined.

Let $p = (x_1, y_1), q = (x_2, y_2) \in E \cup \{\mathcal{O}\}$. Then:

1. If $q = \mathcal{O}$, then $p + \mathcal{O} = p$.
2. If $p = \mathcal{O}$, then $\mathcal{O} + q = p$.
3. If $x_1 \neq x_2$, then $p + q = (x_3, y_3)$, where

$$x_3 = \lambda^2 - x_1 - x_2$$
$$y_3 = \lambda(x_1 - x_3) - y_1.$$

with

$$\lambda = \frac{y_2 - y_1}{x_2 - x_1}.$$

4. If $p = q$ and $y_1 \neq 0$, then $p + q = (x_3, y_3)$, where

$$x_3 = \lambda^2 - x_1 - x_2$$
$$y_3 = \lambda(x_1 - x_3) - y_1$$

with

$$\lambda = \frac{3x_1^2 - a}{2y_1}.$$

5. If $p = q$ and $y_1 = 0$ then $p + q = \mathcal{O}$.

Below is one of the most important results in the entire theory of elliptic curves:

> If E is an elliptic curve, then $E \cup \{\mathcal{O}\}$ forms an abelian group with the operation defined above, with $\mathcal{O}$ being the identity element.

Illustrations of arithmetic with $\mathcal{O}$ are shown in Figure 12.5. The left hand diagram shows that if $p = (x, y)$ and $q = (x, -y)$ then $p + q = \mathcal{O}$; the right hand diagram illustrates $p + \mathcal{O} = p$.

In the left hand diagram, since O is its own reflection, it follows that $P + Q = O$. And in the right hand diagram, the line through O and P meets the curve at R, whose reflection in the x-axis—P again—is the required sum. This shows that O is indeed the identity element of the group, and that the inverse of $P = (x, y)$ is $-P = (x, -y)$.

The only hard aspect of the group laws to prove is associativity; there is no really easy way to do this. There are in fact many different proofs; some algebraic, some geometric, and some based on some very deep and general mathematics from which the associativity of the group law follows as a natural

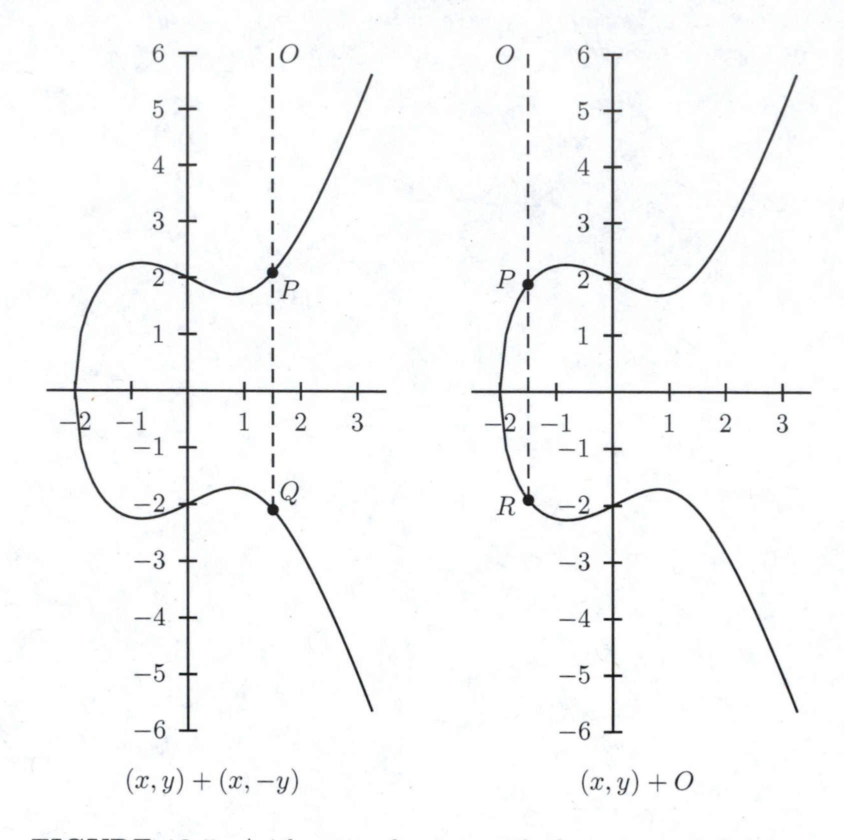

FIGURE 12.5: Arithmetic of points with the point at infinity.

consequence. None of these shall be given here. A very nice approach (using Sage) is given by William Stein [84].

For computation with elliptic curves over finite fields, division in the addition formulas is replaced with inversion in the field. For example, take the curve

$$y^2 = x^3 + 5x + 1$$

over $\mathbb{Z}_{97}$. Suppose $p = (16, 43)$ and $q = (26, 21)$. Then

$$\begin{aligned}\lambda &= (21 - 43)(26 - 16)^{-1} \pmod{97}\\ &= 56.\end{aligned}$$

In Sage:

```
sage: Mod((21-43)/(26-16),97)
  56
```

Then

$$\begin{aligned} x_3 &= \lambda^2 - x_1 - x_2 \\ &= 56^2 - 16 - 26 \pmod{97} \\ &= 87 \end{aligned}$$

and

$$\begin{aligned} y_3 &= \lambda(x_1 - x_3) - y_1 \\ &= 56(16 - 87) - 43 \pmod{97} \\ &= 55. \end{aligned}$$

All this can be easily done in Sage:

```
sage: G = EllipticCurve(GF(97),[5,1])
sage: P = G([16,43])
sage: Q = G([26,21])
sage: P+Q
  (87 : 55 : 1)
```

Note that if a curve E is defined over a finite field, there will be only a finite number of possible points. If $P \in E$, there must be a smallest possible value n for which $nP = \mathcal{O}$. This value n is called the *order* of the point. Many applications of elliptic curves require finding a point whose order is large.

The following results about the elliptic curve group will be useful:

- If the number of points n of the curve E is prime, then the group will be cyclic of order n.
- The order of any point $P \in E$ divides the order of the curve. This is an application of *Lagrange's theorem*, which says that for any group G and subgroup H of G, the order of H must divide the order of G.
- For any curve E, its group is either cyclic, that is, isomorphic to the additive group $\mathbb{Z}_n$, or isomorphic to the group

  $$\mathbb{Z}_n \oplus \mathbb{Z}_m$$

  where n divides m.

For proofs and further discussion, see Washington [92].

The number of points on an elliptic curve

If the curve $y^2 = x^3 + ax + b$ is defined over a finite field, then it will have only a finite number of points. This number is called the *order* of the curve. For a small curve defined over $\mathbb{Z}_p$, the points can be enumerated by a simple procedure:

1. List all values $x^2 \bmod p$ for $x = 0, 1, \ldots, p-1$.

2. List all values $x^3 + ax + b \bmod p$ for $x = 0, 1, \ldots, p-1$.

3. Compare the two lists.

For example, take the curve

$$y^2 = x^3 + 3x + 1 \pmod 7.$$

Consider the table:

x	$x^3 + 3x + 1$	x^2
0	1	0
1	5	1
2	1	4
3	2	2
4	0	2
5	1	4
6	4	1

Comparing the second and third columns, the following points (not at infinity) are immediately obtained:

$$(0,1), (0,6), (2,1), (2,6), (3,3), (3,4), (4,0), (5,1), (5,6), (6,2), (6,5).$$

The easiest way to obtain these points is to look at all possible pairs of equal values in the second and third columns. For each such pair, the corresponding values in the first column give an (x, y) point. For example, the two 4s in rows 5 and 6 provide $(x, y) = (6, 5)$; the two 4s in rows 2 and 6 provide $(x, y) = (6, 2)$.

A famous result from the mathematician Helmut Hasse, and hence known as *Hasse's theorem*, states that for any elliptic curve E over $\mathbb{Z}_p$, the number of points $\#(E)$ satisfies

$$p + 1 - 2\sqrt{p} \leq \#(E) \leq p + 1 + 2\sqrt{p}.$$

Alternatively, this can be written as

$$|\#(E) - (p+1)| \leq 2\sqrt{p}.$$

This can sometimes be used to provide the order of a curve precisely. Suppose that a point $P \in E$ has order m, and there is only one multiple km of m that satisfies

$$p + 1 - 2\sqrt{p} \leq km \leq p + 1 + 2\sqrt{p}.$$

Then the curve must have exactly km points.

There are many families of curves for which the orders can be determined precisely. One simple family is given by curves of the form

$$y^2 = x^3 + kx \pmod{p}$$

where $p = 3 \bmod 4$ and $k \neq 0 \bmod p$. The number of points is easily shown to be $p+1$. To show this, write $p = 4n+3$, so that $(p-1)/2 = 2n+1$. Then

$$\left(\frac{-1}{p}\right) = (-1)^{2n+1} = -1 \pmod{p}$$

and so -1 is not a quadratic residue modulo p. For the same reason, if x is a quadratic residue then $-x$ is not, and vice-versa. This means we can list all the points of the curve starting with $\mathcal{O}$ and $(0,1)$. For every other value $x \bmod p$, since $(-x)^3 + k(-x) = -(x^3 + kx)$; this means that if $x^3 + kx$ is a quadratic residue, then $(-x)^3 + k(-x)$ is not, and vice versa. That means there will be exactly $(p-1)/2$ values of x for which $x^3 + kx$ is a quadratic residue, each of which will have 2 square roots. This provides $p-1$ points, which with $\mathcal{O}$ and $(0,1)$ produces $p+1$ points in total.

A note on projective coordinates

As noted earlier, using Cartesian or affine coordinates leads to several difficulties. The first is that the formulas for addition and doubling require a division or an inversion in a finite field, which is an expensive operation, computationally speaking. The second difficulty is that the point at infinity, although a necessary element in the elliptic curve group, has to be treated as being qualitatively different from the other points in the group.

One way to avoid both these issues is to change the coordinate system. There have been many different coordinate systems proposed for efficient operations on elliptic curves; only projective coordinates will be considered here. In this system, a point is described not by its x and y values, but by the *ratios* between its elements. More particularly, a point in projective coordinates consists of three elements:

$$(X : Y : Z)$$

with

$$(X_1 : Y_1 : Z_1) = (X_2 : Y_2 : Z_2)$$

if

$$\frac{X_1}{Z_1} = \frac{X_2}{Z_2}, \quad \frac{Y_1}{Z_1} = \frac{Y_2}{Z_2}$$

assuming that neither Z_1 or Z_2 are equal to 0. As a convention, we use upper case for projective coordinates. Another way of saying this is that

$$(X : Y : Z) = (\lambda X : \lambda Y : \lambda Z)$$

for all non-zero λ. A projective point

$$(X : Y : Z)$$

with $Z \neq 0$ corresponds to the affine point

$$(X/Z, Y/Z)$$

and an affine point

$$(x, y)$$

corresponds to the equivalence class

$$(\lambda x : \lambda x : \lambda)$$

of projective points. Given that an equivalence class can be represented by any element, in general the Cartesian point

$$(x, y)$$

corresponds to the projective point

$$(x : y : 1).$$

The point at infinity can be represented by the projective point $(0 : 1 : 0)$. Using projective coordinates, addition

$$(X_1 : Y_1 : Z_1) + (X_2 : Y_2 : Z_2) = (X_3 : Y_3 : Z_3)$$

can be performed as follows. Define:

$$u = Y_2Z_1 - Y_1Z_2, \quad v = X_2Z_1 - X_1Z_2, \quad A = u^2Z_1Z_2 - v^3 - 2v^2X_1Z_2.$$

Then

$$X_3 = vA, \quad Y_3 = u(v^2X_1Z_2 - A) - v^3Y_1Z_2, \quad Z_3 = v^3Z_1Z_2.$$

For doubling, $2(X_1 : Y_1 : Z_1) = (X_3 : Y_3 : Z_3)$, define

$$w = aZ_1^2 + 3X_1^2, \quad s = Y_1Z_1, \quad B = X_1Y_1s, \quad h = w^2 - 8B.$$

Then

$$X_3 = 2ha, \quad Y_3 = w(4B - h) - 8Y_1^2s^2, \quad Z_3 = 8s^3.$$

There are other coordinate systems, for example Jacobian projective coordinates, where $(X : Y : Z)$ corresponds to the affine point $(X/Z^2, Y/Z^3)$.

12.3 Background and history

Clearly elliptic curves are not ellipses, and yet their name is no mistake, for their theory and their history are intimately bound up with ellipses. Just as the arc length of the circle can be computed using integrals such as

$$\int_0^s \frac{1}{\sqrt{1-x^2}}\,dx$$

the result of which involves inverse trigonometric functions, the computation of the arc length of an ellipse involves integrals containing square roots of cubic and quartic polynomials, such as

$$\int_0^s \frac{1}{\sqrt{1-x^4}}\,dx$$

which leads to a totally new class of functions. Specifically, if these integrals are inverted, the results are complex-valued functions $f(z)$ called *elliptic functions*, which are distinguished from the trigonometric functions by being *doubly periodic*. This means that there are two complex numbers ω_1 and ω_2 whose ratio is not real, and for which

$$f(z+\omega_1) = f(z)$$
$$f(z+\omega_2) = f(z)$$

for all z. In the middle of the nineteenth century, the German mathematician Karl Weierstrass, in an effort to get to the bottom of these functions, developed what is now known as Weierstrass's $\wp$-function. (The symbol $\wp$ is a sort of cursive p, and the function is spoken of as "Weierstrass's p-function".) This is double periodic, but also has the property that the function and its derivative parameterize an elliptic curve:

$$\wp'(x)^2 = 4\wp(x)^3 + g_1\wp(z) + g_2$$

where g_1 and g_2 are values determined by the function. This is analogous to the function $\sin(x)$ and its derivative $\cos(x)$ parameterizing the unit circle.

The function $\wp(x)$ also has an addition formula:

$$\wp(z+w) = \frac{1}{4}\left(\frac{\wp'(z)-\wp'(w)}{\wp(z)-\wp(w)}\right)^2 - \wp(z) - \wp(w).$$

Given that $\wp'(x)$ and $\wp(x)$ parameterize an elliptic curve, this formula is equivalent (aside from the scaling fraction) to the addition formula developed earlier.

12.4 Multiplication

Just as modular exponentiation is vital to the RSA, Rabin and El Gamal cryptosystems and associated signature schemes, so multiplication is vital to elliptic curve cryptosystems. And in fact the algorithms for point multiplication are exactly the same as for modular exponentiation, with squaring and multiplication replaced by doubling and addition on the elliptic curve.

Using the binary method, the value

$$nP$$

with P being a point on the elliptic curve, and n an integer, can be calculated by first expressing n in binary. Now starting with the point at infinity, just work down the values, for each binary digit b:

$$\text{if } b = \begin{cases} 0, & \text{double the result} \\ 1, & \text{double the result and add } P. \end{cases}$$

For example, calculate

$$19P$$

where $P = (14, 6)$ in the elliptic curve

$$y^2 = x^3 + x + 5$$

over $\mathbb{Z}_{101}$. Use the tabular arrangement; the left-most column is the binary representation of 19, going top to bottom.

$$\begin{array}{lll} 1 & 2\mathcal{O} + (14,6) & = & (14,6) \\ 0 & 2(14,6) & = & (21,46) \\ 0 & 2(21,46) & = & (58,89) \\ 1 & 2(58,89) + (14,6) & = & (91,45) \\ 1 & 2(91,45) + (14,6) & = & (86,7). \end{array}$$

The second method, which doesn't require first computing the binary representation, uses a table of three columns, containing values k_i, a_i and m_i, starting with the values of $k_1 = n$, $a_1 = P$, and $m_0 = \mathcal{O}$. The first column is obtained by dividing the previous value by two and keeping the integer part only, so that

$$k_i = k_{i-1}//2$$

stopping when a value of zero is reached. The second column is obtained by doubling, so that

$$a_i = 2a_{i-1}.$$

For the final column:

$$\text{if } k_i \text{ is } \begin{cases} \text{even, then } m_i = m_{i-1} \\ \text{odd, then } m_i = m_{i-1} + a_i. \end{cases}$$

For the previous example, this produces:

i	k_i	a_i	m_i	
0			$\mathcal{O}$	
1	19	$(14, 6)$	$(14, 6)$	k_1 is odd, so $m_1 = m_0 + a_1$
2	9	$(21, 46)$	$(43, 88)$	k_2 is odd, so $m_2 = m_1 + a_2$
3	4	$(58, 89)$	$(43, 88)$	k_3 is even, so $m_3 = m_2$
4	2	$(81, 60)$	$(43, 88)$	k_4 is even, so $m_4 = m_3$
5	1	$(9, 95)$	$(86, 7)$	k_5 is odd, so $m_5 = m_4 + a_5$
6	0			k_6 is zero, so stop.

The final m_i value is the required result.

12.5 Elliptic curve cryptosystems

Cryptosystems that use elliptic curves are analogues of those cryptosystems discussed in Chapter 6 whose security is based on the discrete logarithm problem. The same problem for elliptic curves, is given points P and Q with

$$Q = nP$$

to determine the value of n. In general this is very difficult; there is no efficient algorithm that will work for all curves and all points on them.

The elliptic curve El Gamal cryptosystem

One standard elliptic curve cryptosystem is based on the El Gamal cryptosystem. Recall how this works: Alice and Bob agree on a prime p and a primitive root a modulo p. Alice's private key is any number $A < p$; her public key is $B = a^A \bmod p$. To send a message m to Alice, Bob chooses a random integer $k < p$ and computes the message key $K = B^k \bmod p$. The ciphertext is the pair $(C_1, C_2) = (a^k, Km) \bmod p$, which Alice can decrypt by $m = C_2(C_1^A)^{-1} \bmod p$.

The elliptic curve system uses exactly the same principles, replacing modular multiplication, inversion and exponentiation with addition, subtraction and multiplication in the elliptic curve. To start, Alice and Bob agree on an elliptic curve E and a point $P \in E$ that has a high order. That is, the value n for which $nP = \mathcal{O}$ is large. Alice chooses her private key $r < n$, and her public key is $Q = rP$.

To send a message m (encoded as a point of E), Bob chooses a random integer k and sends the pair

$$(C_1, C_2) = (kP, kQ + m)$$

as the ciphertext to Alice.

This can be easily decrypted by Alice using

$$m = C_2 - rC_1.$$

To see that this works, note that

$$\begin{aligned} C_2 - rC_1 &= kQ + m - r(kP) \\ &= kQ + m - k(rP) \\ &= kQ + m - kQ \\ &= m. \end{aligned}$$

Figure 12.6 describes this system.

Parameters. An elliptic curve E and a point $P \in E$ of high order n.

Key generation. The private key is any $r < n$. The public key is $Q = rP$.

Encryption. For a message $m \in E$ the ciphertext is $(C_1, C_2) = (kP, kQ + m)$ for a randomly chosen $k < n$.

Decryption. $m = C_2 - rC_1$.

FIGURE 12.6: Elliptic curve El Gamal cryptosystem.

As an example, suppose the curve is

$$y^2 = x^3 + 3x - 2$$

over $\mathbb{Z}_{107}$, and the chosen point is $P = (23, 6)$. The order of p is suitably high:

```
sage: E = EllipticCurve(GF(107),[3,-2])
sage: P = E([23,6])
sage: P.order()
  107
```

Alice chooses a random number as her private key:

```
sage: r = randint(1,107); r
  48
```

and publishes her public key:

```
sage: Q = r*P;Q
  (20 : 51 : 1)
```

Suppose the message to be sent is $m = (86, 25)$. Bob chooses

```
sage: k = randint(1,107);k
  25
```

and computes

```
sage: m = E([86,25])
sage: C = [k*P,k*Q+m];C
  [(54 : 86 : 1), (101 : 37 : 1)]
```

which is the ciphertext he sends to Alice. To recover the plaintext, Alice simply computes:

```
sage: C[1]-r*C[0]
  (86 : 25 : 1)
```

and she is done.

Security and implementation issues

The security of the system depends entirely upon the difficulty of the discrete logarithm problem, Alice's private key r being unobtainable from P and Q. This means that the order of P must have at least one large prime factor; if all its factors are small, then an analogue of the Silver-Pohlig–Hellman algorithm mentioned in Section 6.3 for finding discrete logarithms modulo a prime p can be used.

So here is how Alice and Bob can generate a curve E and a suitable point P.

1. Choose a large prime p for which $(p-1)/2$ is also a prime. (Such a prime is called a *safe prime*.)

2. Generate random values of x, y and a modulo p.

3. Compute $b = y^2 - x^3 - ax$.

4. Check that $4a^3 + 27b^2 \neq 0 \pmod{p}$. Otherwise, generate new values of x, y, a and b.

5. At this stage a curve E has been created, defined by $y^2 = x^3 + ax + b$ and a point $P = (x, y)$ on it.

6. Now the number of points on the E must be determined. There are various methods, but the best one in current use was originally devised by Schoof, and then improved by Elkies and Atkins. It is thus known as the Schoof–Elkies–Atkins or SEA algorithm. If the number of points is p, then the order of P will also be p. If not, start again.

For example, take $p = 227$. In Sage:

```
sage: while true:
    a = mod(randint(1,p),p)
    x = mod(randint(1,p),p)
    y = mod(randint(1,p),p)
    b = y^2-x^3-a*x
    if 4*a^3+27*b^2<>0:
        E = EllipticCurve([a,b])
        if E.cardinality_sea()==p:
            print a,b,x,y
            break
....:
224 201 86 95
```

This has created the elliptic curve E defined by

$$y^2 = x^3 + 224x + 201$$

over the finite field $GF(227)$, with a point $P = (86, 95)$ on it. This can be easily checked:

```
sage: E = EllipticCurve(GF(p),[224,201])
sage: P = E([86,95])
sage: P.order()
  227
```

Since the order of P is equal to the cardinality of E, if follows that the multiples of P will generate all elements of E.

The other problem with the El Gamal system is embedding the message as a point in the curve. With the RSA and modular El Gamal systems, every possible value is a suitable input to the system. But this is not the case for an elliptic curve, where the coordinates of the points take on only a limited number of values. So, given the message m encoded as a single large integer N, there may not be x or y coordinates equal to N. With the curve E above, it is easy to see what values are used neither as x or y coordinates:

```
sage: S = Set(range(227))
sage: for i in E.list():
....:       T = Set([i[0],i[1]])
....:       S = S.difference(T)
```

```
sage: L = list(S); L.sort(); print L
[8, 15, 25, 27, 28, 44, 55, 57, 62, 72, 76, 83, 87, 91, 93,
99, 102, 111, 112, 115, 116, 121, 128, 131, 134, 140, 144,
148, 151, 155, 170, 180, 184, 185, 189, 195, 199, 209, 210,
218, 224, 225]
```

The number 99, for example, is neither an x or a y coordinate.

One probabilistic method, devised by Koblitz [50], is as follows. Suppose that m is encoded as an integer, and that a reasonable probability of failure is 1 out of 2^h. Given a curve E over a field K, suppose there is a one-one mapping between integer values (corresponding to messages) and elements of K. So for the given m, for each $j = 0, 1, 2, \ldots, h$ there is an element of $x \in K$ corresponding to $mh + j$. For such an x, check if the equation

$$y^2 = x^3 + ax + b$$

is solvable for y. If so, embed m as (x, y). If such an (x, y) pair can be found before j gets bigger than h, then m can be recovered with

$$m = \lfloor x/h \rfloor.$$

Since the value x^3+ax+b will be a square roughly half the time, the probability that this method will fail is 2^{-h}.

For example, take the curve E above over $GF(227)$, and let $h = 10$ (so that 10 tries are allowed to embed the message). Since

$$mh + j < p = 227$$

for all $j = 0, 1, 2, \ldots, 10$,

$$(m + 1)h < 227$$

or that $m < 21$. That is, messages higher than 21 cannot be embedded on this curve, with $h = 10$. Suppose that $m = 17$. The first step is to find which of

$$x = mh + j$$

produces a quadratic residue $x^3 + ax + b$ modulo p:

```
sage: f(x) = x^3+a*x+b
sage: m = 17
sage: h = 10
sage: legendre_symbol(f(m*h+0),p)
  -1
sage: legendre_symbol(f(m*h+1),p)
  -1
sage: legendre_symbol(f(m*h+2),p)
  1
```

Thus $x = mh + 2$, so the corresponding y can be determined:

```
sage: x = m*h+2;x
  172
sage: y = Mod(f(x),p).sqrt();y
  90
```

The message $m = 17$ is now embedded as the point $(172, 90)$ on E. To recover the message from this point:

```
sage: x//h
  17
```

The Menezes–Vanstone cryptosystem

This cryptosystem is one for which the plaintext is not embedded as a point on the curve. Instead, the plaintext is *masked* by the curve, in the sense that the values of the plaintext are used as multipliers of coordinates of curve points.

As for El Gamal, Alice and Bob must choose an elliptic curve E and a point $P \in E$ whose order is a large safe prime p. In this system plaintexts m are pairs of integers

$$(m_0, m_1)$$

with each $m_i < p$. Alice chooses as her private key a random integer $A < p$ and publishes $Q = AP$ as her public key. To send m to her, Bob chooses a random integer $k < p$ and computes $(x, y) = kQ$. He then sends the triple

$$(c_0, c_1, c_2) = (kP, xm_0 \bmod p, ym_1 \bmod p)$$

as the ciphertext.

To decrypt, Alice computes

$$\begin{aligned} Ac_0 &= A(kP) \\ &= k(AP) \\ &= kQ \\ &= (x, y). \end{aligned}$$

Then she obtains

$$(m_0, m_1) = (c_1 x^{-1} \bmod p, c_2 y^{-1} \bmod p).$$

Figure 12.7 describes this system.

For a simple example, choose the same $E : y^2 = x^3 + 3x - 2$ over $GF(107)$ and $P = (23, 6)$ as before. Alice creates her private and public keys:

Parameters. An elliptic curve E and a point $P \in E$ of high safe prime order p.

Key generation. The private key is any $A < p$. The public key is $Q = AP$.

Encryption. For a message $m = (m_0, m_1) \in \mathbb{Z}_p^2$ the ciphertext is $(c_0, c_1, c_2) = (kP, xm_0 \bmod p, ym_1 \bmod p)$ for a randomly chosen $k < p$ and where $(x, y) = kQ$.

Decryption. $m = (c_1x^{-1} \bmod p, c_2y^{-1} \bmod p)$.

FIGURE 12.7: Menezes–Vanstone cryptosystem.

```
sage: A = randint(1,p);A
  56
sage: Q = A*P;Q
  (82 : 54 : 1)
```

Suppose Bob wishes to send $m = (100, 50)$ to Alice. He needs to generate a message key and multiply Q by it:

```
sage: k = randint(1,p);k
  89
sage: X = k*Q
sage: (x,y) = (X[0],X[1]);(x,y)
  (83, 71)
```

Now he can create the ciphertext:

```
sage: m = (100,50)
sage: c = (k*P,x*m[0],y*m[1])
sage: c
  ((66 : 66 : 1), 61, 19)
```

To decrypt this, Alice first needs to obtain x and y:

```
sage: X = A*c[0]
sage: (x,y) = (X[0],X[1])
sage: (x,y)
  (83, 71)
```

And now recovering the plaintext is straightforward:

```
sage: m = (c[1]/x,c[2]/y)
sage: m
```

```
(100, 50)
```

12.6 Elliptic curve signature schemes

Just as the El Gamal system above is a direct translation of the modular El Gamal system into elliptic curves, with modular exponentiation replaced by point multiplication, so El Gamal and DSS schemes can be rewritten to use elliptic curves.

Recall El Gamal: there is a prime p and primitive root a. Alice chooses $k < p - 1$ and for which $\gcd(k, p-1) = 1$. To sign a message to Bob using her private key A she computes $r = a^k \bmod p$ and sends

$$(m, r, s) = (m, a^k \bmod p, k^{-1}(m - Ar) \bmod p - 1)$$

to Bob. Bob verifies the signature by confirming that $B^r r^s = a^m \bmod p$, where B is Alice's public key.

To turn this into an elliptic curve scheme, the common element is an elliptic curve E and a point P of suitably high order N. Also required is a function

$$f : E \to \mathbb{Z}.$$

One simple choice of f is to map each point on E to its x-coordinate. Alice chooses a random integer $a < N$ as her private key, and publishes $B = aP$ as her public key. Messages to be signed are represented as integers $m < N$. For large messages, a hash will produce an integer of the appropriate size. To sign a message, Alice

1. Chooses a random integer k with $\gcd(k, N) = 1$, and computes $R = kP$.

2. Computes $s = k^{-1}(m - af(R)) \pmod{N}$.

The signature is (m, R, s). Note that both m and s are integers, but R is a point of E.

To verify the signature, Bob obtains E, P, B and f. He then computes

$$V_1 = f(R)B + sR$$
$$V_2 = mP$$

and accepts the signature if $V_1 = V_2$. This works because:

$$\begin{aligned} V_1 &= f(R)B + sR \\ &= f(R)aP + skP \\ &= f(R)aP + (m - af(R))P \\ &= (f(R)a + m - af(R))P \\ &= mP \\ &= V_2. \end{aligned}$$

12.7 Elliptic curves over binary fields

If the field K has characteristic 2, then the discriminant of

$$x^3 + ax + b$$

is

$$-16(4a^3 + 27b^2) = 0$$

and so the simplified Weierstrass form introduced at the beginning of the chapter cannot be used. Instead, the form

$$y^2 + xy = x^3 + ax^2 + b \tag{12.1}$$

can be used. The addition and doubling formulas are slightly different from those earlier; for details see Hankerson et al. [35]. Curves over binary fields are attractive for cryptography because arithmetic in the field can be made particularly efficient with very low-level operations. This means that operations on the curve are likewise more efficient than operations on curves over $GF(p)$, with p being a large prime.

Here's how such a curve can be defined and used in Sage. First create the field:

```
sage: R.<alpha> = GF(2)[]
sage: F.<X> = GF(2^7,name='X',modulus=alpha^7+alpha+1)
```

Choose $a = X^2 + 1$ and $b = X^2 + X + 1$, and define the curve:

```
sage: E = EllipticCurve(F,[1,X^2+1,0,0,X^2+X+1])
```

Now this curve can be used for an El Gamal cryptosystem. First, find a point with high order. To do this, start by determining the number of points:

```
sage: nE = E.cardinality(); nE
  142
sage: factor(nE)
  2 * 71
```

and so look for a point of order 71:

```
sage: for p in E:
....:       if p.order()==71:
....:           print p;break
....:
(X^2 + X + 1 : X^6 + X^4 + X^3 + 1 : 1)
```

This means the point

```
sage: P = E([X^2 + X + 1 , X^6 + X^4 + X^3 + 1])
```

can be used and the stage is all set for an El Gamal (or Menezes–Vanstone) cryptosystem.

12.8 Pairing-based cryptography

A "pairing" is a mapping that takes elements from two groups, and produces a result in a third group. If the mapping is carefully defined as to have some particular properties, some very powerful cryptographic schemes and protocols can be developed. Although the formal construction of a pairing is somewhat beyond the scope of this text, it is possible to describe the necessary properties of a pairing, discuss how a pairing works, and how it can be used. Pairings were originally investigated as a method for reducing the discrete logarithm problem for elliptic curves to the same problem modulo a prime p. This reduction will be briefly examined.

The basic definition

Although pairings can be defined on any group, this section will consider only mappings of pairs of points from an elliptic curve group to a cyclic group. Suppose that $(G,+)$ is a subgroup of an elliptic curve group of order n, and that $(G_T,\times)$ is the multiplicative group also of order n. Normally n is a prime p, and $(G,+)$ is the subgroup generated by an element $P \in E$ of order p. Then G_T is the multiplicative group generated by multiplication on a finite field $GF(r^k)$ where r is prime and p divides $r^k - 1$. Then G_T is generated by an element $v \in GF(r^k)$ of order p. The value k here is called the *embedding degree.*

Then a *pairing* is a map

$$e : G \times G \to G_T$$

that satisfies the following:

1. e is *bilinear*; that is, it is linear in each variable:

   $$e(P_1+P_2, Q) = e(P_1, Q)e(P_2, Q), \qquad e(P, Q_1+Q_2) = e(P, Q_1)e(P, Q_2).$$

   Sometimes this property is expressed by requiring that

   $$e(aP, bQ) = e(P, Q)^{ab}$$

   for all integers a, b.

2. e is *non-degenerate*:

   $$e(P, P) \neq 1.$$

   In fact, it is reasonable to expect that $e(P, P) = v$.

3. e is efficiently computable.

Supposing that such a pairing has been defined on a curve E, and that P, Q are points of E for which the equation $Q = sP$ is to be solved for s. To do this, first compute

$$u = e(P, P)$$
$$v = e(P, Q).$$

Assume that the discrete logarithm is solvable in G_T so that s can be found such that

$$v = u^s.$$

Then:

$$e(P, Q) = e(P, P)^s = e(P, sP)$$

and so $Q = sP$. Thus the solution of the discrete logarithm problem in G_T produces the solution to the elliptic curve discrete logarithm problem.

Some basic protocols

Tripartite key exchange. One of the simplest protocols involving pairings is Joux's *three-way Diffie Hellman key exchange* [42]. This is a method for three people to agree on a single key to be used between all of them. This can be done without pairings, but requires three steps. In the first step, Alice, Bob and Charlie agree on an elliptic curve E and a point P on it. Each produces a secret integer a, b, c respectively. Alice computes aP, Bob bP and Charlie cP. In the second step, Alice passes her result to Bob, Bob passes his to Charlie, and Charlie passes his to Alice. Then Alice computes $a(cP)$, Bob $b(aP)$ and Charlie $c(bP)$. In the third round, they each pass their new results on as before, and multiply what they receive by their integers: Alice ends up with $a(c(bP))$, Bob with $b(a(cP))$ and Charlie with $(c(b(aP))$—all equal. Figure 12.8 illustrates this protocol.

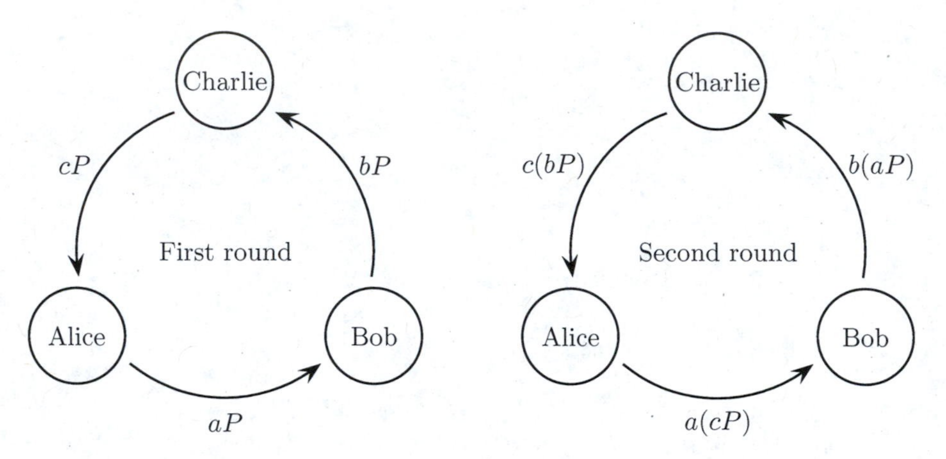

FIGURE 12.8: Diffie–Hellman protocol for three people.

Joux's protocol eliminates one step, by allowing each person to use the other two results simultaneously. As before, P and E are agreed on, and aP, bP and cP are computed. Then Alice computes

$$e(bP, cP)^a = e(P, P)^{abc},$$

Bob computes

$$e(aP, cP)^b = e(P, P)^{abc},$$

and Charlie computes

$$e(aP, bP)^c = e(P, P)^{abc}.$$

As can be seen, each person ends up with the same result.

Identity-based encryption. Recall that one of the problems of public-key cryptography is that an identity does not, itself, authenticate a public key. Alice might have a public key on a database next to her name or email address, but Mallory may have hacked into the database and replaced her public key with his own. Since a public key is just a random looking number, there is no way for Bob to know if the key is really Alice's or not.

One way around this is the idea of *identity-based encryption* (IBE), where the identifier—the name or email address—is itself the public key. At first glance the idea of this seems impossible. In general, it is the private key that is chosen by the user—two large primes for RSA/Rabin systems; a single large integer for El Gamal type systems. Then the public key is constructed from the private key. The mathematics of factoring or discrete logarithms mean that knowledge of the public key provides no information about the private key. But in all these systems it is completely impossible to choose the public key first. For an El Gamal system, this would mean being able to solve the discrete logarithm problem to obtain the corresponding private key, and there would be no security at all.

Amazingly enough, a workable IBE system has been developed using pairings. Although the idea of IBE was proposed in the mid 1980s (by Shamir), the system was not developed until 2001, by Boneh and Franklin [12]. Their basic scheme will be described, treating the pairing as a "black box" computational object, as discussed in Blake et al. [6]. More details on choosing appropriate curves and parameters can be found in Boneh and Franklin's article.

The system requires a "trusted authority" (TA), or a public key infrastructure; the TA shall be named Trent. Trent publishes a suitable elliptic curve E with a group G generated by P, a pairing $e : G \times G \to G_T$, and also creates a "master secret," an integer s. Trent also publishes $Q = sP$. As well, two hash functions are required; H_1 which hashes arbitrarily long strings to elements of G; and H_2 which hashes elements of G_T to binary strings of length n, where n is the message length. So all users on the system will know G, G_T, the pairing e, the points P and Q of G, and the hash functions H_1 and H_2.

For Alice and Bob to exchange encrypted messages, they each have their email address as their public key: `alice@email.org` and `bob@postbox.com` for example. Then for each of these, H_1 will produce an element of G: Q_A for Alice, and Q_B for Bob. With Trent using his secret integer s, Alice and Bob can each be sent their private keys $S_A = sQ_A$ and $S_B = sQ_B$ over a secure channel.

Suppose now that Alice wishes to send a message m to Bob. She performs the following steps:

1. She obtains Bob's public key (his email address), `bob@postbox.com`, and hashes it using H_1 to obtain $Q_B \in G$.

2. With Q_B and Q, Alice computes $u = e(Q_B, Q) \in Q_t$.

3. She creates a random integer t and computes tP.

4. She then computes u^t, hashes it with H_2 and adds it (with XOR) to m.
5. She sends this last result along with tP to Bob.

So the entire ciphertext consists of the pair

$$(C_1, C_2) = (tP, m \oplus H_2(u^t)).$$

To decrypt the ciphertext, Bob performs the following steps:

1. Using his private key S_B sent to him by Trent, he computes $v = e(S_B, C_1)$, and hashes this with H_2
2. He then adds this to C_2 to obtain m: $m = C_2 \oplus H_2(v)$.

To see that this works, note that:

$$\begin{aligned} e(S_B, C_1) &= e(S_B, tP) \\ &= e(sQ_B, tP) \\ &= e(Q_B, P)^{st} \\ &= e(Q_B, sP)^t \\ &= e(Q_B, Q)^t \\ &= u^t \end{aligned}$$

where the bilinearity of the pairing is used to move the indices around. Applying this to the decryption:

$$\begin{aligned} C_2 \oplus H_2(v) &= m \oplus H_2(u^t) \oplus H_2(v) \\ &= m \oplus H_2(u^t) \oplus H_2(e(S_B, C_1)) \\ &= m \oplus H_2(u^t) \oplus H_2(u^t) \quad \text{from above} \\ &= m. \end{aligned}$$

The security of this scheme comes from the hardness of the *computational bilinear Diffie–Hellman problem*: given $P, aP, bP, cP \in G$ for random a, b, c, determine $e(P, P)^{abc}$. This can be reduced to the difficulty of finding discrete logarithms in G; if for example c can be found from P and cP, then $e(P, P)^{abc} = e(aP, bP)^c$.

Short signatures. There are a number of pairing-based signature schemes; and below is one that allows the use of *short signatures*. In this instance a short signature consists of just one group element, as opposed to the two elements required for El Gamal or DSS signatures.

The setup requires an elliptic curve group G generated by P of order n, a hash function H that hashes arbitrarily long strings to elements of G, and a pairing $e : G \times G \to G_T$. Alice chooses as her private key a random integer $a < n - 1$ and produces her public key $A = aP$. To sign a message m, Alice first computes its hash $M = H(m)$, and computes $S = aM$ as the signature. To verify the signature, Bob computes $e(P, S)$ and $e(A, M)$ and checks that they are equal. This works because

$$e(P, S) = e(P, aM) = e(P, M)^a$$

and

$$e(A, M) = e(aP, M) = e(P, M)^a$$

and of course Bob doesn't need to know (and in fact cannot know) the value of Alice's private key a.

12.9 Exploring pairings in Sage

To explore these protocols, it will be necessary to have a simple pairing-like function that satisfies the bilinearity and non-degeneracy properties, but is not efficiently computable. Such a function will require a prime p, and a prime r for which $p|r^k - 1$ for some r. This can easily be done if p is a Mersenne prime, for example $p = 31 = 2^5 - 1$.

This simple pairing is defined as follows: let $Z \in G$ be a group generator, and similarly let $w \in G_T$ be a generator. Then

$$e(P, Q) = w^{\log_Z P \log_Z Q}.$$

This trivially satisfies the bilinearity and non-degeneracy properties, but is not efficiently computable, as it is based on the discrete logarithm problem in G. However, it is quite suitable for small examples.

The next requirement is an elliptic curve of cardinality 31, and one can be found through brute-force search:

```
sage: p = 31
sage: for a in range(p):
    for b in range(p):
        if (4*a^3+27*b^2)%p<>0:
            E=EllipticCurve(GF(p),[a,b])
            if E.cardinality()==p:
                print a,b
```

Of the values produced choose $(a, b) = (4, 4)$, so that:

```
sage: E = EllipticCurve(GF(p),[4,4])
```

Any point of E can be chosen as a suitable generator for the elliptic curve group. The curve can be listed

```
sage: E.list()
```

and from the list of points generated choose

```
sage: Z = E([0,2])
```

Now produce a finite field to use as G_T:

```
sage: R.<alpha> = GF(2)[]
sage: F.<X> = GF(32,name='X',modulus=alpha^5+alpha^2+1)
```

The element `X` is a generator:

```
sage: X.multiplicative_order()
  31
```

Now for the pairing:

```
sage: def e(P,Q):
          return X^(Z.discrete_log(P)*Z.discrete_log(Q))
```

Now for some experiments. First with Joux's tripartite Diffie–Hellman:

```
sage: (a,b,c) = (10,18,26)
sage: e(a*Z,b*Z)^c; e(b*Z,c*Z)^a; e(a*Z,c*Z)^b
  X^4 + X
  X^4 + X
  X^4 + X
```

For the Boneh-Franklin IBE, two hash functions are needed. Using the `str2num` function developed in Chapter 5, H_1 can be defined by taking a string, converting it to a number, and multiplying P by that number:

```
sage: def H1(s): return str2num(s)*P
```

The other hash function H_2 will just be the integer representation of the field element:

```
sage: def H2(z): return int(z.int_repr())
```

Then P, the secret integer s and Q can be chosen and computed:

```
sage: P = E([0,2])
sage: s = randint(1,p);s
  23
sage: Q = s*P;Q
  (17 : 5 : 1)
```

Now Trent can compute Alice and Bob's secret keys S_A and S_B:

```
sage: QA = H1("alice@email.org"); QA
  (29 : 22 : 1)
sage: QB = H1("bob@postbox.com"); QB
  (17 : 5 : 1)
sage: SA = s*QA; SA
  (23 : 24 : 1)
sage: SB = s*QB; SB
  (1 : 28 : 1)
```

Let the message be $m = 25$. Here's how Alice encrypts it (note that XOR is implemented in Sage with a double caret):

```
sage: u = e(QB,Q); u
  X^2
sage: t = randint(1,p); t
  19
sage: C1 = t*P; C1
  (5 : 5 : 1)
sage: C2 = m^^H2(u^t); C2
  13
```

So the ciphertext is the pair

$$(C_1, C_2) = ((5 : 5 : 1), 13).$$

Bob can decrypt this in one step, using

$$m = C_2 \oplus H_2(e(S_B, C_1)).$$

In Sage:

```
sage: C2^^H2(e(SB,C1))
  25
```

For short signatures, using the same G and P, with secret key $a = 17$ and public key

```
sage: a = 17
sage: A = a*P;A
  (2 : 12 : 1)
```

The hash function H required here can just be the hash function H_1 used for the Boneh-Franklin example above. To sign a message:

```
sage: m = "Meet me on Wednesday!"
sage: M = H1(m); M
```

```
  (14 : 13 : 1)
sage: S = a*M; S
  (21 : 7 : 1)
```

To verify this signature, Bob uses the message hash and signature: M, S, and the common point P and Alice's public key A:

```
sage: e(P,S)
  X^3 + X
sage: e(A,M)
  X^3 + X
```

Since they are equal, he accepts the signature as being genuine.

12.10 Glossary

Discrete logarithm problem: Determining the value n for which $P = nQ$ given points P and Q in an elliptic curve.

Elliptic curve: Given a field K, the set of all points $(x, y) \in K^2$ satisfying $y^2 = x^3 + ax + b$ with $4a^3 + 27b^2 \neq 0$.

Elliptic curve group: The group whose elements are points on the curve, and for which the operation is "addition" as defined in Section 12.1.

Embedding: Mapping a plaintext to a particular point in a given curve.

Masking: Using the plaintext to provide a multiplicative value, rather than a point itself.

Pairing: Given a subgroup G of an elliptic curve and a multiplicative subgroup G_T of a finite field, a pairing is a function $e : G \times T \to G_T$ that is linear in each variable.

Point at infinity: A point in $\mathbb{R}^2$, considered projectively, that can be considered to be at the end of every vertical line.

Projective coordinates: A coordinate system where the ratios of values define the points.

Exercises

Review Questions

1. Why are elliptic curves used for cryptographic purposes?
2. In what way does the elliptic curve definition over a binary field differ for the definition over a field of characteristic greater than 3?
3. What are the Weierstrass equations for elliptic curves over both types of fields (binary, and characteristic greater than 3)?
4. What is the identity element of an elliptic curve group?
5. If $P = (x, y)$ is a point on an elliptic curve, what is the inverse P in the curve's group?
6. Under what conditions does Hasse's theorem allow the precise calculation of the number of points on a curve?
7. What is the Diffie–Hellman problem in the context of an elliptic curve?
8. What is a pairing, and what properties does one have?
9. What is "identity-based encryption"?
10. What is a "short signature"?

Beginning Exercises

11. Sketch the graph $y = x^3 - x$, and use that to draw a rough sketch of the elliptic curve

    $$y^2 = x^3 - x$$

    over $\mathbb{R}$.
12. Repeat the above exercise for $y = x^3 - x + 6$.
13. What do these two graphs tell you about the cubic function and the shape of the elliptic curve?
14. Perform the following elliptic curve additions by hand:

    (a) $(0, 1) + (1, 2)$ over $y^2 = x^3 + 2x + 1$
    (b) $(0, 4) + (1, 3)$ over $y^2 = x^3 - 8x + 16$
    (c) $(1, 3) + (2, 2)$ over $y^2 = x^3 - 12x + 20$
    (d) $(1, 2) + (2, 3)$ over $y^2 = x^3 - 2x + 5$

(e) $(0, b) + (1, a)$ over $y^2 = x^3 + (a^2 - b^2 - 1)x + b^2$
(f) $(1, a) + (2, b)$ over $y^2 = x^3 + (b^2 - a^2 - 7)x + (2a^2 - b^2 + 6)$
(g) $3(0, a)$ over $y^2 = x^3 + 2ax + a^2$
(h) $(2, 4) + (3, 7)$ over $y^2 = x^3 + 3x + 2 \pmod{\mathbb{Z}_{11}}$
(i) $(5, 7) + (13, 15)$ over $y^2 = x^3 + x + 4 \pmod{\mathbb{Z}_{17}}$

For the last two exercises you should prepare a table of inverses first.

15. Consider the curve $y^2 = x^3 + 2x \pmod{\mathbb{Z}_{11}}$.

    (a) List all its points.
    (b) Show that for this curve, $2(3, 0) = 2(8, 0) = \mathcal{O}$.
    (c) Use this result to show that there can be no point of order 12. (*Hint:* Suppose there is a point P of order 12. Then the multiples of P must generate all points of E. Let $(3, 0) = nP$ and $(8, 0) = mP$.)

16. Consider the curve $y^2 = x^3 + 3x + 5 \pmod{\mathbb{Z}_7}$.

    (a) List all its points.
    (b) Draw an addition table for the points.
    (c) What is the structure of the curve group?

17. Repeat the above question for the curves

    (a) $y^2 = x^3 + 3x + 3 \pmod{\mathbb{Z}_{11}}$
    (b) $y^2 = x^3 + 6x + 7 \pmod{\mathbb{Z}_{11}}$

18. Repeat the previous question for the curve $y^2 = x^3 + 5x \pmod{\mathbb{Z}_7}$.

19. Show in general that if any elliptic curve E has two distinct points of order 2 then there is no point that generates the entire curve.

20. Suppose that the number of points n of a curve E is known, and that $P \in E$. Show that if $(n/k)P \neq \mathcal{O}$ for all prime divisors k of n, then P has order n.

    (*Hint:* Use the result that the order of a point must divide the order of the curve.)

21. In general, suppose that $nP = \mathcal{O}$ for some n. Show that if $(n/k)P \neq \mathcal{O}$ for all prime divisors k of n, then P has order n.

22. Using Hasse's theorem, prove the result that if $P \in E$ has order m, and there is only one multiple km of m that satisfies

    $$p + 1 - 2\sqrt{p} \leq km \leq p + 1 + 2\sqrt{p}$$

    then the curve must have exactly km points.

Sage Exercises

23. Recall from Chapter 6 that one method of calculating a discrete logarithm over an elliptic curve is the baby step, giant step method. It can also be used for discrete logarithms over elliptic curves. Suppose P has order n in the curve E, and that $s = \lceil \sqrt{n} \rceil$. List all values $Q - bP$ for $b = 1, 2, \ldots, s$ and all values $a(sP)$ for $a = 1, 2, \ldots, s$. By comparing the two lists, find values a' and b' for which $Q - b'P = a'(sP)$. Then the value m for which $Q = mP$ is given by $m = sa' + b'$.

    (a) Show that this method works.

    (b) For the curve $y^2 = x^3 + 1064x + 368 \pmod{\mathbb{Z}_{1367}}$, choose $P = (1179, 40)$ which has order 82, and $Q = (1072, 338)$. Use the baby step giant step method to solve $Q = mP$ for m.

24. Consider the curve E defined by $y^2 = x^3 + 3x + 1 \pmod{661}$, and the point $P = (2, 26) \in E$.

    (a) Use Sage to show that $108P = \mathcal{O}$, but that $(108/2)P \neq \mathcal{O}$ and $(108/3)P \neq \mathcal{O}$.

    (b) Conclude that the order of P is 108.

    (c) Using the result of question 22, determine the order of E.

25. A variation of the method described in question 23 can be used to find the order of a point $P \in E$ over the finite field $\mathbb{Z}_p$.

    (i) Set $Q = (p + 1)P$, and set $m = \lceil p^{1/4} \rceil$.

    (ii) For $0 \leq j, k \leq m$ evaluate $Q + k(2mP)$ and $\pm jP$ until a match is found. If $Q + k'(2mP) = j'P$ then set $N = p + 1 + 2mk' - j'$; if $Q + k'(2mP) = -j'P$ then set $N = p + 1 + 2mk' + j'$.

    (iii) Now $NP = \mathcal{O}$. For each prime factor q of N check if $(N/q)P = \mathcal{O}$. If it does, replace N with N/q, and check the prime factors of this new value of N. Continue until a value N is obtained for which $NP = \mathcal{O}$ but for all prime factors q of N, $(N/q)P \neq \mathcal{O}$. This N will be the required order.

    For this algorithm:

    (a) Show that this method works—why will a match always be found in the second step?

    (b) Demonstrate this method on the points and curves of the previous two questions.

26. Use Koblitz's probabilistic method to embed the plaintext $m = 50$ in the curve E of question 24, using $h = 10$.

27. Using the curve and field defined in section 12.7, work through a full El Gamal encryption and decryption, using parameters of your choice.

28. Repeat the previous question but use the Menezes–Vanstone system.

29. Use the curve and point in question 24 to work through an El Gamal signature. Use the values $m = 50$, $k = 47$, and the private key $a = 47$.

30. An elliptic curve DSS variant can be defined as follows:

    (i) The shared parameters are a curve E over $\mathbb{Z}_p$, and a point $P \in E$ of prime order q. With a private key $a \leq q - 1$, the public key is $Q = aP$.

    (ii) To sign a message represented as an integer m, choose a random $k \leq q$ and compute $(x, y) = kP$. Then the signature is the pair (r, s) where

    $$r = x \pmod q,$$
    $$s = k^{-1}(m + ar) \pmod q.$$

    (iii) The signature is verified by computing the point

    $$(u, v) = (ms^{-1} \bmod q)P + (rs^{-1} \bmod q)Q$$

    and accepting the signature as valid if $r = u \pmod q$.

    For this signature scheme:

    (a) Show that this method is valid.

    (b) Choose as common parameters the curve E defined by $y^2 = x^3 + 5x + 1 \pmod{1367}$, and the point $P = (951, 512)$ which has order 173. With private key $a = 100$, and with $k = 125$, sign the message $m = 1000$, and verify the signature.

31. Demonstrate that pairings could be defined with the multiplicative group $\mathbb{Z}_p$, generated by a primitive root: take the curve $y^2 = x^3 + 4x + x \pmod{31}$ which has order $p = 31$, the point $Z = (0, 2)$, and define a pairing by

    $$e(P, Q) = 3^{\log_Z P \log_Z Q} \pmod p.$$

    Note that 3 is a primitive root modulo p.

    Experiment with Joux's protocol, the Boneh-Franklin cryptosystem, and short signatures, with this pairing.

32. Repeat the above question, with the curve $y^2 = x^3 + 6x + 1 \pmod{1367}$, which has order $p = 1301$. A primitive root modulo p is 2, and a generating point is $Z = (0, 1)$.

Further Exercises

These exercises show how to create an RSA-type system using elliptic curves. More detail can be found in Washington [92].

33. Show that if $p = 2 \pmod 3$ then the elliptic curve

$$y^2 = x^3 + b \pmod p$$

has exactly $p+1$ points. This can be done by letting $t = 3^{-1} \pmod{p-1}$ and showing that if r is a primitive root of p, then so is $s = r^t$. Then if $x = s^k \pmod p$, it follows that $x^3 = r^k \pmod p$. Use this and the number of quadratic residues modulo p to complete the proof.

34. Use the above and Chinese remainder theorem to show that if p and q are both equal to 2 mod 3, then if $N = pq$ the elliptic curve

$$y^2 = x^3 + b \pmod N$$

has exactly $(p+1)(q+1)$ points.

35. An elliptic curve RSA-type system can be created as follows: Alice picks two primes $p, q = 2 \pmod 3$ and a number $e < \mathrm{lcm}(p+1, q+1)$ and relatively prime to $\mathrm{lcm}(p+1, q+1)$, and uses it to produce her private key

$$d = e^{-1} \pmod{\mathrm{lcm}(p+1, q+1)}.$$

She publishes $(N = pq, e)$ as her public key. Messages M are pairs (m_1, m_2) with $0 \le m_1, m_2 \le N-1$. The ciphertext is obtained by

$$C = e(m_1, m_2)$$

with (m_1, m_2) being considered as a point in the elliptic curve E:

$$y^2 = x^3 + b \pmod N$$

with

$$b = m_2^2 - m_1^3 \pmod N.$$

Decryption is obtained by $m = dC$.

Show that the final equation is correct by first showing that by orders of $E_p : y^2 = x^3 + b \pmod p$ and $E_q : y^2 = x^3 + b \pmod q$ it follows that

$$(p+1)M = \infty \pmod p, \quad (q+1)M = \infty \pmod q.$$

Use the fact that $de = 1 + k(\mathrm{lcm}(p+1, q+1))$ for some k, and so $de = 1 + K(p+1)$ for some K. Show that $dC = M \pmod p$ and similarly for q.

Chapter 13

Random numbers and stream ciphers

This chapter explores random numbers and random number generators (RNGs). In particular:

- The importance of random numbers for cryptography.
- True RNGs.
- Pseudo RNGs.
- Cryptographically strong RNGs.
- Linear congruential RNGs.
- RNGs based on "hard" problems such as factorization.
- Stream ciphers, and their relationship with RNGs.
- Some examples of stream ciphers, including Rivest's famous RC4.
- The Blum–Goldwasser probabilistic cryptosystem.

13.1 Introduction

Random numbers sit at the heart of almost all cryptographic and security protocols. For example, they are needed to produce random keys for symmetric encryption, large primes for asymmetric encryption, and initialization vectors for block encryption. For the system to function securely, there needs to be some method of producing, and obtaining, random numbers quickly. Numbers that aren't quite "random" enough can lead to the systems being attacked: if the adversary has enough information to be able to predict at least some of the next number, he or she can use that information against the system. This was the subject of a famous attack against an SSL implementation in an early version of the Netscape web browser.

There are three main classes of random number generators:

1. "True" random number generators (TRNGs). These create random numbers from a digitization of a physical process. These processes may include the timing of keystrokes, the coordinates of a mouse while in use, noise in an electrical component, measurement of radioactive decay, or quantum fluctuations.

2. Pseudo-random number generators (PRNGs). These create random numbers by an algorithm that produces numbers that satisfy statistical tests for randomness. Such numbers are fully determined by the starting value, and hence may be useful for simulation, but useless for cryptography. There are a huge number of different algorithms, with varying strengths and speeds.

3. Cryptographically secure random number generators (CSRNGs). These also create random numbers by a deterministic algorithm, but as well as producing numbers that are statistically random, they should not leak any information about the next number in the sequence. That is, an attacker, even knowing the algorithm and the preceding numbers, should not be able to predict the next number. There are in fact very few such CSRNGs that are trustworthy enough to be used in security-critical applications.

In general a TRNG may be considered to provide the "best" source of randomness. But there are problems here: the process must allow constant access; it must be reliable (that is, it will not break down); it must be secure in that an adversary cannot tinker with the process so as to adjust the output in any way. Many applications that now require random numbers use a number of different physical sources.

For cryptographic purposes, random numbers should satisfy:

- A sequence of random numbers should pass statistical tests. These may include analysis of frequency of output values, runs of values, and of gaps.

- Given a subsequence of random numbers, it should not be possible for an adversary to determine, in a reasonable time, any future or previous values in the sequence.

13.2 Pseudo-random number generators

One very simple PRNG is von Neumann's "middle square" algorithm, proposed by him in 1946 [89]. In a very basic version, it starts with a four-digit number. The next number is obtained by squaring the first number to obtain an eight digit number, and taking the middle four digits. (If the square has less than eight digits, extra digits can be prepended.) This process is repeated as often as needed. So if x_i is one number in the sequence, the next number will be

$$x_{i+1} = \left\lfloor \frac{x_i^2}{100} \right\rfloor \pmod{10000}.$$

If $x_0 = 1234$, then $x_0^2 = 1522756$ and so $x_1 = 5227$. The next few numbers generated are

$$3215, \quad 3362, \quad 3030, \quad 1809, \quad 2724.$$

A major problem with this method is that many numbers produce sequences that start repeating with short periods. Since there are only 10000 different possible values, any sequence will have to repeat sometime! For example, if the sequence produces any of 100, 2500, 3792, 7600, nothing else will be produced, as each of these numbers produce themselves by the middle square method. As another example, any of 2916, 5030, 3009, 540 leads to a period of four:

$$2916 \rightarrow 5030 \rightarrow 3009 \rightarrow 540 \rightarrow 2916.$$

If a small number appears in the sequence, then the sequence immediately dies away to producing nothing but zero. For these reasons, the middle-square method is considered a very unsatisfactory PRNG. However, von Neumann did not intend this method to be statistically sound, but just a quick-and-dirty method (in pre-computing days) for obtaining a few pseudo-random numbers quickly.

An important family of PRNGs that have been used by many computer systems is that of the *linear congruential generators* (LCG). These are all defined by

$$x_{i+1} = ax_i + b \pmod{m}$$

and have been used for many years for simulation purposes. Clearly there are only m different values that can be produced, and after a sequence of at most m numbers, the sequence will repeat. The LCG defined by the equation above is denoted

$$\mathrm{LCG}(m, a, b, x_0).$$

Here are the outputs of a few different LCGs, all with $m = 16$, $x_0 = 0$, and $b = 1$:

$$\begin{array}{ll}
\mathrm{LCG}(16,3,1,0): & 0,1,4,13,8,9,12,5,0,1,4,13,8,9,12,5,0 \\
\mathrm{LCG}(16,5,1,0): & 0,1,6,15,12,13,2,11,8,9,14,7,4,5,10,3,0 \\
\mathrm{LCG}(16,7,1,0): & 0,1,8,9,0,1,8,9,0,1,8,9,0,1,8,9,0 \\
\mathrm{LCG}(16,10,1,0): & 0,1,11,15,7,7,7,7,7,7,7,7,7,7,7,7,7
\end{array}$$

Of these four examples, only $a = 5$ generates a sequence with the maximum period. And in fact for composite m, LCG(m, a, b, x_0) generates a sequence of maximum period m if and only if [49]:

1. $\gcd(b, m) = 1$,
2. $a - 1$ is a multiple of every prime factor of m,
3. If m is a multiple of 4, then $a - 1$ is a multiple of 4.

This means that for $m = 16$ above, a maximal sequence will be generated only for $a = 5, 9, 13$ (the value $a = 1$ generates all values, but the resulting sequence is not "random"). If m is prime, then the maximum length of the sequence will be $m - 1$, and this will be obtained when

1. $\gcd(x_0, m) = 1$,
2. a is a primitive root of m.

One way of measuring the randomness of such generators is to look at the distribution of points (x_i, x_{i+1}) in the plane, or of (x_i, x_{i+1}, x_{i+2}) in three-space. If the points are not distributed uniformly, then the generator is statistically poor.

Consider the two generators

$$x_{i+1} = 85x_i + 1 \pmod{256}$$
$$x_{i+1} = 237x_i + 1 \pmod{256}.$$

Numbers from the first generator can be produced by:

```
sage: a,c,N = 85,1,256
sage: x = [0]
sage: for i in range(N):
....:     x = x+[(a*x[-1]+c)%N]
sage: print x
```

and obtain the sequence

$$0, 1, 86, 143, 124, 45, 242, 91, 56, 153, 206, 103, 52, 69, 234,$$
$$179, 112, 49, 70, 63, 236.$$

The second generator similarly produces

$$0, 1, 238, 87, 140, 157, 90, 83, 216, 249, 134, 15, 228, 21, 114,$$
$$139, 176, 241, 30, 199, 60.$$

Both sequences certainly *look* random, but in fact there's some hidden structure which can be shown by plotting pairs of points on the Cartesian plane. Points obtained from the first generator can be plotted by:

```
sage: pl = [(x[i],x[i+1]) for i in range(N-1)]
sage: list_plot(pl,aspect_ratio=1)
```

Points from the second generator can be plotted by entering the above commands as written, but changing the definition of the LCG first. The results are shown in Figures 13.1 and 13.2.

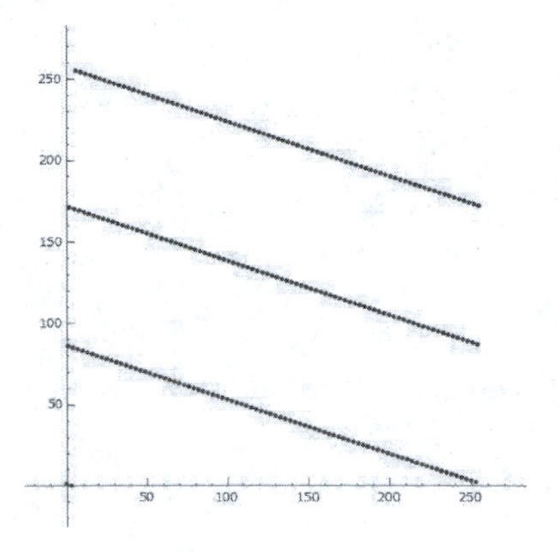

FIGURE 13.1: $x_{i+1} = 85x_i + 1 \pmod{256}$.

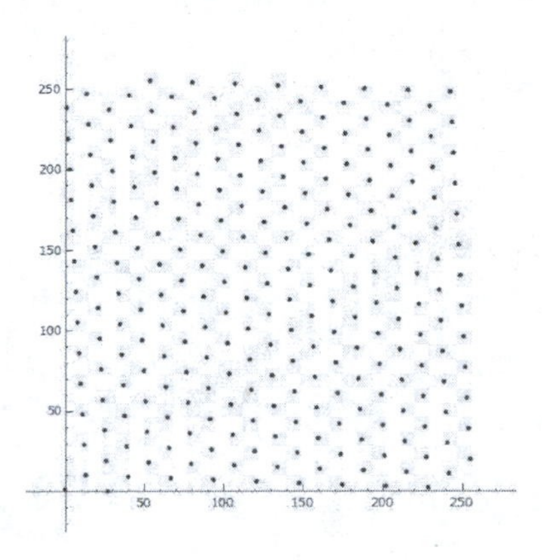

FIGURE 13.2: $x_{i+1} = 237x_i + 1 \pmod{256}$.

Clearly there is much greater correlation between points in the first generator than in the second, so of the two generators, the second one is to be preferred. The linear nature of the generator means that the points plotted will always form a lattice—that is, a collection of points with a periodic structure—and in general the points will form onto hyperplanes in n-space. The best generators are those for which the number of hyperplanes is maximized, and the test that measures the numbers and distances of hyperplanes is called the *spectral test*.

Careful choice of the parameters allows a linear congruential generator to pass many statistical tests—and the spectral test—and they are still used. They are also very fast.

However, a major problem with all LCGs is that they are not cryptographically secure—given a few terms in a sequence, it is possible to determine the parameters, and hence generate any other values. Suppose that $s_i, s_{i+1}, s_{i+2}, s_{i+3}$ are successive terms in the sequence. From the fact that points whose coordinates are successive values fall into planes in 3-space, the determinant

$$\begin{vmatrix} s_i & s_{i+1} & 1 \\ s_{i+1} & s_{i+2} & 1 \\ s_{i+2} & s_{i+3} & 1 \end{vmatrix}$$

will be a multiple of m. For example, suppose that

$$21, 24, 2, 29, 17$$

are terms of a sequence produced by an LCG. Then

$$\begin{vmatrix} 21 & 24 & 1 \\ 24 & 2 & 1 \\ 2 & 29 & 1 \end{vmatrix} = -403 = -(13)(31), \quad \begin{vmatrix} 24 & 2 & 1 \\ 2 & 29 & 1 \\ 29 & 17 & 1 \end{vmatrix} = -465 = -(3)(5)(31).$$

Given the values in the sequence, it seems reasonable to assume that $m = 31$. Taking any three values in the above subsequence, for example the middle three, provides the two equations

$$\begin{aligned} 2 &= 24a + b \pmod{31} \\ 29 &= 2a + b \pmod{31}. \end{aligned}$$

These can be easily solved to produce $a = 3$, $b = 23$.

13.3 Some cryptographically strong generators

This section will introduce some number-theoretic generators all of which can be formally proved to be cryptographically strong if a certain assumption is made. Such assumptions include the intractability of factoring, the discrete logarithm problem, or the quadratic residue problem. Because of the modular arithmetic involved, these generators can be unacceptably slow. However, they have the advantage of having formal security proofs.

Before discussing these generators, the term "cryptographically strong" needs to be formally defined. Of several different definitions, the best is the "next bit test": a PRNG passes the *next bit test* if it generates a sequence of bits b_i, and for any subsequence $b_0, b_1, \ldots, b_k$ there is no way for an adversary to guess the next bit b_{k+1} with a better than 50% accuracy.

The RSA generator

Pick two large primes p and q and set $n = pq$. Choose a value e relatively prime to $\phi(n) = (p-1)(q-1)$, and a starting seed $s_0 < n$. Define

$$s_k = s_{k-1}^e \pmod{n}$$

for all $k \geq 1$. Then the bits produced are the least significant bits (LSBs) of the s_k values:

$$b_k = s_k \pmod 2.$$

For example, suppose $p = 19$, $q = 23$, and $e = 7$. Pick $s_0 = 10$.

```
sage: p,q,e = 19,23,7
sage: n,s = p*q,[10]
sage: for i in range(20):
....:     s + =[power_mod(s[-1],e,n)]
sage: print s
[10, 129, 203, 314, 34, 260, 143, 224, 89, 67, 148, 336,
295, 15, 241, 352, 281, 431, 181, 205, 355, 428, 433, 222]
sage: print [i%2 for i in s]
[0, 1, 1, 0, 0, 0, 1, 0, 1, 1, 0, 0, 1, 1, 1, 0, 1, 1,
1, 1, 1, 0, 1, 0]
```

A formal proof of the security of this generator is somewhat technical, but the intuitive ideas are simple enough. From the equation relating s_k to s_{k-1} it follows that

$$s_{k-1} = s_k^d \pmod n$$

where

$$d = e^{-1} \pmod{\phi(n)}.$$

But finding d means finding $\phi(n)$ which means finding p and q, which means factorizing n...

The Blum–Blum–Shub generator

This generator was developed by Lenore Blum, Manuel Blum and Michael Shub [8], and is known as the BBS generator. It is the first generator to have been formally proven cryptographically strong. It bears the same relation to the RSA generator as the Rabin cryptosystem bears to the RSA cryptosystem.

Choose two large primes p and q each equal to 3 modulo 4, and choose a seed s_0 which is a quadratic residue of $n = pq$. Define

$$s_k = s_{k-1}^2 \pmod n$$

for all $k \geq 1$. Then as for the RSA generator, the bits are the LSBs of this sequence. With the same p, q and $s_0 = 9$ as above:

```
sage: s = [9]
sage: for i in range(20):
```

```
....:        s += [(s[-1]^2)%n]
sage: b = [x%2 for x in s]
sage: print b
[1, 1, 0, 0, 0, 1, 0, 1, 1, 0, 0, 1, 1, 0, 1, 0,
0, 1, 1, 1, 1]
```

The cryptographic strength of this generator is based on the difficulty of deciding whether a given $x < n$ is a quadratic residue of n. But determining this requires the factorization of n. A formal proof is given by Stinson [86].

Note that each value can be obtained directly from the seed, using

$$s_k = (s_0)^{2^k} \pmod{n}$$

and hence

$$b_k = \left((s_0)^{2^k} \pmod{n}\right) \pmod{2}.$$

One trouble with this generator is that is is comparatively slow; only one bit is produced each iteration. This may not matter if only a small number of bits are required, but if a huge number of bits is needed very quickly, for example, producing public/private key pairs for a large public server, or for producing session keys, then the BBS generator may be unacceptably slow. However, it has been shown (see [61]) that it is possible to produce more than one bit at a time. Given $n = pq$, the least significant $\log\log n$ bits may be taken as being cryptographically secure, where the logarithms are taken to base 2.

The discrete logarithm generator

This is in fact one of a family of generators known as the Blum-Micali generators [10]. Suppose p is a large prime, and g is a primitive root. With $s_0 < p$ as the seed, define

$$s_{k+1} = g^{s_k}$$

for all $k \geq 1$. Then the bit sequence b_k is defined by

$$b_k = \begin{cases} 1 & \text{if } s_k > p/2 \\ 0 & \text{if } s_k < p/2 \end{cases}$$

```
sage: p = 97
sage: g = primitive_root(p);g
  5
sage: s = [20]
sage: for i in range(20):
....:        s += [power_mod(g,s[-1],p)]
sage: b = [(x>p//2)*1 for x in s]
```

```
sage: print b
[0, 1, 1, 1, 1, 0, 1, 1, 1, 1, 1, 1, 0, 1, 0, 0,
1, 0, 0, 0, 1]
```

In order to break this system, the adversary would need to be able to solve the discrete logarithm problem modulo p, which is intractable if p is appropriately chosen.

13.4 The shrinking generator

This generator is designed to be both fast and cryptographically strong, and is based on the use of two linear feedback shift registers (LFSRs). A LFSR sequence is generated by a recurrence based on a linear combination of previous elements of the sequence.

For example, suppose a sequence starts with

$$[s_0, s_1, s_2, s_3] = [0, 0, 0, 1]$$

and is continued with the recurrence

$$s_k = s_{k-3} \oplus s_{k-4}.$$

This is easily implemented:

```
sage: s = [0,0,0,1]
sage: for i in range(26):
....:     s += [s[-3]^^s[-4]]
sage: print s
 [0, 0, 0, 1, 0, 0, 1, 1, 0, 1, 0, 1, 1, 1, 1,
  0, 0, 0, 1, 0, 0, 1, 1, 0, 1, 0, 1, 1, 1, 1]
```

Note that this sequence has a period of 15. There is a vital connection between LFSRs and primitive polynomials modulo 2. Consider for example the primitive polynomial

$$p(x) = x^6 + x + 1.$$

With $n = 6$ this polynomial can be written as

$$p(x) = x^n + x^{n-5} + x^{n-6}.$$

The powers provide the recurrence for the sequence:

$$x_k = x_{k-5} \oplus x_{k-6}$$

and since the defining polynomial is primitive, this sequence will have the maximal possible period of $2^6 - 1 = 63$.

```
sage: s = [0,0,0,0,0,1]
sage: for i in range(57):
....:     s += [s[-5]^^s[-6]]
sage: for j in range(7):
....:     print [s[i] for i in range(9*j,9*(j+1))]
[0, 0, 0, 0, 0, 1, 0, 0, 0]
[0, 1, 1, 0, 0, 0, 1, 0, 1]
[0, 0, 1, 1, 1, 1, 0, 1, 0]
[0, 0, 1, 1, 1, 0, 0, 1, 0]
[0, 1, 0, 1, 1, 0, 1, 1, 1]
[0, 1, 1, 0, 0, 1, 1, 0, 1]
[0, 1, 0, 1, 1, 1, 1, 1, 1]
```

If a non-primitive polynomial is used, a sequence will be produced which has a smaller period. For example, suppose $p(x) = x^4 + x^3 + x^2 + x + 1$. This is irreducible, but not primitive. The recurrence obtained from this is

$$s_k = s_{k-1} \oplus s_{k-2} \oplus s_{k-3} \oplus s_{k-4}.$$

```
sage: s = [0,0,0,1]
sage: for i in range(11):
....:     s += [s[-1]^^s[-2]^^s[-3]^^s[-4]]
sage: s
  [0, 0, 0, 1, 1, 0, 0, 0, 1, 1, 0, 0, 0, 1, 1]
```

The period is only 5, as opposed to the sequence above obtained using $s_k = s_{k-3} \oplus s_{k-4}$ which is based on the primitive polynomial $x^4 + x + 1$.

LFSR sequences have many useful properties, and have been used for many years to provide pseudo-random sequences for statistical and communications purposes. They are also very fast and easily implementable in hardware. A classic diagram to describe an LFSR is shown in Figure 13.3.

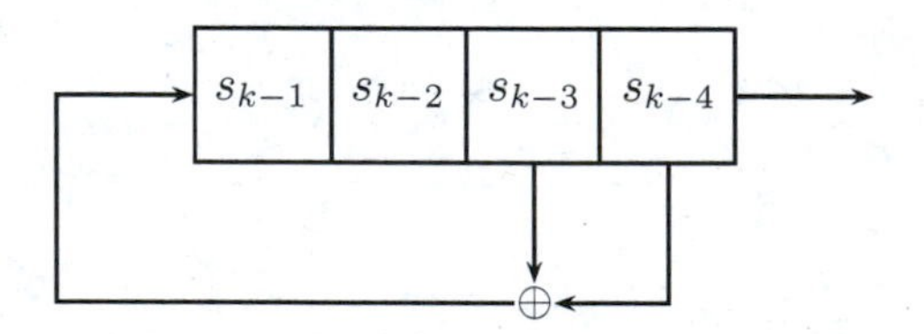

FIGURE 13.3: A linear feedback shift register.

At any state, two elements of the previous four are "tapped", and XOR-ed to produce the new value of s_k, which is pushed onto the left hand end. The other four elements are pushed to the right, and the right-most value, s_{k-4}, is not needed for any further computations.

A single LFSR sequence, being completely deterministic and linear, is however cryptographically very weak. The *shrinking generator* [22] uses the results of two LFSRs, called A and S. The output bits are generated by the A sequence, and their output is controlled by the S sequence. At each stage new bits a_k and s_k are obtained. If s_k is one, then a_k is output as the next generated bit; if s_k is zero, new bits of the A and S sequences are produced. Suppose A is produced by $x^4 + x + 1$ and seeded with $[0, 0, 1, 1]$, and S by $x^5 + x^2 + 1$ seeded with $[0, 0, 0, 1, 1]$. Then:

S:	0	0	0	1	1	0	1	1	1	0	1	0	1	0	0	0	0	1	0	0
A:	0	0	1	1	0	1	0	1	1	1	1	0	0	0	1	0	0	1	1	0
output:				1	0		0	1	1		1		0					1		

The output consists of those bits of A for which the corresponding bits of S are 1.

```
sage: a = [0,0,1,1]
sage: for i in range(50):
....:       a += [a[-3]^^a[-4]]
sage: s = [0,0,0,1,1]
sage: for i in range(50):
....:       s += [s[-3]^^s[-5]]
sage: b = []
sage: for i in range(54):
....:       if s[i]==1:
....:           b += [a[i]]
sage: print b
[1, 0, 0, 1, 1, 1, 0, 1, 1, 1, 1, 0, 0, 0, 1, 0, 0,
1, 1, 1, 1, 0, 0, 1, 0, 1]
```

A variation of the shrinking generator is the self-shrinking generator, which can be applied to any sequence. Given a sequence

$$s_0, s_1, s_2, s_3, \ldots$$

it is considered a sequence of pairs

$$(s_0, s_1), (s_2, s_3), (s_4, s_5) \ldots .$$

Then the output is a sequence b_i defined by

$$b_i = \begin{cases} 1 & \text{if } (s_{2i}, s_{2i+1}) = (1, 0) \text{ or } (1, 1) \\ 0 & \text{otherwise.} \end{cases}$$

For example, suppose s_i is generated by the primitive polynomial $p(x) = x^6 + x + 1$ with the corresponding recurrence

$$x_k = x_{k-5} \oplus x_{k-6}.$$

First create 30 elements of s:

```
sage: s = [0,0,0,1,1,1]
sage: for i in range(24):
....: s += [s[-5]^^s[-6]]
```

and then self-shrink it:

```
sage: b = []
sage: for i in range(len(s)//2):
....: if ([s[2*i],s[2*i+1]]==[1,0]) or ([s[2*i],s[2*i+1]]==[1,1]):
....:     b += [1]
....: else:
....:     b += [0]
sage: b
  [0, 0, 1, 0, 1, 0, 0, 1, 1, 1, 1, 0, 1, 0, 0]
```

It can be shown [60] that any shrinking generator can be implemented as a self-shrinking generator, and conversely that any self-shrinking generator can be implemented as a shrinking generator.

13.5 ISAAC and Fortuna

ISAAC and Fortuna are two RNGs that are designed to be cryptographically secure, although there are some doubts about ISAAC. ISAAC is formally described by its creator Jenkins [41]; the description given below comes from Aumasson [2]. In fact, the given algorithm will be ISAAC+, which has been designed to overcome some perceived deficiencies in ISAAC. The name ISAAC is an acronym for "Indirection, Shift, Accumulate, Add, and Count" which are the operations which comprise it.

ISAAC consists of a number of rounds, which take the following form:

Input: a, b, c, and the internal state s, an array of 256 32-bit words
Output: an array r of 256 32-bit words

$c \leftarrow c+1$
$b \leftarrow b+c$
for $i = 0, \ldots, 255$ **do**
$x \leftarrow s_i$
$a \leftarrow f(a,i) + s_{i+128 \bmod 256}$
$s_i \leftarrow a \oplus b + s_{x \ggg 2 \bmod 256}$
$r_i \leftarrow x + a \oplus s_{s_i \ggg 10 \bmod 256}$
$b \leftarrow r_i$
end for
return r

The symbol $\ggg$ indicates a rotation; the function f is defined as:

$$f(a,i) = \begin{cases} a \lll 3 & \text{if } i = 0 \pmod 4 \\ a \ggg 6 & \text{if } i = 1 \pmod 4 \\ a \lll 2 & \text{if } i = 2 \pmod 4 \\ a \ggg 16 & \text{if } i = 3 \pmod 4). \end{cases}$$

ISAAC can be shown to pass standard statistical tests for randomness, but in Aumasson's words:

> This new generator does not offer much more security guarantees than its brother, and so should not be considered as a proposal for a new pseudo-random generator.

The main problem with ISAAC is the large number of "weak states," which are initial values that cause ISAAC to produce bit streams easily distinguishable from truly random data.

Fortuna has been designed by Ferguson and Schneier [27]. It is designed to be seeded by a TRNG. In fact, to decrease the possibility of such a physical process being attacked, or being unavailable, Fortuna allows the use of many different TRNGs. After that, a pseudo-random process is used to produce an arbitrarily large number of bits, with the initial random seed upgraded as well. To ensure that a good random seed is available even after the computer has been booted, and before a sufficient amount of random seed data is available, data from a previous run of the algorithm is held in storage to be used when required. Thus Fortuna consists of three separate processes:

1. The entropy accumulator. A collection of 32 entropy "pools" collect random data from physical processes.
2. The generator. This produces pseudo-random data from the pools.
3. The seed file manager. The "seed file" provides the initial random value to seed the generator. The manager regulates writing data to this file.

Each of these will now be examined in more detail.

The entropy pools

There are 32 of these pools: $P_0, P_1, \ldots, P_{31}$, and data is written into them uniformly and cyclically from the TRNGs. To ensure the pools don't overflow with data, the data is hashed using SHA-256 so that the pools contain only 32 bytes. The idea of having so may pools is to make it impossible for an attacker to modify all the TRNGs that fill them.

Every now and then—but at least once every 100 milliseconds (also see below)—a "reseed" is performed. This means some pools are emptied and refilled again from the TRNGs. Pool i is reseeded only every 2^i reseeds. So P_0 is reseeded every time; P_1 every second reseed; P_2 every fourth reseed, and so on. Before any pools are emptied and reset to zero, their contents are hashed together using two iterations of SHA-256. The reseeds are designed so that even if an attacker can modify data going into the pools, there will be at least one pool (in practice many) that accumulates enough entropy between reseedings so as to defeat the attacker. So the higher numbered pools are reseeded less often, but accumulate more entropy between reseeds.

Although 32 pools should be enough for most practical purposes (pool P_{31} is reseeded only about every 13 years), the Fortuna algorithm does in fact allow for more pools if huge amounts of random data are required. (If 62 pools were used, then the 62nd pool would not be reseeded in a time equivalent to the age of the Universe).

The generator

This can be any good block cipher such as AES running in CTR mode. To obtain pseudo-random bits, the counter values are used as the plaintexts, and the keys are obtained from the entropy pools. In order to prevent any possible attacks based on periodicity, a reseed is forced after the generation of every 2^{20} bytes.

The seed file manager

When the computer is newly booted, there may not be enough data in the entropy pools to feed to the generator. To overcome this, a "seed file" is maintained, which has random data previously generated. Once the contents of the seed file are used, the seed file is immediately replaced with new seed data. This data is obtained from the entropy pools. The creators recommend that new seed file contents be generated every 10 minutes.

For an example of an implementation of Fortuna, see McEvoy et al. [59].

13.6 Stream ciphers

Stream ciphers are intimately bound up with the theory of random number generation. To a large extent, a stream cipher works by XOR-ing the plaintext bits with the key stream, the key stream itself being produced by some pseudo-random process. Recall from Chapter 8 that some encryption modes work as stream ciphers, in that the encryption algorithm is used to generate the key stream, and the decryption algorithm is never used.

However, the creation of the key stream requires great care. Here is a case where the need for cryptographic strength in the PRNG is vital. As an example, consider the use of an LFSR as a key stream, with a known plaintext attack. Given a key stream x, and plaintext and ciphertext p and c respectively, they are related by $c = p \oplus x$ (where the operation is understood to be performed bit-wise), so that $x = c \oplus p$. That is, an attacker can be assumed to be able to obtain at least as much of the key stream as has already been used.

Suppose that an LFSR has been used to create the key stream. Then

$$x_k = \sum_{i=1}^{n} c_i x_{k-i} \pmod 2.$$

For example, this equation for generating an LFSR

$$x_k = x_{k-3} \oplus x_{k-4}$$

can be written

$$x_k = 0x_{k-1} + 0x_{k-2} + 1x_{k-3} + 1x_{k-4}$$

and so

$$[c_4, c_3, c_2, c_1] = [0, 0, 1, 1].$$

If these c_i values can be obtained, the system will be broken, because all further key stream values are now known to the attacker. Suppose that $n = 4$ is known. Then

$$\begin{aligned} x_4 &= c_4x_0 + c_2x_1 + c_2x_2 + c_1x_3 \\ x_5 &= c_4x_1 + c_2x_2 + c_2x_3 + c_1x_4 \\ x_6 &= c_4x_2 + c_2x_3 + c_2x_4 + c_1x_5 \\ x_7 &= c_4x_3 + c_3x_4 + c_2x_5 + c_1x_6 \end{aligned}$$

or in matrix form

$$\begin{bmatrix} x_4 \\ x_5 \\ x_6 \\ x_7 \end{bmatrix} = \begin{bmatrix} x_0 & x_1 & x_2 & x_3 \\ x_1 & x_2 & x_3 & x_4 \\ x_2 & x_3 & x_4 & x_5 \\ x_3 & x_4 & x_5 & x_6 \end{bmatrix} \begin{bmatrix} c_4 \\ c_3 \\ c_2 \\ c_1 \end{bmatrix}$$

so that

$$\begin{bmatrix} c_4 \\ c_3 \\ c_2 \\ c_1 \end{bmatrix} = \begin{bmatrix} x_0 & x_1 & x_2 & x_3 \\ x_1 & x_2 & x_3 & x_4 \\ x_2 & x_3 & x_4 & x_5 \\ x_3 & x_4 & x_5 & x_6 \end{bmatrix}^{-1} \begin{bmatrix} x_4 \\ x_5 \\ x_6 \\ x_7 \end{bmatrix}.$$

Using the first eight values from the stream above

x_0	x_1	x_2	x_3	x_4	x_5	x_6	x_7
0	0	0	1	0	0	1	1

Now the computations can be done in Sage:

```
sage: n = 4
sage: M = Matrix(GF(2),[x[i:i+n] for i in range(n)])
sage: X = vector(GF(2),x[n:2*n])
sage: c = M.inverse()*X
sage: c
  (1, 1, 0, 0)
```

which are the correct values of c_i.

13.7 RC4

The stream cipher RC4 was developed by Ronald Rivest (the "R" of "RSA") and trademarked as a protected algorithm in 1987 ("RC" stands for "Ron's code" or "Rivest Cipher.") In 1994 its details were leaked onto the Internet by an anonymous attacker, and since then the algorithm has been publicly known. It has been the most intensively researched stream cipher since 1994, and it has been extensively used. It suffers from some defects, some of which can be ameliorated by careful use. It has the delightful quality—almost uniquely among modern bit-based ciphers—of being extraordinarily easy to describe and to implement.

It has two parts. First is the *key scheduling algorithm* (KSA) in which the key is used to produce a permutation of all values 0 to 255. Second is the *pseudo-random stream generation* (PRSG) where the result of the first step is used to generate random values in the range 0 to 255. RC4 can be implemented as operating on bytes, or on integers $0 \dots 255$, or a mixture of both.

Given a key k consisting of a sequence k_i of bytes (or integers less than 256) of length n, the KSA is

for $\mathtt{i} = 0, \dots, 255$ **do**
    $\mathtt{S}[\mathtt{i}] \leftarrow \mathtt{i}$
end for
$\mathtt{j} \leftarrow 0$
for $\mathtt{i} = 0, \dots, 255$ **do**
    $\mathtt{j} \leftarrow (\mathtt{j} + \mathtt{S}[\mathtt{i}] + \mathtt{k}[\mathtt{i} \bmod \mathtt{n}]) \bmod 256$
    swap $\mathtt{S}[\mathtt{i}]$, $\mathtt{S}[\mathtt{j}]$
end for

Having created the initial array, the PRSG is

$\mathtt{i} \leftarrow 0$

$\texttt{j} \leftarrow 0$
while not end of plaintext:
  $\texttt{i} \leftarrow \texttt{i} + 1 \bmod 256$
  $\texttt{j} \leftarrow \texttt{j} + \texttt{S}[\texttt{i}] \bmod 256$
  swap $\texttt{S}[\texttt{i}]$, $\texttt{S}[\texttt{j}]$
  next byte is $\texttt{S}[\texttt{S}[\texttt{i}] + \texttt{S}[\texttt{j}] \bmod 256]$

To get a feel of RC4, first experiment with a baby version, using only the integers $0 \dots 7$.

```
sage: S = [0,1,2,3,4,5,6,7]
sage: k = [3,5,7]
sage: j = 0
sage: for i in range(8):
....: j = (j+S[i]+k[i%3])%8;print i,j,S
....: S[i],S[j]=S[j],S[i]
0 3 [0, 1, 2, 3, 4, 5, 6, 7]
1 1 [3, 1, 2, 0, 4, 5, 6, 7]
2 2 [3, 1, 2, 0, 4, 5, 6, 7]
3 5 [3, 1, 2, 0, 4, 5, 6, 7]
4 6 [3, 1, 2, 5, 4, 0, 6, 7]
5 5 [3, 1, 2, 5, 6, 0, 4, 7]
6 4 [3, 1, 2, 5, 6, 0, 4, 7]
7 0 [3, 1, 2, 5, 4, 0, 6, 7]
sage: S
[7, 1, 2, 5, 4, 0, 6, 3]
```

Even if the key is set to zeros, the final result is still a "random" looking permutation of the S values. And to generate 10 values of the stream:

```
sage: i,j,ks = 0,0,[]
sage: for k in range(10):
....: i = (i+1)%8
....: j = (j+S[i])%8
....: S[i],S[j] = S[j],S[i]
....: ks += [S[(S[i]+S[j])%8]]
sage: ks
[2, 3, 5, 2, 5, 5, 3, 6, 3, 6]
```

It is easy to write programs in Sage to implement "full" RC4. The first step is to ensure that each integer $0 \dots 255$ is represented as two hexadecimal digits, and `zfill` will do this:

```
sage: hex(11).zfill(2)
  '0b'
sage: hex(123).zfill(2)
  '7b'
```

Now for the complete RC4 program, which is simply the above programs for key scheduling and stream generation, put together using 256 instead of 8:

```
def rc4(Key,pl):
    key = map(ord,Key)
    kl = len(key)
    S = range(256)
    j = 0
    for i in range(256):
        j = (j+S[i]+key[i%kl])%256
        S[i],S[j] = S[j],S[i]
    i = 0
    j = 0
    pln = map(ord,pl)
    out = ''
    for k in pln:
        i = (i+1)%256
        j = (j+S[i])%256
        S[i],S[j] = S[j],S[i]
        c = S[(S[i]+S[j])%256]
        out += hex(Integer(k^^c)).zfill(2)
    return out.upper()
```

And to test it out:

```
sage: rc4("Key","Plaintext")
  'BBF316E8D940AFA0D3'
sage: rc4("Another key","...and another plaintext!")
  'A3A26C9F918C8CA6BCFAE6211D719AE2F66F799CFAD3A07493'
```

A problem with RC4 is that the first bytes produced have been shown to be non-random, and indeed leak information about the key. To avoid this, the first bytes are dropped; one recommendation is to drop the first 3072 bytes.

For other, newer stream ciphers, look through the references to "The eSTREAM Portfolio" [3].

13.8 The Blum–Goldwasser cryptosystem

This is a remarkable public-key cryptosystem with the efficiency of a stream cipher, which was developed by Blum and Goldwasser in 1984 [9]. It works by producing a BBS random stream as the key stream, and encrypting by XOR-ing the key stream with the plaintext. However, if $n = pq$ is the public key, and used to generate the key stream, then the use of p and q as the private key can be used to determine the seed. The use of this system requires that both p and q are equal to 3 modulo 4.

For example, consider the example given previously, with $p = 19$, $q = 23$ and the seed $s_0 = 9$. Suppose that

$$s_{20} = 101.$$

Recall from the Rabin cryptosystem that a square root of m of $a \pmod{n}$ can be found by the following sequence of steps:

1. Find the values s and t for which
$$sp + tq = 1.$$
2. Compute
$$a_p = c^{(p+1)/4} \pmod{p}$$
$$a_q = c^{(q+1)/4} \pmod{q}.$$
3. Then the value m for which $m^2 = a \pmod{n}$ is
$$spa_q + tqa_p \pmod{n}.$$

A slight variation of this enables the computation of 2^k-th roots:

1. Find the values s and t for which
$$sp + tq = 1.$$
2. Compute
$$A_p = ((p+1)/4)^{k+1} \pmod{p-1}$$
$$A_q = ((q+1)/4)^{k+1} \pmod{q-1}.$$
3. Compute
$$a_p = a^{A_p} \pmod{p}$$
$$a_q = a^{A_q} \pmod{p}.$$

4. Then the value m for which $m^{2^k} = a \pmod{n}$ is

$$spa_q + tqa_p \pmod{n}.$$

Below is a test, with the value $s_{20} = 101$ above.

```
sage: k = 19
sage: Ap = power_mod((p+1)//4,k+1,p-1);Ap
  7
sage: Aq = power_mod((q+1)//4,k+1,q-1);Aq
  12
sage: ap = power_mod(101,Ap,p);ap
  9
sage: aq = power_mod(101,Aq,q);aq
  9
sage: r = mod(s*p*aq+t*q*ap,n);r
  9
```

To show that this works, recall from Chapter 5 that if y is a quadratic residue modulo p, then

$$y^{(p-1)/2} = 1 \pmod{p}.$$

For any two consecutive values $x = s_k$ and $y = s_{k+1}$ produced by the BBS generator,

$$x^2 = y \pmod{n},$$

which means that y is a quadratic residue modulo n and so

$$\begin{aligned} x^2 &= y \pmod{p}, \\ x^2 &= y \pmod{q}. \end{aligned}$$

Then

$$\begin{aligned} \left(y^{(p+1)/4)}\right)^2 &= y^{(p+1)/2} \pmod{p} \\ &= y^{(p+1)/2} \pmod{p} \\ &= y^{(p-1)/2}y \pmod{p} \\ &= y \pmod{p}. \end{aligned}$$

Thus

$$y^{(p+1)/4)}$$

is a square root of s_k and is also a quadratic residue (so its square root can be computed, and so on). A square root that is itself a quadratic residue is called a *principal square root.*

In general,

$$\left(y^{(p+1)/4)}\right)^{h+1} \pmod p$$

will be the principal 2^{h+1}-th root of y. Since by Fermat's theorem, $z^{p-1} = 1 \pmod p$ for any z, the exponent can be reduced

$$((p+1)/4)^{h+1}$$

by $p - 1$ before raising y to that power. That is, the principal 2^{h+1}-th root of y modulo p can be obtained by

$$y^{((p+1)/4)^{k+1} \bmod p-1} \pmod p.$$

This is easiest broken down into two steps:

1. $A_p = ((p+1)/4)^{k+1} \bmod p - 1.$
2. $a_p = y^{A_p} \pmod p.$

The principal 2^{h+1}-th root modulo n is obtained by performing these operations for both p and q, and applying the Chinese remainder theorem to obtain a result modulo n. But that is exactly the sequence of steps given above.

Having set up the mathematical groundwork, the cryptosystem can now be defined. Alice chooses two large primes p and q as her private key, and publishes their product $n = pq$ as her public key. To encrypt a plaintext message m to her, the message is treated as a string of bits m_i of length N. Bob chooses a random seed $s_0 < n$ and creates a BBS key stream

$$b_1, b_2, \ldots, b_N$$

where

$$\begin{aligned} s_{k+1} &= s_k^2 \pmod n \\ b_k &= s_k \pmod 2 \end{aligned}$$

for all $1 \leq k \leq N$. For each k a value c_k is obtained by

$$c_k = b_k \oplus m_k.$$

Then the ciphertext is

$$(c_1, c_2, \ldots, c_N, s_{N+1}).$$

That is, Alice is sent all the bits of the plaintext XOR-ed with the BBS key stream, along with the next value in the BBS sequence.

Parameters. Two large primes p and q, and their product $n = pq$.

Key generation. The private key is the two primes p and q. The public key is their product n.

Encryption. To encrypt a message m consisting of N bits m_i create a BBS key stream $b_i = s_i \pmod 2$ where $s_i = s_{i-1}^2 \pmod n$, and $c_i = m_i \oplus b_i$. Then the ciphertext is $(c_1, c_2, \ldots, c_N, s_{N+1})$.

Decryption. Use the factorization of n to compute square roots of s_{N+1} and previous values, and so reconstruct the key stream.

FIGURE 13.4: The Blum–Goldwasser cryptosystem.

To decrypt, Alice applies the method above of finding the principal 2^{N+1}-th root of s_{N+1} which will give her the starting seed s_0 of the BBS key stream. She can then create the key stream, and XOR it to the ciphertext values c_i to recover the plaintext. The system is described in Figure 13.4.

For a complete example, start with

```
sage: p = next_prime(10^6); p; p%4
  1000003
  3
sage: q = next_prime(2^20); q; q%4
  1048583
  3
```

Now Alice can publish her public key:

```
sage: n = p*q; n
  1048586145749
```

Suppose Bob wishes to send Alice the message 11110000 which is of length $N = 8$. In order for the decryption to work, Bob must choose a seed that is a quadratic residue. He can do this by choosing a random integer less than n, and squaring it modulo n:

```
sage: s = [randint(1,n)^2%n];s
  [336695326782]
```

Now Bob needs to enter the plaintext, and create enough s values and their least significant bits:

```
sage: m = [1,1,1,1,0,0,0,0]
sage: N = len(m)
sage: for i in range(N):
....: s += [(s[-1]^2)%n]
```

```
sage: b = [x%2 for x in s[:N]]
```

Now for the encryption:

```
sage: c = [i^^j for i,j in zip(b,m)]
  [1, 0, 0, 1, 0, 0, 0, 1]
```

Now the ciphertext is obtained by appending the last s value:

```
sage: c += [s[N]]; c
  [0, 0, 1, 0, 0, 0, 1, 1, 368189415063]
```

To decrypt, Alice applies the steps from above. First the extended Euclidean algorithm:

```
sage: [g,ss,tt] = xgcd(p,q)
```

and then obtaining the initial seed:

```
sage: k = N
sage: Ap = power_mod((p+1)//4,k,p-1); Ap
  60547
sage: Aq = power_mod((q+1)//4,k,q-1); Aq
  698372
sage: aq = power_mod(s[k+1],Aq,q); aq
  568397
sage: ap = power_mod(s[k+1],Ap,p); ap
  316700
sage: r = mod(ss*p*aq+tt*q*ap,n); r
  336695326782
```

Now Alice can create the key stream as Bob did, and recover the plaintext:

```
sage: [i^^j for i,j in zip(b,c[:N])]
  [1, 1, 1, 1, 0, 0, 0, 0]
```

The system can be sped up by taking not just one bit each time to form the key stream, but $t = \log\log n$ bits each time. In this case the message is broken into blocks of t bits each. For details see [61].

13.9 Glossary

Cryptographically secure random number generator (CSRNG.) An RNG for which an attacker is unable to determine the next or previous outputs from any list of outputs.

Linear congruential generator. A PRNG created by an affine map over a modulus.

Linear feedback shift register (LFSR). A sequence generated recursively by a linear combination (XORs) of previous terms.

Pseudo random number generator (PRNG.) A deterministic algorithm whose output cannot be statistically differentiated from a true RNG.

Stream cipher. A cipher that encrypts one bit at a time by XOR-ing the plaintext with bits obtained from a PRNG.

True random number generator (TRNG.) An RNG that uses an unpredictable physical process to generate output.

Exercises

Review Questions

1. Why are LCGs not considered to be cryptographically secure?
2. What are the three processes which comprise Fortuna?
3. How does Fortuna seed its entropy pools?
4. What is a known weakness in RC4 and how can that weakness be overcome?

Beginning Exercises

5. A one-time pad can be considered as a Vigenère cipher with a keyword as long as the plaintext. Decrypt the ciphertext `XNLURWGLQZXARZ` with the key `EGHTXDVHZWPXJG`.
6. Create a pseudo-random Vigenère key stream as follows: use the linear congruential generator

   $$x_{i+1} = 237x_i + 1 \pmod{256}$$

   with $x_0 = 11$ to create a list of x_i values. Reduce each value modulo 26, and turn each new value into a letter. What are the first 10 letters of the key stream?
7. Using the key stream from the previous question, decrypt `MNRPZSBXTG`.
8. The RSA-generated values given on page 339 have been used as a key stream to produce the ciphertext:

```
[0, 0, 1, 1, 0, 1, 1, 0, 1, 0, 0, 0, 1, 0, 1, 1, 1,
0, 1, 1, 0, 1, 0, 0]
```

(a) Decrypt this ciphertext to find the plaintext.

(b) Turn this numeric plaintext back into characters by treating each 8 bits as the binary expansion of an ASCII value (with the least significant bit on the right).

9. For the linear congruential generator $x_{i+1} = 17x_i + 1 \pmod{32}$:

   (a) Start with $x_0 = 0$ and list the next 30 values.

   (b) Plot the pairs (x_i, x_{i+1}) for all $0 \le i \le 19$.

   (c) Comment on the resulting plot.

10. Repeat the above question with the generator $x_{i+1} = 13x_i + 1 \pmod{32}$.

11. The linear congruential generator RANDU, which was used on IBM mainframes in the 1960s, is defined by

$$x_{i+1} = 65539x_i \pmod{2^{31}}.$$

Show that [62]

$$x_{i+2} = 6x_{i+1} + 9x_i \pmod{2^{31}}$$

for all i. (It is this high correlation between successive values—among other things—that has caused Knuth to describe this LCG as being "really horrible.")

12. Suppose the RSA generator is to be used to simulate the rolls of a fair die by taking blocks of 5 bits of the output, and for each block $[b_0, b_1, b_2, b_3, b_4]$, defining a roll as

$$1 + b_0 + b_1 + b_2 + b_3 + b_4.$$

Show that this is not a good simulation method: what are the probabilities of obtaining each of the rolls?

13. Find a fairer way of simulating dice rolls with RSA output.

Sage Exercises

14. List the points generated by RANDU as defined in question 11 in $\mathbb{R}^3$ as follows:

```
sage: N = 2^31
sage: rnd = [2^30-1]
sage: for i in range(1002): rnd += [(65539*rnd[i])%N]
```

```
sage: rf = [float(i/(2^31)) for i in rnd]
sage: list_plot([(rf[i],rf[i+1],rf[i+2])\
....: for i in range(1000)],size=1)
```

Move the resulting cube of points about with the mouse. What do you notice?

15. The following consecutive values have been produced with an LCG:

    $$116, 95, 109, 14, 163.$$

    Use the method given on page 338 to find the recurrence used. Use the command `solve_mod` to solve the simultaneous congruences.

16. The RSA generator can be sensitive to the starting value. Set $p, q, e = 19, 23, 7$ as before but set $s_0 = 11$. List the first 20 s_i values. What do you notice?

17. Repeat the last question but use $e = 5$ and $s_0 = 7$.

18. Now use the Blum-Micali discrete log generator with $p, g = 97, 5$ but with $s_0 = 37$.

19. Repeat the above question but for $p, g = 53, 2$ and $s_0 = 9$.

20. The following key stream

    $$[0, 1, 0, 1, 0, 1, 1, 1, 1, 1, 1, 0]$$

    has been obtained with an order 6 LFSR. Determine the polynomial and recurrence used.

21. Repeat the above question, but using

    $$[1, 1, 0, 0, 0, 1, 0, 1, 1, 0, 1, 0, 0, 1, 0, 0]$$

    which has been generated with an order 8 LFSR.

22. Test the RC4 cipher with the following test values:

Plaintext	Key	Ciphertext
0123456789ABCDEF	0123456789ABCDEF	75B7878099E0C596
0000000000000000	0000000000000000	DE188941A3375D3A
Attack at dawn	Secret	45A01F645FC35B38355 2544B9BF5

Note that the top two plaintexts and keys are given as hexadecimal, the bottom in ASCII.

23. Using the parameters for the Blum–Goldwasser system given on page 354, decrypt the ciphertext

```
[1, 1, 1, 0, 0, 1, 1, 1, 997692783568]
```

Further Exercises

24. For the RSA generator:

    (a) Show that $s_k = s_0^{(e^n)} \pmod{n}$.

    (b) What does this tell you about the upper bound for the period of the s_k list?

25. For the discrete log generator:

    (a) Show that $s_k = g^{(s_0^k)} \pmod{p}$.

    (b) Show that if $a^m = 1 \pmod{p-1}$ then the list generated by the discrete log generator modulo p with starting value $s_0 = a$ will repeat with period m.

Chapter 14

Advanced applications and protocols

Previous chapters have been concerned with the basic building blocks of cryptography: information integrity and authenticity, digital signatures, hash functions and their uses, and the associated mathematics and algorithms. This chapter shows how those building blocks can be used for more advanced purposes. In particular:

1. Secure multi-party communication, where different parties collude to create a shared result, with each party keeping its input secret from the others.

2. Zero knowledge proofs, where a "prover" must provide a proof of some knowledge to a "verifier," but without actually giving away any of that knowledge. As an example, the prover might try to prove the existence of the factorization of a large integer, but without providing any information that will enable the verifier to find out its factors.

3. Oblivious transfer, when a "sender" provides information to a "receiver," but is unable to learn exactly what the receiver obtained.

4. Digital cash, which like "real" cash can be spent, loaned, but leaves no audit trail.

5. Electronic voting, by which a voter, after registering a vote, has a receipt that enables the voter to verify the tallying, and ensures that the vote is counted exactly once.

14.1 Secure multi-party computation

The general problem is this: suppose that a function of n variables

$$f(x_1, x_2, \ldots, x_n)$$

is to be computed by n people, each person supplying one variable. However, no person wants their variable's value to be revealed. How can the function f be computed without each person giving away their values?

This problem was originally posed by Andrew Yao [98] in 1982, as the "millionaires problem." Here Alice and Bob, both millionaires, wish to establish which one of them is richer, without either of them revealing their wealth. In other words, they wish to compute the Boolean-valued function $f(x, y)$ which returns True if $x > y$ and False otherwise.

Here is Yao's solution. It assumes that Alice and Bob's wealth can be measured as i and j millions respectively, where $1 \leq i, j \leq 10$. The protocol can be used with any public-key system; for ease of explanation the RSA system will be used. Suppose $n = pq$ is an N-bit integer, and suppose that $e < n$ is Alice's public key.

1. Bob starts by choosing a random N-bit integer x and computes $k = x^e \pmod{n}$.

2. Bob sends to Alice the value $k - j + 1$. Note that this value gives away nothing to Alice, as she doesn't know either x (and hence k) or j.

3. Alice computes the values

   $$y_u = (k - j + u)^d \pmod{n}$$

   for all $u = 1, 2, \ldots, 10$ using her private key d. She keeps all these values private. Note that since $v = k - j + 1$ is the value Bob sent to Alice, this can be expressed as Alice computing

   $$(v + s)^d \pmod{n}$$

   for all $s = 0, 1, 2, \ldots, 9$.

4. She now chooses a random prime r of $N/2$ bits, and computes

   $$z_u = y_u \pmod{r}$$

   for each u. She must obtain values of z which differ by at least two. If two values differ by only 1, then she must choose another prime and compute the values again.

5. She sends the prime r to Bob, along with the values

   $$z_1, z_2, \ldots, z_i, z_{i+1} + 1, \ldots, z_{10} + 1$$

   (all computed modulo p). That is, she sends the first i values unchanged, but adds one to all values from $i + 1$ onwards. Recall that i is Alice's wealth.

6. Bob looks at this last list of 10 integers, and finds the j-th value. If it is equal to $x \pmod{r}$ he deduces that $i \geq j$ (so Alice is at least as rich as him), and $i < j$ otherwise.

To see why this works, suppose that $i = 10$, and so Alice sends

$$((k - j + u)^d \bmod n) \bmod r$$

to Bob for all $u = 1, 2, 3, \ldots, 10$. Since $k = x^e \pmod n$ and since Bob's wealth is j, the j-value in Alice's list will be

$$\begin{aligned} ((k - j + j)^d \bmod n) \bmod r &= (k^d \bmod n) \bmod r \\ &= x \pmod r \end{aligned}$$

since d and e are the RSA keys. In general, the j-value z_j will only satisfy $z_j = x \pmod r$ if Alice has not added one before sending it. But this will only occur if her wealth i is at least as great as Bob's wealth j.

Here is an example. Suppose Alice's wealth is $i = 3$ and Bob's wealth is $j = 7$. Alice chooses $p = 991$, $q = 631$ so that $n = 625321$. This is a 20-bit number, so $N = 20$. Alice chooses 107551 as her public key, and computes

$$d = e^{-1} \pmod{(p-1)(q-1)} = 281251$$

as her private key.

To start the protocol, Bob chooses $x = 568479$ which has 20 bits, and computes

$$k = x^e \pmod n = 233640.$$

He then sends $233640 - 7 + 1 = 233634$ to Alice.

Now Alice needs to compute

$$\begin{aligned} y_1 &= 233634^d \pmod n, \\ y_2 &= 233635^d \pmod n, \\ y_3 &= 233636^d \pmod n, \\ &\vdots \\ y_{10} &= 233643^d \pmod n. \end{aligned}$$

If she is sensible, she will use Sage (assuming that d and n have already been entered):

```
sage: y = [power_mod(233634+u,d,n) for u in range(10)]
sage: y
[209850, 268589, 420577, 554655, 5938, 344918,
 568479, 283124, 58855, 68721]
```

Now she needs to choose a random prime r of $N/2$ bits and reduce all the y values modulo r:

```
sage: r = next_prime(randint(2^9,2^10))
sage: r
757
sage: z = [s%r for s in y];z
[161, 611, 442, 531, 639, 483, 729, 6, 566, 591]
```

All the values here differ by at least 2:

```
sage: diffs = [[abs(z[i]-z[j]) for j in range(i+1,10)]\
....: for i in range(9)]
sage: min(flatten(diffs))
20
```

Using her own wealth $i = 3$ she sends r to Bob, along with values

$$z_1, z_2, z_3, z_4 + 1, z_5 + 1, \ldots, z_{10} + 1.$$

```
sage: z1 = z[:i]+[s+1 for s in z[i:]]; z1
[161, 611, 442, 532, 640, 484, 730, 7, 567, 592]
```

Bob now checks the sixth value (recall that Bob's wealth is $j = 6$) against $x \pmod{r}$

```
sage: z1[5]==x%r
False
```

He conclude that he is richer.

Suppose the values were reversed; that $i = 7$ and $j = 3$. He sends $k-j+1 = 233638$ to Alice, and she computes

```
sage: y = [power_mod(233638+u,d,n) for u in range(10)]; y
[5938, 344918, 568479, 283124, 58855, 68721, 80563,
 542835, 371835, 310942]
```

Using the same r as above:

```
sage: z = [s%r for s in y];z
[639, 483, 729, 6, 566, 591, 321, 66, 148, 572]
```

and as above all these values differ by at least two. Now the values are sent to Bob: p as before, and

```
sage: z1 = z[:i]+[s+1 for s in z[i:]]; z1
[639, 483, 729, 6, 566, 591, 321, 66, 149, 573]
```

Bob's check:

```
sage: z1[j-1]==x%r
True
```

He concludes that Alice is at least as rich as him.

Another classic is Chaum's "dining cryptographers problem" [15]. Here is the problem with its solution, as given by Chaum:

> Three cryptographers are sitting down to dinner at their favorite three-star restaurant. Their waiter informs them that arrangements have been made with the maitre d'hotel for the bill to be paid anonymously. One of the cryptographers might be paying for the dinner, or it might have been NSA (U.S. National Security Agency). The three cryptographers respect each other's right to make an anonymous payment, but they wonder if NSA is paying. They resolve their uncertainty fairly by carrying out the following protocol:
>
> Each cryptographer flips an unbiased coin behind his menu, between him and the cryptographer on his right, so that only the two of them can see the outcome. Each cryptographer then states aloud whether the two coins he can see—the one he flipped and the one his left-hand neighbor flipped—fell on the same side or on different sides. If one of the cryptographers is the payer, he states the opposite of what he sees. An odd number of differences uttered at the table indicates that a cryptographer is paying; an even number indicates that NSA is paying (assuming that the dinner was paid for only once). Yet if a cryptographer is paying, neither of the other two learns anything from the utterances about which cryptographer it is.

To see how this works, note that as three coins are tossed, either they will all be the same, or two will be the same and the other different. Both possibilities are illustrated in Figure 14.1.

Suppose the NSA is paying, so that each cryptographer calls out what he or she sees. On the left, the calls will be "same, same, same", on the right "same" from Bob and "different" from Alice and Carol. Either way there will be an even number of "different" calls.

Suppose that one of the cryptographers is paying. Take the left scenario, and suppose that payer is Alice; she will then call "different," for a total of two "sames" and one "different." On the right, again suppose Alice is paying. She will call "same" for a total again of two "sames" and one "different." If Bob is paying, he will call "different" for a total of three "different" calls. Either way, there will be an *odd* number of "different" calls.

This protocol may be considered as a means of computing the Boolean function

$$f(x, y, z) = x\overline{y}\,\overline{z} + \overline{x}y\overline{z} + \overline{x}\,\overline{y}z$$

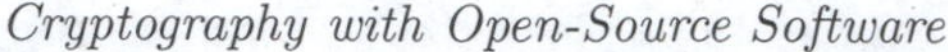

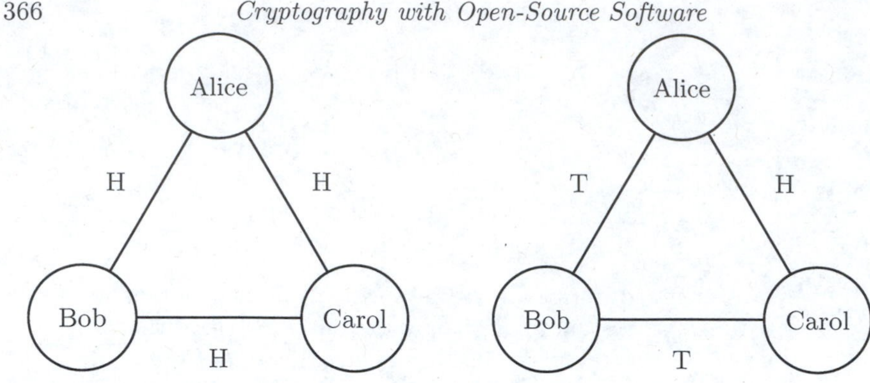

FIGURE 14.1: The dining cryptographers.

where the operations are to be interpreted in their Boolean sense, and a result of True means that one of the cryptographers is paying, and False means that none of them are.

The protocol can be generalized to any number of cryptographers, and instead of tossing coins they may produce a random bit, and state either the XOR of the two bits on either side if they did not pay, and the opposite if they did. The results can all be added (modulo 2), so that a result of 0 indicates nobody paid, and a result of 1 means one of them did.

14.2 Zero knowledge proofs

The idea of a zero knowledge proof is for Peggy, the *prover*, to be able to prove to Victor, the *verifier*, that she is in possession of some useful knowledge, but without revealing any of that knowledge to Victor.

For example, suppose Peggy and Victor are working on separate research teams investigating factorization. Peggy's team has succeeded in factorizing a large composite number, and she would like to communicate this to Victor. She does not want him to know the factors she has found, so how can she convince him that she has produced a successful factorization?

A delightful example of a zero knowledge proof is provided by Naor et al. [65]. Here Peggy is faced with the task of proving to Victor that she has solved one of the visual puzzles in the "Where's Wally?" books (known as "Where's Waldo" in the US and Canada). The puzzle is to find the figure of Wally (Waldo) in a busy, intricate picture containing hundreds of other characters and masses of confusing detail. How does Peggy convince Victor that she knows Wally's whereabouts, without revealing it to Victor? One "low-tech" solution given is this: Peggy takes a very large sheet of cardboard and cuts a Wally-shaped hole in it. She then places her cardboard over the book so that only Wally is seen through the hole. Victor can certainly see Wally—so he knows that Peggy has found him—but as the rest of the puzzle is hidden it does not help Victor find Wally himself.

One of the major uses of a zero knowledge proof is for *identification*, where Peggy wishes to identify herself (to Victor, say), but without giving Victor any information that would allow him to impersonate Peggy later on.

An early identification protocol was given by Fiat and Shamir [28]; it involves the use of a trusted third party, Trent, who chooses two large primes p and q and publishes their product $n = pq$. Peggy creates a secret s so that $1 \le s \le n-1$ and publishes $v = s^2 \bmod n$ as her public key. To prove herself to Victor, Peggy "commits" to a random $r \le n-1$; this means that she may reveal r to Victor after the computation to show that she did not cheat, and computes $x = r^2 \bmod n$ which she sends to Victor. Victor chooses at random $e \in \{0, 1\}$ and sends his choice to Peggy. Peggy then sends back

$$y = \begin{cases} r & \text{if } e = 0, \\ rs \bmod n & \text{if } e = 1. \end{cases}$$

(In other words, Peggy sends $rs^e \bmod n$.) Victor verifies the proof if $y^2 = xv^e \bmod n$. In fact the complete protocol uses this just as a building block; a more complete version has Peggy producing several secret keys $s_1, s_2, \ldots, s_k$ and corresponding public keys v_i. There will be t challenges, and for each challenge Peggy starts by committing to a random value r and sending to Victor $x = r^2 \bmod n$. Victor sends Peggy a random binary vector

$$[e_1, e_2, \ldots, e_k]$$

to which Peggy responds with

$$y = r \prod_{e_j = 1} s_j^{-1} \pmod{n}$$

and Victor verifies that

$$x = y^2 \prod_{e_j = 1} v_j \pmod{n}.$$

Note that y and x can be equivalently written as

$$y = r \prod_{j=1}^{k} s_j^{-e_j}$$

and

$$x = y^2 \prod_{j=1}^{k} v_j^{e_j}.$$

Victor accepts Peggy's proof of identity only if all t checks are successful. Victor is convinced because

$$\begin{aligned} y^2 \prod_{e_j=1} v_j &= \left(r_i \prod_{e_j=1} s_j^{-1}\right)^2 \prod_{e_j=1} v_j, \\ &= r^2 \prod_{e_j=1} ((s_j^{-1})^2 v_j) \\ &= r^2 \\ &= x \end{aligned}$$

where all operations are performed modulo n.

To see this in operation, use the values $(p, q, n) = (991, 631, 625321)$. With $k = 6$, Peggy can create values for s and v:

```
sage: k = 6
sage: s = []
sage: v = []
sage: for i in range(k):
....:     t = mod(randint(1,n),n)
....:     s += [t]
....:     v += [t^2]
....:
sage: s
[571081, 247560, 166434, 193662, 556140, 294482]
sage: v
[467616, 118353, 432019, 92627, 429148, 132044]
```

To identify herself to Victor, Peggy first commits to a value r and sends $x = r^2$ to Victor:

```
sage: r = randint(1,n)
sage: x = r^2%n
sage: x
229852
```

Now Victor sends a random binary vector to Peggy:

```
sage: e = [randint(0,1) for i in range(k)]; e
[1, 1, 0, 1, 1, 1]
```

Peggy responds by returning the value y:

```
sage: y = r*prod((1/s[j])^e[j] for j in range(k))
sage: y
301252
```

Now Victor can make the final check:

```
sage: y^2*prod(v[i]^e[i] for i in range(k)) == x
True
```

In this protocol, Peggy is actually proving to Victor that she knows her secret s_i numbers, without revealing them. Because of the difficulty of computing modular square roots, Victor cannot obtain these secrets, and so he cannot later impersonate Peggy.

Suppose also that an eavesdropper, Eve, tries to convince Victor that she is Peggy. She does not have Peggy's secret numbers s_i, so she has to guess a suitable binary vector $[a_1, a_2, \ldots, a_k]$ for which

$$x = y^2 \prod_{a_j=1} v_j \pmod{n}.$$

She might try by sending

$$x = r^2 \prod_{a_j=1} v_j \pmod{n}, \text{ and } y = r$$

at the beginning, so when Victor sends his challenge vector, Eve is ready for it. But the probability of Eve choosing acceptable values a_i is small, and becomes vanishingly small after more challenges.

This protocol is not quite *zero* knowledge though, as one bit has been leaked: the knowledge that the prover is actually Peggy. To eliminate this leaked bit, the protocol should determine not that the prover *is* Peggy, but that the prover can *identify* whether or not she is Peggy. The Feige-Fiat-Shamir protocol [25] "tidies up" the original Fiat-Shamir protocol by being fully zero knowledge, as well as more efficient in practice.

As before, Peggy chooses random values s_j and publishes the values $v_j = s_j^{-2} \bmod n$. A challenge then consists of

1. Peggy choosing r at random, and sending $x = \pm r^2 \pmod{n}$ to Victor.

2. Victor choosing a random k-length binary vector $[e_j]$ and sending it to Peggy.

3. Peggy sending

$$y = r \prod_{e_j=1} s_j \pmod{n}$$

to Victor.

4. Victor checking that

$$x = \pm y^2 \prod_{e_j=1} v_j \pmod{n}.$$

The introduction of the sign eliminates the extra bit leakage, and provides a fully zero knowledge protocol. Full proofs of all assertions can be found in the original paper.

Back to the problem posed at the beginning of this section: how can Peggy prove to Victor that she can factorize n without revealing any of the factors?

One approach to a protocol is this: Victor picks a random $s < n$ and sends $x = s^2 \bmod n$ to Peggy. If Peggy can return s, that shows she can factorize n. But this technique can be used to advantage by Victor: since n is a product of two odd primes there will be four different square roots of x: s and $-s$, and two others t and $-t$. If Peggy returns t or $-t$, then $\gcd(s + t, n)$ will be a proper factor. So this approach does not satisfy the zero knowledge property.

A better protocol [90] is this:

1. With n being a k-bit integer, Peggy chooses random a of at least $k/2$ bits and sends $A = a^2 \bmod n$ to Victor.

2. At the same time, Victor chooses a random b of at least $k/2$ bits and sends $B = b^2 \bmod n$ to Peggy.

3. Since A and B are both quadratic residues by construction, so will AB, and so Peggy can use her knowledge of the factorization of n to find square roots of $AB \bmod n$. She picks any one, x.

4. Victor sends Peggy a random bit e.

5. If $e = 0$ Peggy sends a to Victor, and if $e = 1$ Peggy sends x to Victor.

6. Victor can now check: if $e = 0$ he checks that $a^2 = A \bmod n$ (recall he was sent A in step 1); if $e = 1$ he checks that $x^2 = AB \bmod n$.

This is repeated as many times as required (20 or 30 for example), and if Peggy "passes" every test Victor can be convinced. This is a better protocol than the previous naive approach for two reasons: Peggy cannot fake square roots if she cannot factor n—the best she can do is send a when $e = 0$, but the probability of e being always zero (for uniform distribution) becomes vanishingly small for a large number of trials. Also, for each challenge Victor does not get enough information to enable him to factor n. He gets at most one square root, not the two required.

In general, a zero knowledge proof should satisfy three criteria (which are here given informally):

Completeness. If Peggy really does have the information, and if both Peggy and Victor follow the protocol, then Victor will accept the result.

Soundness. If Peggy is lying about having the information, then Victor will reject the result with probability p close to 1. This means that although Peggy may be able to lie once or twice, she will not be able to lie enough to convince Victor in the long term.

Zero knowledge. No information about the actual result is leaked during the protocol.

An excellent introduction to the theory of zero knowledge systems is given by Goldreich [34].

14.3 Oblivious transfer

Oblivious transfer has much in common with zero knowledge proofs. However, here the concern is in information being sent from a "sender" to a "receiver," with the sender being unable to ascertain what the receiver actually obtains. The first protocol was developed by Rabin [74].

For Rabin's protocol, the sender (Bob) uses an RSA encryption, and so chooses two large primes p and q, calculates $n = pq$ and chooses e relatively prime to $(p-1)(q-1)$. The steps of the protocol are

1. The message m to be sent is encrypted as $c = m^e \bmod n$, the values n, e and c are sent to the receiver (Alice).

2. Alice chooses a random $x < n$ and sends $y = x^2 \bmod n$ to Bob. If p and q are large, the probability that p or q is a divisor of x is negligible, and so it may be assumed that there are four square roots of $y \bmod n$; they are x and $-x$, x' and $-x'$.

3. Bob finds a square root z of $y \bmod n$ and sends z to Alice.

If z is equal to x' or $-x'$ then Alice can factor n and so determine $d = e^{-1} \bmod (p-1)(q-1)$ and thus decrypt c to obtain m. If however z is equal to x or $-x$ Alice can do nothing. With this protocol Bob cannot be sure that Alice has obtained the message; she can obtain the message only with probability $1/2$.

This protocol has been shown to have some vulnerabilities; for example Alice can cheat in step 2 by sending a value y for which she does not actually know any square roots. When Bob sends back a square root, Alice will be able to factor n. This vulnerability can be overcome by requiring Alice to provide a zero knowledge proof that she has a square root of y.

This protocol, and others like it, once strengthened, can be used as a building block for secure multi-party computation [56].

A similar protocol to Rabin's was developed by Even, Goldreich, and Lempel [24]. With this protocol, Bob will send two messages to Alice, but be unable to tell which message Alice receives. The protocol can of course be engineered to work with any public-key system, but for ease of explanation here it is with RSA:

1. Bob generates RSA parameters: modulus $n = pq$, the public key e and corresponding private key d. He also generates two random values, x_0 and x_1, and sends them to Alice along with his public values n and e.

2. Alice picks r at random to be either 0 or 1, and also generates a random value k. She then blinds x_r by computing

$$v = x_r + (k^e \bmod n)$$

which she sends to Bob.

3. Bob does not know which of x_0 and x_1 Alice chose, so he attempts to unblind with both of his random messages and comes up with two possible values for k: $k_0 = (v - x_0)^d \bmod n$ and $k_1 = (v - x_1)^d \bmod n$. One of these will be equal to k since it will correctly decrypt, while the other will produce another random value that does not reveal any information about k.

   He then chooses s to be a random bit, and sends the three values

$$x'_0 = x_0 + k_s,\ x'_1 = x_1 + k_{1-s},\ s$$

   to Alice.

4. Alice sets $c = r + s \bmod 2$ and so now is able to compute exactly one of the messages $x_c = x'_c - k$.

This can be shown to work by enumerating all possible choices of r and s. Suppose for example that $r = 1$ and $s = 0$. Then

$$v = x_1 + (k^e \bmod n).$$

Then the values of k_i are

$$\begin{aligned}
[k_0, k_1] &= [(v - x_0)^d \bmod n, (v - x_1)^d \bmod n] \\
&= [(x_1 + k^e - x_0)^d \bmod n, (x_1 + k^e - x_1)^d \bmod n] \\
&= [(x_1 - x_0 + k^e)^d \bmod n, (k^e)^d \bmod n] \\
&= [(x_1 - x_0 + k^e)^d \bmod n, k],
\end{aligned}$$

and the x'_i values are

$$\begin{aligned}
[x'_0, x'_1] &= [x_0 + k_0, x_1 + k_1] \\
&= [x_0 + (x_1 - x_0 + k^e)^d \bmod n, x_1 + k].
\end{aligned}$$

Since $c = 1$, Alice computes

$$x'_1 - k = x_1$$

for her message. If $s = 1$, then

$$\begin{aligned}[x_0', x_1'] &= [x_0 + k_1, x_1 + k_0] \\ &= [x_0 + k, x_1 + (x_1 - x_0 + k^e)^d \bmod n].\end{aligned}$$

And since $c = 0$ now, Alice computes

$$x_0' - k = x_0.$$

Here is a little example, with $n = 625321$ as before, and public and private keys $e = 286147$ and $d = 555283$. Bob starts by creating random messages for Alice:

```
sage: x = [mod(randint(1,n),n),mod(randint(1,n),n)]
sage: x
[45885, 443990]
```

Now Alice chooses a random bit and value k:

```
sage: r = randint(0,1)
sage: k = mod(randint(1,n),n)
sage: k
602353
```

Now she blinds x_r:

```
sage: v = x[r]+k^e
sage: v
479713
```

Bob now attempts to unblind v:

```
sage: K = [(v-z)^d for z in x]
sage: K
[602353, 120593]
```

Note that one of these is indeed k; however Bob does not know which one it is. Bob now continues:

```
sage: s = randint(0,1)
sage: X = [x[0]+K[s],x[1]+K[1-s]]
sage: X
[22917, 564583]
```

All these values are sent to Alice. Alice can now recover a message:

```
sage: c = (r+s)%2
sage: X[c]-k
45885
```

and this value is indeed a message. In fact, it is message x_0:

```
sage: x[0]
45885
```

14.4 Digital cash

The idea of digital cash is to provide some sort of money transfer that is anonymous and untraceable. Digital cash should satisfy these properties [63]:

Unforgeability. This is the most basic and fundamental requirement. There should be no easy method by which cash can be forged in a reasonable time.

Untraceability. Once digital cash is issued by the bank, it cannot be traced either by the bank or by any third parties.

Anonymity. The owner of the cash should not be identifiable.

Double spending. Digital cash should be able to be spent only once. Attempts at double spending or of coin duplication should be detectable by the bank.

Fairness. This is an optional property, required only for "fair cash" schemes, in which anonymity is only conditional, and a trusted third party can detect the owner if the cash is used illegally.

There are several digital cash schemes, of which one of the most comprehensive and efficient has been designed by Brands [13]. There are three parties: the *Bank*, the *User* (also known as the *Client* or the *Spender*), and the *Shop* (or *Shopkeeper*, also known as the *Merchant* or *Vendor*). For ease of description, it will be assumed that all digital withdrawals and spendings have the same currency value, and a single unit of digital cash will be called a "coin."

There are five phases to the digital cash scheme, of which the first two have to be done only once, and the last three as many times as required:

1. The Bank sets itself up with all necessary parameters and public keys.
2. The User opens an account.
3. The User withdraws a coin from the Bank.
4. The User spends the coin at the Shop.
5. The Shop deposits the coin with the Bank.

Brand's scheme requires many variables and steps, but all the steps are themselves simple. The complexity arises from the necessity of providing anonymity of the owner, while at the same time protecting the coin.

The Bank's initial setup

The following parameters must be known to all users of the system: Bank, Users, and Shops.

- Large primes p and q for which $q|(p-1)$.
- Two hash functions:

  $$\mathcal{H} : \mathbb{Z} \times \mathbb{Z} \times \mathbb{Z} \times \mathbb{Z} \times \mathbb{Z} \to \mathbb{Z}_q$$

  and

  $$\mathcal{H}_0 : \mathbb{Z} \times \mathbb{Z} \times \mathbb{Z} \times \mathbb{Z} \to \mathbb{Z}_q.$$

- Three elements g, g_1, g_2 of $\mathbb{Z}_p$ all of which have multiplicative order q. These can be obtained as follows: let α be any primitive root modulo p, and define

  $$g = \alpha^{(p-1)/q} \pmod{p}.$$

  Let k_1, k_2 be randomly chosen elements of $\mathbb{Z}_q$ and define

  $$g_1 = g^{k_1}, \; g_2 = g^{k_2} \pmod{p}.$$

  The values k_1, k_2 should be deleted, as knowledge of them would compromise the security of the system.

The Bank starts by choosing a random $x \in \mathbb{Z}_q$ and publishing

$$h = g^x \pmod{p}$$

as its public key.

Opening an account

The User chooses $U \in \mathbb{Z}_q$ at random and computes

$$I = g_1^U \pmod{p}$$

and checks that

$$g_1^U g_2 \neq 1 \pmod{p}.$$

This value I is the User's identity, which is sent to the Bank. The Bank can store I and use it to identify the User for all transactions. The Bank also computes

$$z = (Ig_2)^x \pmod{p}$$

which is sent back to the User.

Withdrawing a coin

If the User wishes to withdraw a coin, the following steps must be followed.

Step 1. The Bank first chooses a random $w \in \mathbb{Z}_q$ and computes

$$a = g^w \pmod p$$
$$b = (Ig_2)^w \pmod p$$

which are sent to the User. A new w must be chosen for each transaction.

Step 2. The User then chooses at random $s, x_1, x_2, u, v \in \mathbb{Z}_q$ (these should be freshly chosen for each new coin) and computes

$$A = (Ig_2)^s$$
$$B = g_1^{x_1} g_2^{x_2}$$
$$z' = z^s$$
$$a' = a^u g^v$$
$$b' = b^{su} A^v$$

where all arithmetic is done modulo p. The value $A = 1$ is not permitted: this would assume either a value $s = 0 (\mathrm{mod} q)$, or that $Ig_2 = 1 (\mathrm{mod} p)$ which means that the User has managed to find U to solve the discrete logarithm problem $g_1^U = g_2^{-1} (\bmod p)$. The User next computes the hash value

$$H = \mathcal{H}(A, B, z', a', b')$$

and then

$$c = H/u \pmod q$$

and sends this value to the Bank.

Step 3. The Bank then computes

$$r = cx + w \pmod q$$

which is sent to the User.

Step 4. The User now checks the following:

$$g^r = h^c a \pmod p$$
$$(Ig_2)^r = z^c b \pmod p.$$

If both of these check out correctly, the User now computes

$$r' = ru + v \pmod q$$

to obtain the coin

$$(A, B, z', a', b', r')$$

and the amount of the coin is deducted from the User's account.

To see that the checking steps are valid, note that

$$\begin{aligned} g^r &= g^{cx+w} \\ &= (g^x)^c g^w \\ &= h^c a \end{aligned}$$

and similarly

$$\begin{aligned} (Ig_2)^r &= (Ig_2)^{cx+w} \\ &= ((Ig_2)^x)^c (Ig_2)^w \\ &= z^c b. \end{aligned}$$

In Brands' original description, the pair (A, B) constituted the coin, and the quadruple (z', a', b', r') its signature.

Spending the coin

The User starts by sending the coin to the Shop, and then the following steps are followed.

Step 1. The Shop first computes the hash H for himself, and then checks that

$$\begin{aligned} g^{r'} &= h^H a' \pmod{p} \\ A^{r'} &= (z')^H b' \pmod{p}. \end{aligned}$$

This ensures the validity of the coin, and ensures the Shop that both the Bank and the User have followed the protocol properly. In Brands' original description, this check was done right at the end (as part of Step 3 below), but it makes sense to perform it at the beginning.

These equations can be proved by noting that $H = cu \bmod q$ and so

$$\begin{aligned} g^{r'} &= g^{ru+v} \\ &= g^{(cx+w)u+v} \\ &= g^{wu+v} g^{cxu} \\ &= a^u g^v h^{cu} \\ &= a' h^H. \end{aligned}$$

The Shop next computes

$$d = \mathcal{H}_0(A, B, S, t)$$

where S is the Shop's identity (analogous to the user's value I), and t is a number representing the date and time of the transaction. This value d is sent to the User.

Step 2. The User next computes

$$r_1 = dUs + x_1 \pmod q$$
$$r_2 = ds + x_2 \pmod q$$

where U is the number originally chosen by the User to establish his[1] identity, and s, x_1, x_2 were random numbers chosen as part of the protocol for withdrawing the coin. The User then sends r_1 and r_2 to the Shop.

Step 3. The Shop checks that

$$g_1^{r_1} g_2^{r_2} = A^d B \pmod p.$$

If this is true, the Shop accepts the coin. This equation should be true:

$$\begin{aligned} A^d B &= (I g_2)^{sd} g_1^{x_1} g_2^{x_2} \\ &= (g_1^U g_2)^{sd} g_1^{x_1} g_2^{x_2} \\ &= g_1^{sdU + x_1} g_2^{sd + x_2} \\ &= g_1^{r_1} g_2^{r_2}. \end{aligned}$$

Depositing the coin

To deposit the coin in the Bank, the Shop tenders the coin

$$(A, B, z', a', b', r')$$

and the triple (d, r_1, r_2). The Bank first checks that the coin has not been deposited before, and then makes the same checks on the coin as the Shop did:

$$\begin{aligned} g^{r'} &= h^H a' \pmod p \\ A^{r'} &= (z')^H b' \pmod p \\ g_1^{r_1} g_2^{r_2} &= A^d B \pmod p. \end{aligned}$$

If all these equations hold true, the Bank credits the amount of the coin to the Shop's account.

An example

Here is an example with small numbers.

[1]Since English does not have a generic third person singular pronoun, I will use "he" to refer to the User when necessary. If this seems sexist, I will use "her" for the generic person in the next section.

Setup. First the setup—it can be verified that $q|(p-1)$:

```
sage: p = 10000019; q = 1523
sage: alpha = primitive_root(p)
sage: g = mod(alpha,p)^((p-1)//q); g
8407386
sage: def rq(): return randint(1,q)
sage: k1,k2 = rq(),rq(); g1,g2 = g^k1,g^k2; g1,g2
(2219954, 6695507)
sage: x = rq(); h = g^x; h
1177538
```

Opening an account. Next, opening an account (with the check that $g_i^U g_2 \neq 1$):

```
sage: U = rq(); g1^U*g2 != 1
True
sage: I = g1^U; I
6481318
sage: z = (I*g2)^x; z
4477168
```

Withdrawing a coin. Now a coin can be created for withdrawal by the User:

```
sage: w = rq();a,b = g^w,(I*g2)^w; w,a,b
(417, 3335316, 6369540)
sage: A,B,zd,ad = (I*g2)^s,g1^x1*g2^x2,z^s,a^u*g^v
sage: bd = b^(s*u)*A^v
```

Rather than create a hash function, here is a value to use:

```
sage: H = mod(1000,q)
```

Continuing with the last few steps:

```
sage: c = H/u; r = c*x+w; c,r
(123, 62)
```

Now the checks can be done:

```
sage: g^r == h^c*a
True
sage: (I*g2)^r == z^c*b
True
```

Finally the coin can be completed:

```
sage: rd = r*u+v
sage: A,B,zd,ad,bd,rd
(8158746, 4371050, 9839354, 1595842, 2728069, 376)
```

Spending the coin. Now the coin can be spent, first with the Shop's check:

```
sage: g^rd == h^H*ad
True
sage: A^rd == zd^H*bd
True
```

As before, choose a value for d:

```
sage: d = mod(500,q)
```

The User's values:

```
sage: r1,r2 = d*U*s+x1,d*s+x2
```

and the Shop's check:

```
g1^r1*g2^r2 == A^d*B
True
```

Why the protocol works

This protocol can be shown to satisfy the properties given at the beginning of this section. Here are some skeleton arguments (rather than full proofs) of the assertions:

Unforgeability. First note that the coin must be signed to be valid, and that the signature requires input from both the Bank and the User. The User cannot forge a coin alone, because the user cannot determine the value of r using his own input. The value of r depends on x which only the Bank knows. The User has access to the Bank's public key h, but determining x would involve solving a discrete logarithm problem.

Similarly the Bank cannot forge a coin without the User. In fact, all that the Bank has are x, w, and later on c. These are not enough to create a full signature for the coin. The Bank may try to forge a coin that satisfies $g^{r'} = h^H a$ but with the User's value U not known to the Bank, the Bank cannot create an appropriate value of r_1 to use for spending. In fact, the Bank could create an appropriate value of r_1 only if $s = 0$, which is a forbidden value in the coin creation protocol.

Untraceability. If the coin were to be traceable, the User's identity must somehow be coded into the coin's signature. But even though the User requires his identity I to create the coin (in particular the number A), his identity is hidden unless the Bank or any other third party can find the User's random value s, which is itself hidden by the discrete logarithm problem.

Anonymity. The Shop has the six values that comprise the coin, as well as the triple d, r_1, r_2. But none of these values enable the Shop to identify the User: the values I and U cannot be obtained as they are safely hidden by virtue of the discrete logarithm problem. Also, the values that comprise the coin aren't seen by the Bank until the Shop makes a deposit, which happens only after the coin has been spent by the User.

Double spending. Suppose the User tries to spend the coin twice, once with the Shop, and a second time with a Merchant. This will require first the creation of (d, r_1, r_2) with the Shopkeeper, and also the creation of (d', r'_1, r'_2) with the Merchant. (The Merchant will create d' using his own identity M in place of the Shop's S.)

When the coin is tendered to the Bank as a deposit, the Bank will see the values (A, B, z', a', b', r') twice, and also have the two triples (d, r_1, r_2) and (d', r'_1, r'_2). Recall from above that the coin and a single triple does not make it possible to identify the User. However, in this case the equations for the r_i and r'_i values can be used to determine U:

$$\begin{aligned} r_1 - r'_1 &= dUs - d'Us \\ &= Us(d - d') \end{aligned}$$

(all modulo q), and

$$\begin{aligned} r_2 - r'_2 &= ds - d's \\ &= s(d - d'). \end{aligned}$$

Dividing the last equations then yields

$$U = (r_1 - r'_1)/(r_2 - r'_2) \pmod q$$

which enables the Bank to identify the User. Since the only other way U can be obtained is by solving a discrete logarithm problem, the Bank has proof that double spending has occurred.

For a more complete discussion of the security of this protocol, see the original paper [13] or the text of Trappe and Washington [88].

14.5 Voting protocols

The following requirements for a secure voting scheme are among those listed in an Internet Policy Institute report [40]:

Eligibility. Only authorized voters can vote.

Uniqueness. No one can vote more than once.

Privacy. No one can determine for whom anyone else voted.

Integrity. Votes should not be able to be modified, forged, or deleted without detection.

Non-coercibility. Voters should not be able to prove how they voted (if they were able to, this would allow vote selling and coercion).

Verifiability. Every voter can make sure that her[2] vote has been taken into account in the final tally.

A number of secure voting schemes have been proposed, all of which aim to satisfy at least those requirements, and possibly some others as well (such as convenience and ease of use; and robustness and reliability, where the system can work in spite of errors and breakdowns in some of its component parts). Most voting systems get alarmingly complicated, depending on the amount of security required. A reasonable balance between usability and security can be obtained by using a hybrid system, with some sort of paper-based ballots, and cryptographic methods used for secure tallying and verification. Two such recent schemes are Prêt à Voter [75] and Scantegrity [16].

Use of homomorphic encryption

A *homomorphic cryptosystem* is one where given two ciphertexts c_1 and c_2 corresponding to plaintexts p_1 and p_2, it is possible to produce the ciphertext of some function of of p_1 and p_2 without decryption. For example, in the RSA system:

$$c_1 = (p_1)^e \pmod{n}$$
$$c_2 = (p_2)^e \pmod{n}.$$

It follows that

$$c_1c_2 = (p_1p_2)^e \pmod{n}$$

so that the product c_1c_2 is the ciphertext corresponding to p_1p_2.

[2] I told you I would!

For many modern voting systems, the cryptosystem of choice is one developed by Pascal Paillier in 1999 [68]. For Paillier's cryptosystem, start with two large primes p and q and let $n = pq$. Let g be an integer whose multiplicative order modulo n^2 is a multiple of n. (This can be obtained easily by setting $g = kn + 1$ for any $1 \leq k < n$; then the multiplicative order modulo n^2 will be n.) Define $\lambda = \text{lcm}((p-1),(q-1))$. Then the public key is (g, n) and the private key is λ. Messages are elements $m \in \mathbb{Z}_n$. To encrypt such a message, choose $x \in \mathbb{Z}_n^*$ at random, and compute the ciphertext

$$c = g^m x^n \pmod{n^2}.$$

Define the function $L(u)$ by

$$L(u) = \frac{u-1}{n}$$

where the division is an integer division, and not a modular division, and where the domain of L is the set of all integers equal to 1 modulo n. Then the ciphertext can be decrypted by

$$m = L(c^\lambda \bmod n^2)/L(g^\lambda \bmod n^2) \pmod{n}.$$

Note that the denominator is independent of c and so can be computed at the beginning to be used in all decryptions. This cryptosystem is particularly suitable because it is strongly homomorphic: if c_1 and c_2 are the ciphertexts of plaintexts m_1 and m_2 respectively, then $c_1 c_2$ is the ciphertext of message $m_1 + m_2$. Full proofs of the correctness of this system can be found in the original paper.

Figure 14.2 describes this system.

Parameters. Two large primes p qnd q, and an integer g of multiplicative order kn modulo n^2 for some k.

Key generation. The private key is $\lambda = \text{lcm}((p-1),(q-1))$. The public key is (g, n).

Encryption. For a message $m < n$ the ciphertext is $c = g^m x^n \pmod{n^2}$ for random $x \in \mathbb{Z}_n^*$.

Decryption. $m = L(c^\lambda \bmod n^2)/L(g^\lambda \bmod n^2) \pmod{n}$ where $L(u) = (u-1)/n$.

FIGURE 14.2: Paillier cryptosystem.

A basic electronic system (that is, without paper ballots), has been described by Baudron et al. [4]. Their scheme proposes a "1-out-of-p" election, where one candidate is to be elected among p others. A brief description of this scheme follows.

The setup. Election authorities consist of a hierarchy of local, regional and national authorities whose public keys are denoted pk, pk_i and $pk_{i,j}$ respectively. There is nothing special about a three-level hierarchy; more or fewer levels would also work. If l is the number of voters, then M is the smallest power of 2 which is larger then l.

Voting. This consists of the following three steps.

Step 1. In order to cast a vote for candidate m, the voter downloads from the online voting center the three public keys $pk = (n, g)$, $pk_i = (n, g_i)$ and $pk_{i,j} = (n, g_{i,j})$ for her zone. She then encrypts the integer M^m using the Paillier system using each public key and generates three ciphertexts C_n, C_r, C_l.

This means that

$$C_n = g^{M^k} x^n \pmod{n^2}$$
$$C_r = g_i^{M^k} x^n \pmod{n^2}$$
$$C_l = g_{i,j}^{M^k} x^n \pmod{n^2}.$$

Step 2. She generates zero knowledge proofs to convince any verifier that each ciphertext encrypts a valid vote. This can be done as follows.

1. She chooses $x \in \mathbb{Z}_n$ at random and $x \in \mathbb{Z}_n^*$, and computes $u = g^x s^n \pmod{n^2}$ and commits to u.
2. The verifier chooses a challenge e which is sent to the voter.
3. The voter then computes

$$v = x - em \pmod{n}$$
$$w = sr^{-e} g^{\lfloor (x-em)/n \rfloor} \pmod{n}$$

and sends those values to the verifier.

4. The verifier checks that $g^v c^e w^N = u \pmod{n^2}$.

Step 3. She generates a proof to convince any verifier that all three ciphertexts encrypt the same vote. This can be done by using zero knowledge proofs of the equality of the plaintexts.

Tallying. This is the business of collecting all the votes together, and adding up the votes for each candidate. Again there is an online voting center where all voters can log in, and and access their votes. (The original authors speak of a "bulletin board"; anyway it will consist of some sort of shared system where each voter can read and sign her vote, but nobody else's.) The bulletin board is shown in Table 14.1.

Each voter signs the data in her row, along with her name and a time

Name	C_l	C_r	C_n
Voter 1	$g_{i,j}^{v_{i,j,1}}$	$g_i^{v_{i,j,1}}$	$g^{v_{i,j,1}}$
Voter 2	$g_{i,j}^{v_{i,j,2}}$	$g_i^{v_{i,j,2}}$	$g^{v_{i,j,2}}$
⋮			
Voter k	$g_{i,j}^{v_{i,j,k}}$	$g_i^{v_{i,j,k}}$	$g^{v_{i,j,k}}$
⋮			
Voter l	$g_{i,j}^{v_{i,j,l}}$	$g_i^{v_{i,j,l}}$	$g^{v_{i,j,l}}$

TABLE 14.1: The local authority's Bulletin Board

stamp. The proofs and signatures are verified by the authorities, and the local authority $A_{i,j}$ computes the products of all the correct votes in the three vote columns. Because the Paillier cryptosystem has the property that

$$E(m_1 + m_2) = E(m_1)E(m_2)$$

(where E is an encryption) it follows that in general

$$E\left(\sum_1^l m_i\right) = \prod_1^l E(m_i).$$

The regional authority can thus verify that the product in the C_r column decrypts to the sum of the local tallies obtained from the local authorities.

The regional authority's Bulletin Board is shown in Table 14.2.

Finally the national authority computes the product of the elements in its second column, decrypts it and verifies that the decryption is equal to the sum of elements from the first column. This first column consists of tallies obtained from the regional authorities. This is shown in Table 14.3.

Having a multi-layer system (three in this example) means that there are multiple opportunities for verification. Each voter can verify that her vote has been used in the local tally. She can also determine that the local tally has been correctly added to the regional tally, and that her vote has been added to the national tally.

This scheme is in many ways very elegant, but it is not particularly easy to administer. More recent voting systems have taken the general idea of Paillier style tallying, and added to it some sort of paper-based ballot.

Prêt à Voter

Prêt à Voter (PAV) obtains part of its security from the randomization of candidates on a ballot paper. A ballot paper consists of two parts: one part

Local Authorities	$V_{j,k}$	$\prod_{j,k} C_r$	$\prod_{j,k} C_n$
Local Auth 1	$\sum_k v_{i,1,k}$	$\prod_k g_i^{v_{i,1,k}}$	$\prod_k g^{v_{i,1,k}}$
Local Auth 2	$\sum_k v_{i,2,k}$	$\prod_k g_i^{v_{i,2,k}}$	$\prod_k g^{v_{i,2,k}}$
$\vdots$			
Local Auth j	$\sum_k v_{i,j,k}$	$\prod_k g_i^{v_{i,j,k}}$	$\prod_k g^{v_{i,j,k}}$
$\vdots$			
Local Auth l	$\sum_k v_{i,l,k}$	$\prod_k g_i^{v_{i,l,k}}$	$\prod_k g^{v_{i,l,k}}$
Sum of A_i	$\sum_{j,k} v_{i,j,k}$	$\prod_{j,k} g_i^{v_{i,j,k}}$ $\Downarrow$ decrypts $\sum_{j,k} v_{i,j,k}$	$\prod_{j,k} g^{v_{i,j,k}}$

TABLE 14.2: The regional Bulletin Board

Regional Authorities	$V_{i,j,k}$	$\prod_{i,j,k} C_n$
Regional Auth 1	$\sum_{j,k} v_{1,j,k}$	$\prod_{j,k} g_i^{v_{1,j,k}}$
Regional Auth 2	$\sum_{j,k} v_{2,j,k}$	$\prod_{j,k} g_i^{v_{2,j,k}}$
$\vdots$		
Regional Auth i	$\sum_{j,k} v_{i,j,k}$	$\prod_{j,k} g_i^{v_{i,j,k}}$
$\vdots$		
Regional Auth l	$\sum_{j,k} v_{l,j,k}$	$\prod_{j,k} g_i^{v_{l,j,k}}$
Sum of A	$\sum_{i,j,k} v_{i,j,k}$	$\prod_{i,j,k} g_i^{v_{i,j,k}}$ $\Downarrow$ decrypts $\sum_{i,j,k} v_{i,j,k}$

TABLE 14.3: The national Bulletin Board

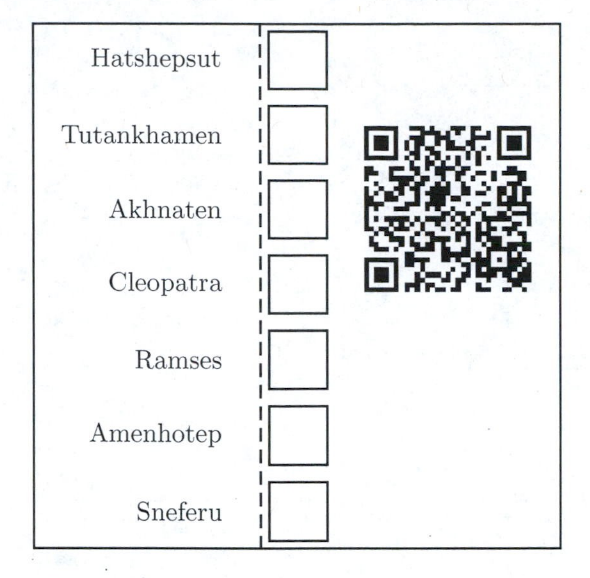

FIGURE 14.3: A Prêt à voter ballot.

has just a randomized list of all candidates; and the other part has boxes for marking the voter's choice as well as some sort of optical code which encodes the candidates' order. The two parts are joined by a perforated strip and so can be separated after the marks have been made. Figure 14.3 shows an example.

The optical code must encode the order of the candidates in such a way that no single party can obtain the order from the code. This code will contain multiply-nested encryptions, and is called an "onion."

To cast a vote, the voter receives a ballot, marks the boxes according to the type of vote (a single cross, or a list of numbers), and then separates the ballot into its two halves. The list of names must be destroyed (by shredding, for example), and the part with the code and mark can be scanned by an optical reader and its results recorded. After recording the voter keeps the ballot as a receipt. (In some versions the optical reader prints a receipt for the voter; this may contain an alphanumeric sequence which the voter can use later to verify her vote.) Voting results may be listed online for verification.

Having voted, the vote then passes through several "mixnets" which shuffle and encrypt the votes at each stage, thus ensuring no possible connection between the votes and the voters. The nature again of the Paillier cryptosystem means that multiple encryptions can be easily decrypted.

More details are given by Xia [96].

Scantegrity

Scantegrity is another system designed to provide verifiability with anonymity which uses a mixnet system to provide anonymity. It uses a different ballot design from PAV: the ballots are to be marked with a special pen, which also displays a hidden code in the mark space. Each ballot has its own set of invisible marks, and the voter can write down the revealed code to use for verification later on.

The entire system is quite complex, but once in place is supposedly easy both to administer and to run. And it has been used for at least one election. See the web page `http://www.scantegrity.org` for more information, white papers and diagrams.

14.6 Glossary

Challenge-response system. A family of protocols, of which zero knowledge proofs are one type, that involves the generation of challenges by the verifier, and responses from the prover.

Completeness, soundness, zero knowledge. Formal requirements of a zero knowledge proof.

Digital cash. Unforgeable, untraceable and anonymous digital money.

Homomorphic encryption. A cryptosystem for which if (p_i, c_i) and (p_j, c_j) are plaintext, ciphertext pairs, then $(p_i p_j, c_i, c_j)$ is another legitimate plaintext, ciphertext pair.

Mixnet. A system that shuffles and encrypts values before sending them on.

Oblivious transfer. Sending information without being able to know what was received.

Secure multi-party computation. The computation of a function of multiple variables, where no holder of any one variable has a knowledge of any other.

Zero knowledge proofs. A proof where the verifier is convinced of its proof without obtaining any knowledge of the context.

Exercises

Review Questions

1. What is the "millionaires problem," and how is it solved?

2. What is the "dining cryptographers problem," and how is it solved?

3. Why is Peggy's consistently finding square roots not a zero knowledge proof of factorization?

4. What is "double spending," and how does Brands' digital cash scheme guard against it?

5. What are the three parties required for a digital cash scheme?

6. List the requirements for a secure voting protocol.

7. In general terms, how does the use of a homomorphic cryptosystem make a secure voting scheme possible?

Beginning Exercises

8. Here is a simple multi-party computation for finding the sum of several different values (incomes for example). Alice picks a random number r, adds her income a to it and sends $r + a$ to Bob. Bob adds his income b and sends $r + a + b$ to Carol. Carol adds c and sends $r + a + b + c$ back to Alice. Alice subtracts r and the sum is obtained.

   (a) Suppose Alice cheats by sending a to Bob. He does not know that a is Alice's income, so he computes $a + b$. If Alice can obtain this value, she now knows Bob's income.
   How can this attack by Alice be prevented?

   (b) How can this protocol be adjusted to stop Alice from using r at the beginning but subtracting another value s at the end?

   (c) Describe a version of this protocol that uses encrypted transfers for extra security.

9. Show how the dining cryptographers problem can be generalized to any number.

10. A simple example that better explains zero knowledge proofs is given by Quisquater et al. [73]. Imagine a loop-shaped tunnel as shown in Figure 14.4 with one entrance and a door hidden from the entrance.

    Peggy wants to prove to Victor that she can pass through the door, without giving any information to Victor how she can do it—even which way (left to right, right to left, or both) she does it. Here is the protocol:

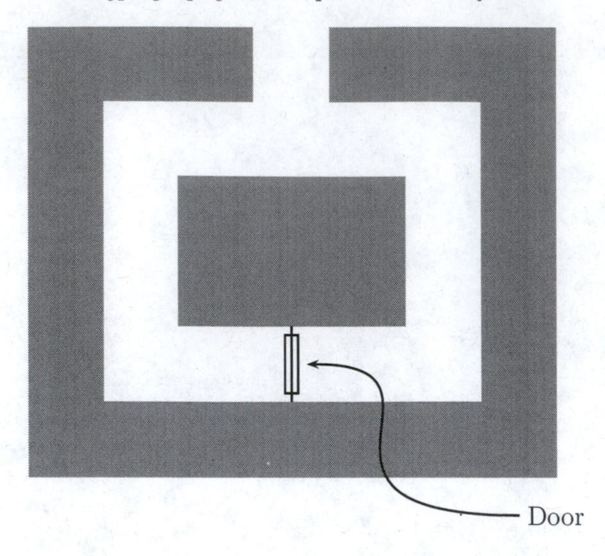

FIGURE 14.4: The tunnel for a zero knowledge proof.

(i) Peggy enters the entrance, and tosses a coin. If it shows heads she goes to the right side of the tunnel; if tails she goes to the left.

(ii) Victor stands in the entrance and also tosses a coin. If it shows heads, he requests Peggy to exit from the right hand side; if tails he asks Peggy to exit from the left.

Clearly if both coins show heads then Peggy can just disappear out of sight and then reappear. So if she is lying, she has only a 50% chance of fooling Victor.

(a) Suppose Eve is watching carefully from the entrance. Can she gain any useful information that will enable her to fool Victor later?

(b) Show that this protocol satisfies the three properties (completeness, soundness and zero knowledge) of a zero knowledge proof.

11. Here is a zero knowledge proof of a discrete logarithm. Peggy needs to keep an El Gamal key A safe, and her public key (p, a, B) is published, where $B = a^A \pmod{p}$. But Peggy may need to prove her identity to Victor, without revealing A:

    (i) She chooses at random $1 \leq r \leq p - 1$ and sends $h = a^r \pmod{p}$ to Victor.

    (ii) Victor sends back a random bit b.

    (iii) Peggy sends $s = r + bA \pmod{p-1}$ back to Victor.

    (iv) Victor computes $a^s = a^{r+bA} = hB^b \pmod{p}$.

Note that this last step is in fact a verification; that $a^s = hB^b \pmod{p}$.

(a) Show that this satisfies the three properties of a zero knowledge proof.

(b) Can Eve cheat and try to trick Victor to believe that she is Peggy?

12. One protocol for voting is given by Schneier [77]. Assume the existence of a Central Tabulating Facility (CTF) which collects and counts the votes. Here is the protocol:

    (a) Each voter encrypts his or her vote with the public key of the CTF.

    (b) The voter than sends that result to the CTF.

    (c) The CTF decrypts the votes (with its private key), counts the votes, and announces the outcome.

    Now, which of the requirements given at the beginning of Section 14.5 does this protocol satisfy?

13. Here is another protocol:

    (a) All voters sign their votes with their private keys.

    (b) All signed votes are then encrypted with the public key of the CTF.

    (c) These votes are then sent to the CTF.

    (d) The CTF decrypts the votes (with its private key), checks the signatures, counts the votes, and announces the outcome.

    Now, which of the requirements does this protocol satisfy?

14. Show that $(kn+1)^n = 1 \pmod{n^2}$. (*Hint*: Expand the left hand side using the binomial theorem.)

Sage Exercises

15. Here is a simple hash that can used in the digital cash system. It takes a list of values, turns them into a string, hashes the string with SHA256, and returns a value in $\mathbb{Z}_q$:

```
sage: from hashlib import sha256
sage: def Hash(x,q):
....:     xs = join(map(str,x),'')
....:     xh = sha256(xs).hexdigest()
....:     return mod(ZZ('0x'+xh),q)
....:
```

A time value (in seconds from January 1, 1970) can be found with:

```
sage: import time
sage: time.time()
```

(This particular time is called "UNIX time.") Use these to build a complete Sage system for digital cash.

16. Let $p = 5051$ and $q = 5987$. Let $n = pq$ and $g = 4321n + 1$.
    (a) Verify that $g^n = 1 \pmod{n^2}$.
    (b) Encrypt the message $m = 10^7$ with the Paillier cryptosystem to the ciphertext c, using $x = 12345678$.
    (c) Decrypt the result.
17. Using the parameters in the previous question, encrypt $m' = 10^6$ to the ciphertext c', and verify that cc' is the encryption of $m + m'$.
18. Suppose Peggy wants to prove to Victor that she has factorized $n = 2^{67} - 1$. Here is how Sage can be used for multiple challenges:
    (a) Peggy to Victor:

```
sage: a = [randint(floor(sqrt(n)),n) for i in range(10)]
sage: A = [mod(x,n)^2 for x in a]
```

    (b) Victor to Peggy:

```
sage: b = [randint(floor(sqrt(n)),n) for i in range(10)]
sage: B = [mod(x,n)^2 for x in b]
```

    (c) Peggy's square roots (this is where she would use her knowledge of the factorization):

```
sage: x = [sqrt(mod(u*v,n)) for u,v in zip(A,B)]
```

    (d) Victor's challenge:

```
sage: e = [randint(0,1) for i in range(10)]
```

    (e) Peggy's response:

```
sage: X=[]
sage: for i in range(10):
....:     if e[i]==0:
....:         X+=[a[i]]
```

```
....:     else:
....:         X+=[x[i]]
....:
```

(f) Victor's check:

```
sage: for i in range(10):
....:     if e[i]==0:
....:         C+=[a[i]^2==A[i]]
....:     else:
....:         C+=[x[i]^2==A[i]*B[i]]
....:
```

Work through this scenario. What is the final value of C?

19. Use a similar scheme to provide multiple instances of the discrete logarithm proof described in question 11.

Further Exercises

20. Describe a version of Baudron's voting scheme where there is no hierarchy of authorities, just the voters and one authority who tallies the votes.

    (a) What security is possible with this system?

    (b) How much verification is possible?

    (c) Set up and experiment with such a system when there are only three voters and two candidates.

Appendix A

Introduction to Sage

This appendix provides a very brief introduction to Sage: where to get it, how to install it, how to use it, and how to program with it.

A.1 Obtaining and installing Sage

Sage's home on the web is at `http://www.sagemath.org`, and there are various mirror servers. Sage can be downloaded from any of them. The installation depends on the operating system. Sage's base system is Linux, and many of the tools which make up Sage are Linux-based. This means that if you use Linux, installing Sage locally is easy, but if you run any other system it can be a little more difficult. In brief:

Linux. You can download a binary for your distribution, or you can download the source and compile it. This last can take some time on older machines, but is actually quite fool-proof, and easier than it sounds.

Windows. To run Sage under Windows, you have to fool Windows into thinking it is Linux. This can be done in several ways:

- Install a virtualization system such as VMWare or VirtualBox, and install Sage into that.
- You can install a version of Linux that runs within Windows, such as Wubi, and install Sage into that.

Full instructions can be found on the Sage web page.

MacOS. This is easier than for Windows, because of the Unix-based nature of the operating system. Sage binaries exist for OSX 10.4 onwards.

An alternative is to simply create an account at either `http://www.sagenb.org` or `http://sagenb.kaist.ac.kr/` and use Sage within your browser. This requires absolutely no setup, but is less convenient, in the long run, than having Sage running locally.

A.2 Starting with Sage

Sage can run either in a console, or in a browser. For cryptography, a console is quite sufficient, although some people may prefer the typeset output and layout of the browser-based notebook. Figure A.1 shows some RSA computations in a console.

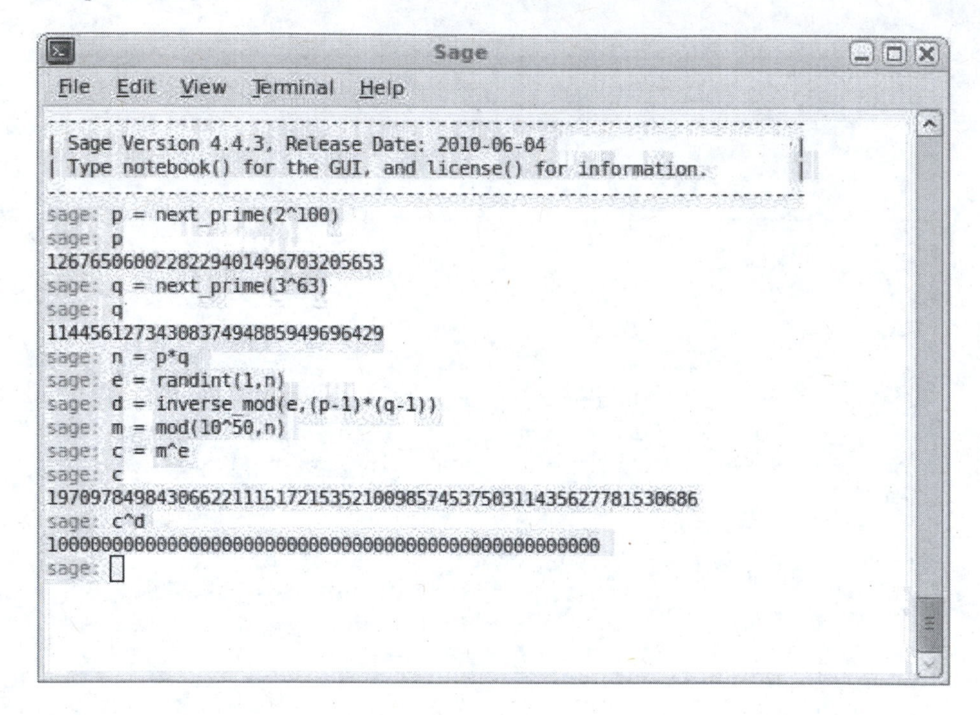

FIGURE A.1: Sage running in a console.

Figure A.2 shows similar RSA computations in the notebook.

To save space, all Sage computations will be presented in console mode.

A.3 Basic usage

Arithmetic. Sage supports arbitrary precision arithmetic, as well as real arithmetic to arbitrary precision:

```
sage: 17^100
  1108899372780783641306111715875094966436017167649879524402 7
```

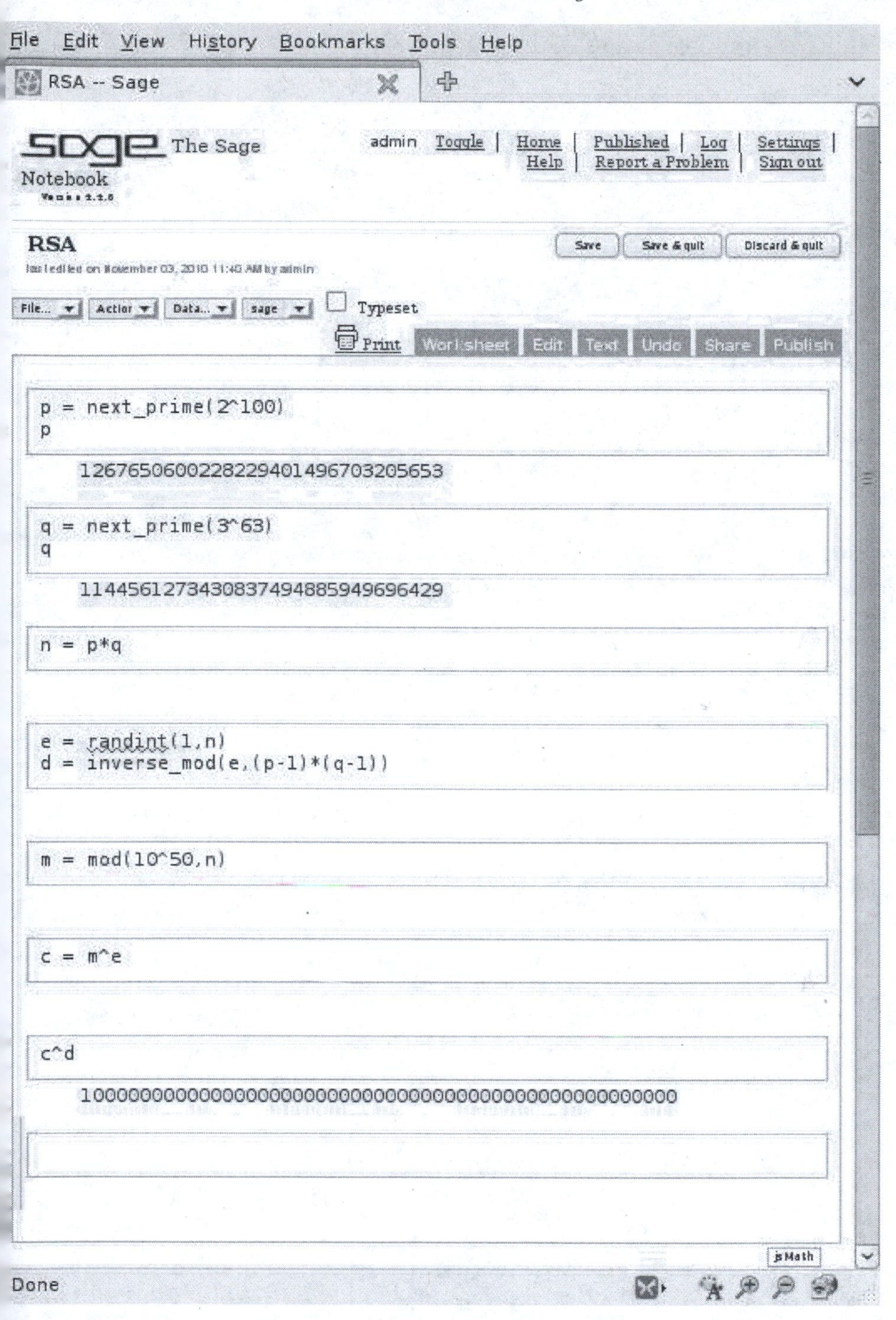

FIGURE A.2: Sage running in the notebook.

```
6984127888758050136669771242469425600509358924845150306839760
8001
sage: sqrt(2).n(digits=100)
  1.414213562373095048801688724209698078569671875376948073176
679737990732478462107038850387534327641573
```

All the standard arithmetic operations are supported, as well as many others; for example logarithms:

```
sage: log(3.0)
  1.09861228866811
sage: log(64,2)
  6
sage: log(63,2)
  log(63)/log(2)
sage: log(63.0,2)
  5.97727992349992
```

Note that a logarithm without a second value is assumed to be the natural logarithm. Also note that outputs are given in closed form unless otherwise requested: by given a floating point input, by using `n()` as above, or by the `float` command:

```
sage: float(log(2,10))
  0.30102999566398114
```

Variables. Variables can be defined either by giving a value, or introducing them with `var`:

```
sage: a = 64
sage: a^(1/3)
  4
sage: var('x')
sage: sage: var('x')
  x
sage: solve(x^2+2*x+3==0,x)
  [x == -I*sqrt(2) - 1, x == I*sqrt(2) - 1]
sage: solve
sage: solve(y^2+3*y-2==0,y)
---------------------------------------------------------------
NameError                         Traceback (most recent call last)

/home/Sage/<ipython console> in <module>()

NameError: name 'y' is not defined
```

Note that the definition of a variable (such as in the first line above) does not display the variable's value. A variable can be deleted with `del`:

```
sage: del(y)
```

This has the effect of removing the variable and all its values from the workspace. The command:

```
sage: var('y')
```

will remove any current values or definitions of the variable, but will maintain its presence.

The `parent` command will produce the variable's type:

```
sage: a = 64
sage: parent(a)
  Integer Ring
sage: b = a^(1/3)
sage: parent(b)
  Rational Field
```

Note that although `a` was considered an integer, `b` is not. However, `b` can be coerced (that is, have its type changed) to an integer as follows:

```
sage: b = ZZ(b)
sage: parent(b)
  Integer Ring
```

There are two different division operators whose output can be differentiated using `parent`: a single slash produces a rational output, and a double slash produces an integer output:

```
sage: a = 6/3; a
  2
sage: parent(a)
  Rational Field
sage: b = 6//3; b
  2
sage: parent(b)
  Integer Ring
```

Multiple variables can be defined simultaneously:

```
sage: a,b,c = 2,3,4
```

The convenient C-style "`+=`" abbreviation is supported, so that

```
i += 1
```

abbreviates

```
i = i+1
```

Ranges and lists. A range of values can be obtained with the `range` command. This has several variations:

`range(n)`:	Lists all values from 0 to $n-1$.
`range(m,n)`:	Lists all values from m to $n-1$.
`range(m,n,s)`:	Lists all values starting from m going in steps of size s to below n.

For example:

```
sage: range(5)
  [0, 1, 2, 3, 4]
sage: range(3,8)
  [3, 4, 5, 6, 7]
sage: range(3,11,2)
  [3, 5, 7, 9]
sage: range(10,1,-2)
  [10, 8, 6, 4, 2]
```

The output of a `range` command is a list. Lists can be constructed by iterating over a range, or over another list:

```
sage: [x^2 for x in range(10)]
  [0, 1, 4, 9, 16, 25, 36, 49, 64, 81]
sage: [floor(sqrt(x)) for x in range(10,100,10)]
  [3, 4, 5, 6, 7, 7, 8, 8, 9]
sage: L = [floor(x^(1/3)) for x in range(100,1000,50)]
sage: L
  [4, 5, 5, 6, 6, 7, 7, 7, 7, 8, 8, 8, 8, 9, 9, 9, 9, 9]
sage: [(x>7)*1 for x in L]
  [0, 0, 0, 0, 0, 0, 0, 0, 0, 1, 1, 1, 1, 1, 1, 1, 1, 1]
```

Note that in the last command that `(x>7)*1` gives a numeric value to a logical output: 1 for `True` and 0 for `False`.

Lists can contain any definable Sage objects, which includes other lists, so lists can be multiply nested:

```
sage: [chr(x) for x in range(80,130,5)]
  ['P', 'U', 'Z', '_', 'd', 'i', 'n', 's', 'x', '}']
sage: var('x')
sage: [x^i for i in range(1,11)]
  [x, x^2, x^3, x^4, x^5, x^6, x^7, x^8, x^9, x^10]
sage: var('i,j')
```

```
sage: [[i-j for j in range(i+1)] for i in range(5)]
  [[0], [1, 0], [2, 1, 0], [3, 2, 1, 0], [4, 3, 2, 1, 0]]
```

List elements are indexed with square brackets. For example

```
sage: P = [nth_prime(x) for x in range(1,11)]
sage: P
  [2, 3, 5, 7, 11, 13, 17, 19, 23, 29]
sage: P[0]
  2
sage: P[5]
  13
```

A negative index indicates counting indices from the right:

```
sage: P[-1]
  29
```

There are other commands that will also produce a list of primes:

```
sage: primes_first_n(10)
  [2, 3, 5, 7, 11, 13, 17, 19, 23, 29]
sage: prime_range(2,30)
  [2, 3, 5, 7, 11, 13, 17, 19, 23, 29]
```

Ranges of elements from a list can be obtained using semicolons. If `L` is a list, then:

`L[n]`:	Produces the n-th value (starting from 0).
`L[m:n]`:	Lists all values from the m-th value to the $(n-1)$-th value.
`L[m:n:s]`:	Lists all values starting at the m-th going in steps of size s to below n.

For example:

```
sage: L = [(x^2+x)//2 for x in range(1,11)]
sage: L
  [1, 3, 6, 10, 15, 21, 28, 36, 45, 55]
sage: L[0:3]
  [1, 3, 6]
sage: L[1:6:2]
  [3, 10, 21]
```

The variations `L[:n]` and `L[n:]` list the first n values, and everything but the first n values respectively:

```
sage: L[:4]
  [1, 3, 6, 10]
sage: L[4:]
  [15, 21, 28, 36, 45, 55]
```

This form of listing can also be used to reverse a list:

```
sage: L[::-1]
  [29, 23, 19, 17, 13, 11, 7, 5, 3, 2]
```

Operations on two lists together can be performed by using `zip`. This requires the lists be of equal length:

```
sage: [x+y for x,y in zip(P,L)]
  [3, 6, 11, 17, 26, 34, 45, 55, 68, 84]
```

Lists can be concatenated using `+`:

```
sage: P[:5]+L[:5]
  [2, 3, 5, 7, 11, 1, 3, 6, 10, 15]
```

Lists can also use the "`+=`" abbreviation:

```
sage: P += [31]
sage: P
  [2, 3, 5, 7, 11, 13, 17, 19, 23, 29, 31]
```

A.4 Tab completion and help

Tab completion means that if the first few characters of a command are entered, followed by the Tab key, then Sage will return all commands beginning with those characters:

```
sage: fa Tab
factor         factorization      farey          fast_float
factor_base    falling_factorial  fast_arith
factorial      false              fast_callable
```

```
sage: exp Tab
exp        experimental_packages   exponential_integral_1
exp_int    explain_pickle          expovariate
expand     expnums
```

But there is more than this. Given an object `X`, then `X.` Tab will list all methods associated with the class (or type) of that object. For example, with the list `L` from above:

```
sage: L. Tab
L.append    L.extend    L.insert    L.remove    L.sort
L.count     L.index     L.pop       L.reverse
```

Here is the same thing with some permutations:

```
sage: PM = Permutations(6)
sage: PM. Tab
PM.Element      PM.element_class PM.list           PM.rank
PM.cardinality  PM.filter        PM.map            PM.rename
PM.category     PM.first         PM.n              PM.reset_name
PM.count        PM.identity      PM.next           PM.save
PM.db           PM.is_finite     PM.previous       PM.union
PM.dump         PM.iterator      PM.random         PM.unrank
PM.dumps        PM.last          PM.random_element PM.version
```

Help is obtained with the question mark. The output should contain a description of the command, the arguments for which it is defined, an enumeration of all its parameters, and copious examples. Sometimes these are quite short (not the entire output is shown here):

```
sage: ferrers_diagram?
Docstring:
       Return the Ferrers diagram of pi.

       INPUT:

       * ``pi``- a partition, given as a list of integers.

       EXAMPLES:

          sage: print Partition([5,5,2,1]).ferrers_diagram()
          *****
          *****
          **
          *
```

Other help results can be very long—the help for `factor`, for example, runs to many screens with discussion of keywords and examples. Help is also given for methods defined on a given class:

```
sage: L.index?
```

```
Type:         builtin_function_or_method
Base Class:   <type 'builtin_function_or_method'>
String Form:  <built-in method index of list object at 0x5cd95a8>
Namespace:    Interactive
Docstring:
    L.index(value, [start, [stop]]) -> integer -- return first
    index of value. Raises ValueError if the value is not present.

Class Docstring:
    <attribute '__doc__' of 'builtin_function_or_method' objects>
```

and (displaying only the DocString):

```
sage: PM.filter?
Docstring:
     Returns the combinatorial subclass of f which consists
     of the elements x of self such that f(x) is True.

      EXAMPLES:

     sage: P = Permutations(3).filter(lambda x: x.avoids([1,2]))
     sage: P.list()
     [[3, 2, 1]]
```

A.5 Basic programming

Sage is based on Python, and thus inherits all of Python's programming syntax and style. The main interest here is with loops and branches. Python is distinguished from languages based on C in that programming blocks are not delimited by brackets, but by indentation.

Loops. The `for` statement provides basic loops:

```
for i in I:
    first statement in the loop block
    ...
    last statement in the loop block
```

Here `I` is any structure that allows iteration, such as lists. Here is an example that loops over a list of primes p, and displays the largest prime factor of $p-1$:

```
sage: for p in primes(3,30):
```

```
....:     f = [q for q,e in factor(p-1)]
....:     print p, f[-1]
....:
3 2
5 2
7 3
11 5
13 3
17 2
19 3
23 11
29 7
```

The next program lists subsets:

```
sage: for s in subsets([1,2,3]):
....:     print s
....:
[]
[1]
[2]
[1, 2]
[3]
[1, 3]
[2, 3]
[1, 2, 3]
```

The `while` loop continues looping until a particular criterion is satisfied. Here is an example: finding the smallest power of two that contains all ten digits (such a number is called *pandigital*). Start with noting that if n is any integer, then

```
set([x for x in str(n)])
```

will produce a set containing the string versions of all integers in n:

```
sage: n = 2^20
sage: set([x for x in str(n)])
  set(['1', '0', '5', '4', '7', '6', '8'])
```

Then

```
sage: alldigs = set([x for x in '0123456789'])
```

will produce a set of all digits. To obtain 10 digits, the starting power of two must be 2^{29}. Now for the loop:

```
sage: e = 28
sage: n = 2^e
sage: while true:
....:     n,e = n*2, e+1
....:     if set([x for x in str(n)]) == alldigs:
....:         print e, n
....:         break
....:
  68 295147905179352825856
```

The `break` statement simply exits the loop when the criterion is obtained.

Branching. The basic statement here is `if`—already seen in the above example—which has the following general form:

```
if (statement):
    indented
    block
    of code
else:
    another
    indented
    block
```

The `else` part is optional. Notice that for both `for` and `if` there is no need for any "end" statement, as the block is delimited by its indentation.

For example, here is how to list a few Mersenne primes, primes of the form $2^n - 1$ (which requires that n is also prime):

```
sage: for i in prime_range(100):
....:     p = 2^i-1
....:     if is_prime(p):
....:         print i,p
....:
2 3
3 7
5 31
7 127
13 8191
17 131071
19 524287
31 2147483647
61 2305843009213693951
89 618970019642690137449562111
```

Functions. A function may be considered a program that produces an output for a given input. A function has the following form:

```
def ftn(x):
    blocks of code
    implementing the
    function, all
    appropriately
    indented
    ...
    return output
```

The `def` and `return` are reserved words, and should not be used for any other purpose. The value of `output` is generated by the function, and this is the value that is returned.

A.6 A programming example

This section will, as an example, explore the "$3x + 1$" or Collatz problem. Starting with any integer $n = n_0$, recursively generate $n_1, n_2, \ldots$ with:

$$n_k = \begin{cases} n_{k-1}/2 & \text{if } n_{k-1} \text{ is even,} \\ 3n_{k-1} + 1 & \text{if } n_{k-1} \text{ is odd.} \end{cases}$$

The problem (as yet unsolved) is to show that no matter what starting value, the sequence will always eventually end up as $4, 2, 1, 4, 2, 1, \ldots$ or alternatively, that there will always be a k for which $n_k = 1$.

To implement this, a function will produce, as output, the sequence n_k, stopping whenever a 1 is reached:

```
def collatz(n):
    seq = [n]
    s = seq[-1]
    while s != 1:
        if s%2 == 0:
            t = s//2
        else:
            t = 3*s+1
        seq += [t]
        s = t
    return seq
```

Note the use of a `while` loop, and the layers of indentation. This program can be stored in a file called (say) `collatz.sage` and then read into Sage with

```
sage: load collatz.sage
```

Now it can be used:

```
sage: collatz(17)
  [17, 52, 26, 13, 40, 20, 10, 5, 16, 8, 4, 2, 1]
sage: collatz(27)
  [27, 82, 41, 124, 62, 31, 94, 47, 142, 71, 214, 107, 322, 161,
   484, 242, 121, 364, 182, 91, 274, 137, 412, 206, 103, 310,
   155, 466, 233, 700, 350, 175, 526, 263, 790, 395, 1186, 593,
   1780, 890, 445, 1336, 668, 334, 167, 502, 251, 754, 377, 1132,
   566, 283, 850, 425, 1276, 638, 319, 958, 479, 1438, 719, 2158,
   1079, 3238, 1619, 4858, 2429, 7288, 3644, 1822, 911, 2734,
   1367, 4102, 2051, 6154, 3077, 9232, 4616, 2308, 1154, 577,
   1732, 866, 433, 1300, 650, 325, 976, 488, 244, 122, 61, 184,
   92, 46, 23, 70, 35, 106, 53, 160, 80, 40, 20, 10, 5, 16, 8,
   4, 2, 1]
```

Try `collatz(837799)` and see what happens!

Exercises

1. Find the smallest integer n for which

$$\sum_{k=1}^{n} \frac{1}{k} > 10.$$

2. Explore the commands `factor`, `binomial`, `fibonacci`.

3. Look up the command "`prime_range`" and find out what it does. Also look up the command "`len`." Use these two commands to find the number of

   (a) Three digit prime numbers.

   (b) Four digit prime numbers.

   (c) Five digit prime numbers.

   (d) The number of primes less than one million.

4. Look up the command "`prime_pi`" and use it to find

   (a) The number of primes less than one million.

(b) The number of primes between one million and one billion (10^9).

5. What happens if you try to use the method of question 3 to answer question 4(b)?

6. Create the list of all values $n < 1000$ for which $3^n + 2$ is prime.

7. Find the smallest factorial that contains all ten digits.

8. Using the `collatz` function, create a list of Collatz "records," numbers for which the length of the Collatz list is the longest so far.

9. Look up the command `randint`, and see if you can find a prime number that contains all ten digits.

10. *Twin primes* are those that differ by 2, such as 17 and 19, or 41 and 43.

    (a) List all twin primes beneath 100.

    (b) Find the number of twin prime pairs less than (i) 1000, (ii) 1000000.

11. *Sexy primes* (so called after the Latin "sex" for six) are pairs of primes that differ by 6, and a *sexy triplet* consists of three primes p, q, r with a common difference of 6, such as 11, 17, 23.

    Find the number of sexy triplets (a) less than 1000, (b) less than 1000000.

12. Find the longest arithmetic progression of primes where all primes are (a) less than 200, (b) less than 1000, (c) less than 3000.

Appendix B

Advanced computational number theory

This appendix provides an introduction to some more advanced number theoretical algorithms. In particular:

- The quadratic sieve method for factorization.
- The AKS method of primality testing.
- The Pollard rho and the Pohlig–Hellman methods for discrete logarithms.

B.1 The quadratic sieve

The quadratic sieve is a relatively recent algorithm, which can be used to factor numbers up to about 100 digits. After that a more sophisticated algorithm—the "number field sieve"—which is beyond the scope of this text, needs to be used. The idea is to find congruences of the form

$$x^2 = y^2 \pmod{n} \tag{B.1}$$

and then testing to see if either of $\gcd(x+y,n)$ or $\gcd(x-y,n)$ are proper factors. To find such congruences, a small set of primes $B = \{p_1, p_2, \ldots p_k\}$ called a "factor base" is chosen. Then the aim is to find numbers close to n whose remainder modulo n can be factored entirely over the factor base. That is, congruences of the form

$$w = p_1^{a_1} \cdot p_2^{a_2} \cdots p_k^{a_k} \pmod{n}$$

where $\{p_1, p_2, \ldots, p_k\}$ is the factor base.

This can be done by setting

$$s = \lfloor \sqrt{n} \rfloor$$

and factoring $w = (i+s)^2 - n$ for increasing values of i. A quadratic congruence can then be found by multiplying appropriate values. To see how this works with a small example, take $n = 91489$, and the factor base $B = \{2, 3, 5\}$. Here

$$s = \lfloor \sqrt{n} \rfloor = 302.$$

Using trial and error, the following congruences can be found:

$$\begin{aligned}
(1+s)^2 &= 2^6 \cdot 5 \pmod{n} \\
(3+s)^2 &= 2^9 \cdot 3 \pmod{n} \\
(6+s)^2 &= 3^3 \cdot 5^3 \pmod{n} \\
(11+s)^2 &= 2^4 \cdot 3^4 \cdot 5 \pmod{n} \\
(15+s)^2 &= 2^3 \cdot 3^2 \cdot 5^3 \pmod{n}.
\end{aligned}$$

If the first and fourth congruences are multiplied, then

$$(1+s)^2(11+s)^2 = 2^{10} \cdot 3^4 \cdot 5^2 \pmod{n}.$$

Taking square roots of both sides of this congruence produces

$$\begin{aligned}
x &= (1+s)(11+s) \\
&= 3350 \pmod{n} \\
y &= 2^5 \cdot 3^2 \cdot 5 \\
&= 1440 \pmod{n}.
\end{aligned}$$

Then

$$\begin{aligned}
\gcd(x-y, n) &= 191 \\
\gcd(x+y, n) &= 479
\end{aligned}$$

and the factors of n have been found.

To find the quadratic congruence from the set of congruences

$$(i+s)^2 = p_1^{a_1} \cdot p_2^{a_2} \cdots p_k^{a_k} \pmod{n}$$

it is convenient to invoke some linear algebra. Consider the above list of congruences. The exponents of the primes in the factor base can be placed in an array indexed by the factor base and the values of i as follows:

	2	3	5
1	6	0	1
3	9	1	0
6	0	3	3
11	4	4	1
15	3	2	3

and a quadratic congruence can be obtained by finding rows whose sum contains only even numbers. To simplify this, all arithmetic can be done modulo

2, and rows found whose sum is zero. The array of exponents modulo 2 is then

$$\begin{bmatrix} 0 & 0 & 1 \\ 1 & 1 & 0 \\ 0 & 1 & 1 \\ 0 & 0 & 1 \\ 1 & 0 & 1 \end{bmatrix}$$

and the nullspace of this matrix will produce the required rows. In Sage

```
sage: A=matrix(GF(2),[[0,0,1],[1,1,0],[0,1,1],[0,0,1],[1,0,1]])
sage: A.kernel().basis_matrix()

  [1 0 0 1 0]
  [0 1 1 0 1]
```

This shows that there are two possible sums: rows 1 and 4, or rows 2, 3 and 5.

A number whose prime factors are all contained in the factor base B is said to *split* over B. This can be tested in Sage with a little function:

```
def splits(n,B):
   f = list(factor(n))
   f1 = set(a[0] for a in f)
   return f1.issubset(B)
```

For the number $n = 814159$ to be factored, with $s = 902$, first enter a factor base, and then test values for splitting:

```
sage: n=814159;s=int(sqrt(n))
sage: B=set([2,3,5,11,19])
sage: for i in list(1..70):
    if splits((i+s)^2-n,B):
        print i
....:
1
15
26
40
51
70
```

Factoring each of $(i + s)^2 - n$ for the values of i just produced, leads to the following array:

	2	3	5	11	19
1	1	0	4	0	0
15	1	5	1	1	0
26	0	2	2	1	1
40	0	0	1	4	0
51	1	2	2	1	1
70	0	0	4	1	1

which can be entered into Sage:

```
sage: A=matrix(GF(2),[[1,0,4,0,0],[1,5,1,1,0],[0,2,2,1,1],\
[0,0,1,4,0],[1,2,2,1,1],[0,0,4,1,1]]); A

  [1 0 0 0 0]
  [1 1 1 1 0]
  [0 0 0 1 1]
  [0 0 1 0 0]
  [1 0 0 1 1]
  [0 0 0 1 1]

sage: A.kernel().basis_matrix()

  [1 0 0 0 1 1]
  [0 0 1 0 0 1]
```

If (from the first row here) rows 1, 5 and 6 are used, the congruence is

$$(1+s)^2(51+s)^2(70+s)^2 = 2^2 \cdot 3^2 \cdot 5^{10} \cdot 11^2 \cdot 19^2 \pmod{n}$$

and so, taking square roots of both sides:

```
sage: x = (1+s)*(51+s)*(70+s)%n; x
  322055
sage: y = 2*3*5^5*11*19%n; y
  662114
sage: gcd(x-y,n)
  431
sage: gcd(x+y,n)
  1889
```

and n has been factored.

B.2 The AKS primality test

This test, published by computer scientists Manindra Agrawal, Neeraj Kayal, and Nitin Saxena in 2004 [1], has the advantage of being fully deterministic, and running in polynomial time. Although it is not as fast as the Miller–Rabin test, and so is still not suitable for the testing of very large primes, it is a very exciting development in the field of primality testing. Its starting point is the result that all the binomial coefficients

$$\binom{n}{k}, k = 0, 1, 2, \ldots n$$

are divisible by n if and only if n is prime. This means that if $\gcd(a, n) = 1$, then

$$(X + a)^n = X^n + a \pmod{n}.$$

For example:

```
sage: R.<X> = Zmod(13)[]
sage: (X+2)^13
  X^13 + 2
sage: (X+5)^13
  X^13 + 5
```

but for a composite n:

```
sage: R.<X> = Zmod(14)[]
sage: (X+2)^14
  X^14 + 4*X^7 + 4
```

By itself, this observation does not produce a very efficient test, because in the worst case n coefficients need to be examined. The insight of Agrawal, Kayal and Saxena was to reduce the polynomial by a further polynomial $X^r - 1$ for some r:

$$(X + a)^n = X^n + a \bmod (X^r - 1, n).$$

This equation is satisfied for all primes n and for any $a < n$, and also for some composites. They showed that if this equation was satisfied by all a below a certain limit, then n must be a prime power.

Using the notation

$$\mathrm{ord}_r n$$

for the smallest power k for which

$$n^k = 1 \pmod{r}$$

the AKS algorithm is as follows (with logarithms taken to base 2):

1. If $n = a^b$ for some $a, b > 1$, then output COMPOSITE and stop.
2. Find the smallest r such that $\text{ord}_r n > \log^2 n$.
3. If $1 < \gcd(a, n) < r$ for some $a \leq r$, then output COMPOSITE and stop.
4. If $n \leq r$, then output PRIME and stop.
5. For a from 1 to $\lfloor \sqrt{\phi(r)} \log n \rfloor$ do:

   if $(X + a)^n \neq X^n + a \bmod (X^r - 1, n)$,
   then output COMPOSITE and stop.

6. Output PRIME.

Here is an example in Sage, with $n = 4673$. The first job is to check that n is not a perfect power:

```
sage: n = 4673
sage: n.is_perfect_power()
  False
```

The next job is to find r. First compute $\log^2 n$:

```
sage: log2 = (log(n,2)^2).n()
sage: log2
  148.599350404839
```

and then find r using a loop, starting at $\lfloor \log^2 n \rfloor$ and working up:

```
sage: r = floor(log2)
sage: while True:
....:      r = r+1
....:      if Mod(n,r).multiplicative_order() > log2:
....:          break
sage: r
  163
```

Now to check all the gcds, as per step 3:

```
sage: for a in list(1..r):
....:      if gcd(a,n)>1 & gcd(a,n)<r:
....:          print a
```

This produces nothing, so n has not yet been shown to be composite. Step 4 is relevant only for $n < 5,690,034$, and in this example $n > r$.

For step 5, the upper limit can be calculated:

```
sage: upper = floor(sqrt(euler_phi(r))*log(n,2))
sage: upper
  155
```

and then the loop:

```
sage: R.<X> = Zmod(n)[]
sage: for a in list(1..upper):
....:     if ((X+a)^n-X^n-a).quo_rem(X^r-1)[1]<>0:
....:         print a
```

There is no output, so n can be classified as definitely prime.

The version described here is from the original paper; since that time several improvements have been suggested, that speed up the algorithm.

B.3 Methods of computing discrete logarithms

Recall from Chapter 6 that the baby step, giant step method for computing $\log_a b \pmod{p}$ has running time approximately $\sqrt{p}$. However, if p is a prime for which $p-1$ has small factors, so that

$$p - 1 = q_1^{a_1} q_2^{a_2} \dots q_k^{a_k}$$

where each q_i is a small prime, then it is faster to determine the values of

$$\log_a b \pmod{q_i^{a_i}}$$

for each q_i, and combine the results with the Chinese remainder theorem. This is the basis for the Pohlig–Hellman algorithm [71] (sometimes also called the Silver-Pohlig–Hellman algorithm).

To introduce the algorithm, consider first the special case where each $a_i = 1$. For example, take the prime $p = 2311$ for which

$$p - 1 = 2 \cdot 3 \cdot 5 \cdot 7 \cdot 11.$$

Then, for example, the problem

$$\log_3 2295 \pmod{p}$$

requires finding x for which

$$3^x = 2295 \pmod{p}.$$

Consider the factor $q = 5$ of $p - 1$. From the previous equation, it follows that

$$(3^x)^{(p-1)/5} = 2295^{(p-1)/5} \pmod{p}.$$

Suppose that $x = 5m + y$. Then from the left hand side of the previous equation

$$\begin{aligned}(3^x)^{(p-1)/5} &= (3^{5m+y})^{(p-1)/5} \\ &= 3^{p-1}3^{y(p-1)/5} \\ &= 3^{y(p-1)/5}\end{aligned}$$

where the last equality follows from Fermat's theorem. In order to find y then, test all values $y = 0, 1, 2, 3, 4$ until a value is found that is equal to $2295^{(p-1)/5} \pmod{p}$.

A quick computation reveals that $y = 3$:

```
sage: A = mod(2295,p)^((p-1)/5)
sage: A
  2006
sage: [mod(3,p)^(y*(p-1)/5) for y in range(5)]
  [1, 585, 197, 2006, 1833]
```

This can be easily implemented with the `index` method:

```
sage: [mod(3,p)^(x*(p-1)/5) for x in range(5)].index(A)
  3
```

Since by definition $y = x \pmod{5}$ this result shows that $x = 3 \pmod{5}$. By the same means

$$\begin{aligned} x &= 1 \pmod{2} \\ x &= 2 \pmod{3} \\ x &= 6 \pmod{7} \\ x &= 7 \pmod{11}. \end{aligned}$$

Using the Chinese remainder theorem:

```
sage: crt([1,2,3,6,7],[2,3,5,7,11])
  293
```

produces $x = 293$ as the solution to the original logarithm problem.

This approach can be easily extended to exponents $a_i > 1$. To introduce the general method, consider first the exponent $a_i = 2$. Suppose the problem is to determine

$$x = \log_6 504858 \pmod{p}$$

where $p = 5336101$ and

$$p - 1 = 2^2 \cdot 3^2 \cdot 5^2 \cdot 7^2 \cdot 11^2.$$

As above, if $y = x \pmod{5^2}$, then $y = y_0 + 5y_1$, or $x = 25m + 5y_1 + y_0$ for some m, and y_0 can be found as before by enumerating all possible values until the equation:

$$6^{y_0(p-1)/5} = 504858^{(p-1)/5)}$$

is satisfied.

```
sage: p = 5336101
sage: a  = mod(6,p)
sage: A = 504858^((p-1)/5)
sage: [a^(y*(p-1)/5) for y in range(5)].index(A)
  3
```

and so $y_0 = 3$.

To find y_1, start with

$$6^{25m+5y_1+y_0} = 504858 \pmod{p}$$

from which

$$6^{25+5y_1} = 504858 \cdot 6^{-y_0} \pmod{p}.$$

Raising each side to the power of $(p-1)/25$ and using Fermat's theorem produces

$$6^{y_1(p-1)/5} = (504858 \cdot 6^{-y_0})^{(p-1)/25}.$$

Since y_0 is known, the right hand side can be computed:

```
sage: B = (504858*a^(-3))^((p-1)/25); B
  2865518
```

and now y_1 can again be found by trial and error:

```
sage: [a^(y*(p-1)/5) for y in range(5)].index(B)
  2
```

This means that

$$\begin{aligned} x &= y_0 + 5y_1 \pmod{25} \\ &= 13 \pmod{25}. \end{aligned}$$

Similarly:

$$\begin{aligned} x &= 1 \pmod{2^2} \\ x &= 5 \pmod{3^2} \\ x &= 31 \pmod{7^2} \\ x &= 101 \pmod{11^2} \end{aligned}$$

leading to the solution:

```
sage: crt([1,5,13,31,101],[4,9,25,49,121])
  4401113
```

or $x = 4401113$ as the required logarithm.

In general, suppose the problem is to determine

$$x = \log_a A \pmod{p}$$

where

$$p - 1 = q_1^{a_1} q_2^{a_2} \dots q_k^{a_k}.$$

To compute $x \pmod{q_i^{a_i}}$ proceed as follows. The explanation will be simplified by writing $q = q_i$ and $a = a_i$

$$x \pmod{q^a} = y = y_{a-1}q^{a-1} + \cdots + y_2q^2 + y_1q + y_0.$$

Then y_0 is first found by solving

$$a^{y_0(p-1)/q} = A^{(p-1)/q} \pmod{p}.$$

This is done by enumerating the left hand side for all values $y_0 = 0, 1, 2, \dots, q-1$ until the right hand side is obtained. This shows also why $p - 1$ is required to have small factors, for if any of the prime factors q were large, this method of enumeration would be unworkable.

Next, y_1 is found by solving

$$a^{y_1(p-1)/q} = (Aa^{-y_0})^{(p-1)/q^2} \pmod{p}.$$

This can be simplified by first computing

$$A_1 = Aa^{-y_0} \pmod{p}$$

and then solving

$$a^{y_1(p-1)/q} = (A_0)^{(p-1)/q^2} \pmod{p}.$$

Similarly, writing

$$A_2 = A_1a^{-y_1q}$$

then y_2 can be found by solving

$$a^{y_2(p-1)/q} = (A_2)^{(p-1)/q^3} \pmod{p}.$$

And so on. In general, if $A_0 = A$, then the values A_k and y_k are determined alternatively by

$$A_{k+1} = A_k a^{-y_k q^k}$$

and

$$a^{y_k(p-1)/q} = (A_k)^{(p-1)/q^{k+1}} \pmod{p}.$$

Further details can be found in Yan [97] or Crandall and Pomerance [23].

There is also a Pollard rho method, similar in style to the method for factoring. To find

$$\log_a y \pmod{n}$$

where n is not necessarily prime, the method generates pseudo-random triples $[x, r, s]$ for which

$$x = a^r y^s \pmod{n}$$

until two values are found that have equal x values:

$$a^{r_i} y^{s_i} = a^{r_j} y^{s_j} \pmod{n}.$$

In that case the logarithm is

$$(r_i - r_j)/(s_i - s_j) \pmod{n-1}$$

assuming that the inverse exists. To find equal values, the following "random" function:

$$\psi(x, a, y) = \begin{cases} (x^2 \bmod n, 2r \bmod n-1, 2s \bmod n-1) & \text{if } x = 0 \pmod{3} \\ (ax \bmod n, r+1 \bmod n-1, s \bmod n-1) & \text{if } x = 1 \pmod{3} \\ (yx \bmod n, r \bmod n-1, s+1 \bmod n-1) & \text{if } x = 2 \pmod{3} \end{cases}$$

is used. This function is devised so that if

$$a^r y^s = x \pmod{n}$$

and if $(X, R, S) = \psi(x, r, s)$ then

$$a^R y^S = X \pmod{n}.$$

Start with

$$X_0 = [1, 0, 0]$$

and

$$Y_0 = X_0.$$

Then using the function ψ above compute

$$X_k = \psi(X_{k-1})$$

and

$$Y_k = \psi(\psi(Y_{k-1}))$$

until k is found for which the first values of X_k and Y_k are equal.

As an experiment, consider the problem

$$x = \log_7 23 \pmod{71}$$

the values are $a = 7$, $y = 23$ and $n = 71$. First enter the values and set up the function ψ:

```
sage: a,y,n = 7,12,71
sage: def psi(X):
....:     x,r,s=X[0],X[1],X[2]
....:     if x%3==0:
....:         return [(x^2)%n,(2*r)%(n-1),(2*s)%(n-1)]
....:     if x%3==1:
....:         return [(a*x)%n,(r+1)%(n-1),s]
....:     if x%3==2:
....:         return [(y*x)%n,r,(s+1)%(n-1)]
```

And next calculate values of X and Y:

```
sage: for i in range(1,11):
    X = psi(X); Y = psi(psi(Y))
    print i,X,Y,X[0]==Y[0]
....:
1 [7, 1, 0] [49, 2, 0] False
2 [49, 2, 0] [8, 3, 1] False
3 [59, 3, 0] [60, 6, 4] False
4 [8, 3, 1] [14, 12, 9] False
5 [42, 3, 2] [22, 12, 11] False
6 [60, 6, 4] [2, 26, 22] False
7 [50, 12, 8] [38, 27, 23] False
8 [14, 12, 9] [12, 28, 24] False
9 [38, 12, 10] [46, 56, 49] False
10 [22, 12, 11] [22, 57, 50] True
```

The values in the last row should satisfy

$$a^{12}y^{11} = 22 \pmod{71}$$

and

$$a^{57}y^{50} = 22 \pmod{71}.$$

Since

$$a^{12}y^{11} = a^{57}y^{50} \pmod{71}$$

it follows that

$$a^{12-57} = y^{50-11} \pmod{71}$$

and so

$$x = (12 - 57)/(50 - 11) \pmod{70}$$

by Fermat's theorem.

```
sage: (12-57)/mod(50-11,70)
  15
```

This is the required logarithm, as can be quickly checked:

```
sage: mod(7,71)^15
  23
```

Again, Crandall and Pomerance [23] provide full details.

Exercises

1. Use the quadratic sieve with a factor base

   $$B = (2, 3, 5, 7, 11, 13, 17, 19)$$

   to factor $n = 2167069$. Use values of i up to 15 to find splits.

2. Use the quadratic sieve with a factor base

   $$B = (2, 3, 5, 7, 11, 13, 17, 19, 23, 29)$$

   to factor $n = 22201423$. Use values of i up to 10.

3. Using as a factor base all primes less than 40, factor $n = 961562969$ with values of i up to 70.

4. Using as a factor base all primes less than 100, factor $n = 1491089161$ with values of i up to 170.

5. Using as a factor base all primes less than 200, factor $n = 55736054986393$ with values of i up to 130.

6. Using as a factor base all primes less than 200, factor $n = 77436508049557$ with values of i up to 180.

7. Use the AKS algorithm to show that the following numbers are prime:

   (i) 14321, (ii) 69761, (iii) 96893.

8. Use the Pohlig–Hellman algorithm to evaluate the following discrete logarithms, where in each case $p-1$ is a product of unique small primes (that is, $p-1$ is square-free):

   (a) $\log_2 2275 \pmod{3571}$,
   (b) $\log_2 2433 \pmod{9283}$,
   (c) $\log_3 1705 \pmod{2731}$,
   (d) $\log_7 21228 \pmod{51871}$,
   (e) $\log_{18} 411435 \pmod{461891}$.

   Compare your answers with the results of the `discrete_log` command.

9. Use the Pohlig–Hellman algorithm to evaluate the following discrete logarithms, where in each case $p-1$ is a product of powers of small primes:

   (a) $\log_3 140794 \pmod{404251}$,
   (b) $\log_{17} 113881 \pmod{159721}$,
   (c) $\log_3 25892 \pmod{28901}$,
   (d) $\log_5 846573 \pmod{3516553}$.

10. Use Pollard's rho method to compute the discrete logarithms in questions 8 and 9.

11. Use either method to compute the discrete logarithms given in the multiplicative Merkle–Hellman knapsack example given on page 134.

Bibliography

[1] Mahindra Agrawal, Neeraj Kayal, and Nitin Saxena. PRIMES is in P. *Annals of Mathematics*, 160(2):781–793, 2004.

[2] Jean-Philippe Aumasson. On the pseudo-random generator ISAAC, 2006. `http://eprint.iacr.org/2006/438.pdf`.

[3] Steve Babbage et al. The eSTREAM portfolio, 2008. `http://www.ecrypt.eu.org/stream/portfolio.pdf`.

[4] Olivier Baudron, Pierre-Alain Fouque, David Pointcheval, Guillaume Poupard, and Jacques Stern. Practical multi-candidate election system. In *Proceedings of the ACM Symposium on Principles of Distributed Computing*, pages 274–283. ACM Press, 2001.

[5] Jem Berkes and David Hopwood. RFC Errata: RFC1319, "The MD2 Message-Digest Algorithm", April 1992, 2002. Available at `http://www.ietf.org/rfc/rfc1319.txt`.

[6] Ian F. Blake, Gadiel Seroussi, Nigel P. Smart, and N. J. Hitchin. *Advances in Elliptic Curve Cryptography.* Cambridge, 2005.

[7] Daniel Bleichenbacher. Generating ElGamal signatures without knowing the secret key. In *EUROCRYPT: Advances in Cryptology: Proceedings of EUROCRYPT*, pages 10–18, 1996.

[8] Lenore Blum, Manuel Blum, and Mike Shub. A simple unpredictable pseudo-random number generator. *SIAM J. Comput.*, 15(2):364–383, 1986.

[9] Manuel Blum and Shafi Goldwasser. An efficient probabilistic public key encryption scheme which hides all partial information. In G. R. Blakley and D. C. Chaum, editors, *Proc. CRYPTO 84*, volume 196 of *Lecture Notes in Computer Science*, pages 289–299. Springer, 1985.

[10] Manuel Blum and Silvio Micali. How to generate cryptographically strong sequences of pseudo-random bits. *SIAM J. Comput.*, 13(4):850–864, 1984.

[11] Dan Boneh. Twenty years of attacks on the RSA cryptosystem. *Notices of the AMS*, 46:203–213, 1999.

[12] Daniel Boneh and Matthew Franklin. Identity-based encryption from the Weil pairing. In Joe Kilian, editor, *Advances in Cryptology–CRYPTO '01*, volume 1514 of *Lecture Notes in Computer Science*, pages 213–229. Springer, 2001.

[13] Stefan Brands. Untraceable off-line cash in wallet with observers (extended abstract). In Douglas R. Stinson, editor, *Advances in Cryptology—CRYPTO '93*, volume 773 of *Lecture Notes in Computer Science*, pages 302–318. Springer-Verlag, 1993.

[14] Johannes Buchmann. *Introduction to Cryptography.* Springer, 2002.

[15] David Chaum. The dining cryptographers problem: Unconditional sender and recipient untraceability. *J. Cryptology*, 1(1):65–75, 1988.

[16] David Chaum, Richard T. Carback, Jeremy Clark, Aleksander Essex, Stefan Popoveniuc, Ronald L. Rivest, Peter Y. A. Ryan, Emily Huei-Yi Shen, Alan T. Sherman, and Poorvi L. Vora. Scantegrity II: End-to-end verifiability by voters of optical scan elections through confirmation codes. *IEEE Transactions on Information Forensics and Security*, 4(4):611–627, 2009.

[17] David Chaum, Eugène van Heijst, and Birgit Pfitzmann. Cryptographically strong undeniable signatures, unconditionally secure for the signer. In Joan Feigenbaum, editor, *Proceedings of Crypto '91*, volume 576 of *Lecture Notes in Computer Science*, pages 470–484. Springer, 1991.

[18] Benny Chor and Ronald L. Rivest. A knapsack-type public key cryptosystem based on arithmetic in finite fields. *IEEE Transactions on Information Theory*, 34(5):901–909, 1988.

[19] C. Cid, S. Murphy, and M. Robshaw. Small scale variants of the AES. In *Proceedings of Fast Software Encryption*, volume 3557 of *LNCS*. Springer, 2005. Available at `http://www.isg.rhul.ac.uk/~sean/smallAES-fse05.pdf`.

[20] John H. Conway. *On Numbers and Games.* Academic Press, New York, 1976.

[21] Don Coppersmith. The Data Encryption Standard (DES) and its strength against attacks. *IBM Journal of Research and Development*, 38(3):243–250, May 1994.

[22] Don Coppersmith, Hugo Krawczyk, and Yishay Mansour. The shrinking generator. In Douglas R. Stinson, editor, *Proceedings of Advances in Cryptology—CRYPTO '93, 13th Annual International Cryptology Conference*, volume 773 of *Lecture Notes in Computer Science*, pages 22–39, 1993.

[23] Richard Crandall and Carl Pomerance. *Prime Numbers: A Computational Perspective.* Springer, 2005.

[24] Shimon Even, Oded Goldreich, and Abraham Lempel. A randomized protocol for signing contracts. *Communications of the ACM*, 28(6):637–647, 1985.

[25] Uriel Feige, Amos Fiat, and Adi Shamir. Zero knowledge proofs of identity. In *Proceedings of the Nineteenth Annual ACM Symposium on Theory of Computing*, pages 210–217. ACM, 1987.

[26] Michael Fellows and Neal Koblitz. Kid krypto. In Ernest Brickell, editor, *Advances in Cryptology CRYPTO 92*, volume 740 of *Lecture Notes in Computer Science*, pages 371–389. Springer, 1993.

[27] Niels Ferguson and Bruce Schneier. *Practical Cryptography.* John Wiley & Sons, 2003.

[28] Amos Fiat and Adi Shamir. How to prove yourself: Practical solutions to identification and signature problems. In Andrew M. Odlyzko, editor, *Advances in Cryptology—CRYPTO '86, Santa Barbara, California, USA*, volume 263 of *Lecture Notes in Computer Science*, pages 186–194, 1986.

[29] Helen Fouché Gaines. *Cryptanalysis: A study of ciphers and their solution.* Dover Publications, Inc., 1944.

[30] Praveen Gauvaram, Lars Knudsen, Krystian Matusiewicz, Florian Mendel, Christan Rechberger, Martin Schläffer, and Søren Thomsen. Grøstl—a sha-3 candidate. *Technical Report*, 2008. Available at `http://www.groestl.info/Groestl.pdf`.

[31] Praveen Gauvaram, Lars Knudsen, Krystian Matusiewicz, Florian Mendel, Christan Rechberger, Martin Schläffer, and Søren Thomsen. Grøstl addendum. *Technical Report*, 2009. Available at `http://www.groestl.info/Groestl-addendum.pdf`.

[32] J. K. Gibson. Discrete logarithm hash function that is collision free and one way. *IEE Proceedings-E*, pages 407–410, 1991. No. 138.

[33] John Gilmore. *Cracking DES: Secrets of Encryption Research, Wiretap Politics & Chip Design.* Electronic Frontier Foundation, San Francisco, 1998.

[34] Oded Goldreich. *Foundations of Cryptography: Basic Tools.* Cambridge Univerrity Press, 2004.

[35] Darrel Hankerson, Alfred Menezes, and Scott Vanstone. *Guide to Elliptic Curve Cryptography.* Springer, 2004.

[36] Lester S. Hill. Cryptography in an algebraic alphabet. *American Mathematical Monthly*, 36(6):306–312, 1929.

[37] Lester S. Hill. Concerning certain linear transformation apparatus of cryptography. *American Mathematical Monthly*, 38(3):135–154, 1931.

[38] Jeffrey Hoffstein, Jill Pipher, and Joseph H. Silverman. *An Introduction to Mathematical Cryptography.* Springer, 2008.

[39] Thomas R. Hungerford. *Abstract Algebra: An Introduction.* Brooks Cole, 2nd edition, 1996.

[40] Internet Policy Institute (IPI). Report of the National Workshop on Internet Voting: Issues and research agenda. *Technical Report*, National Science Foundation (NSF), March 2001.

[41] Robert J. Jenkins. ISAAC. In Dieter Gollman, editor, *Fast Software Encryption: 3rd International Workshop*, volume 1039 of *Lecture Notes in Computer Science*, pages 41–50. Springer, 1996.

[42] Antoine Joux. A one round protocol for tripartite Diffie-Hellman. In Wieb Bosma, editor, *Proceedings of the ANTS-IV Conference*, volume 1838 of *Lecture Notes in Computer Science*, pages 385–394, 2000.

[43] D. Kahn. *The Codebreakers.* Weidenfeld and Nicolson, 1974.

[44] Burton Kaliski, Ronald Rivest, and Alan Sherman. Is the Data Encryption Standard a group? *Journal of Cryptology*, 1:3–36, 1988.

[45] Burton S. Kaliski. *The MD2 Message-Digest Algorithm.* Available at `http://www.ietf.org/rfc/rfc1319.txt`.

[46] John Kelsey, Bruce Schneier, and David Wagner. Related-key cryptanalysis of 3-WAY, Biham-DES,CAST, DES-X, NewDES, RC2, and TEA. In Yongfei Han, Tatsuaki Okamoto, and Sihan Qing, editors, *Information and Communications Security*, volume 1334 of *Lecture Notes in Computer Science*, pages 233–246. Springer, 1997.

[47] Joe Kilian and Phillip Rogaway. How to protect DES against exhaustive key search (an analysis of DESX). *J. Cryptology*, 14(1):17–35, 2001.

[48] Israel Kleiner. The evolution of group theory: A brief survey. *Mathematics Magazine*, pages 195–215, 1986.

[49] Donald E. Knuth. *Seminumerical Algorithms*, volume 2 of *The Art of Computer Programming.* Addison-Wesley, Reading, MA, 2nd edition, 1981.

[50] Neal Koblitz. *A Course in Number Theory and Cryptography.* Springer, 2nd edition, 1994.

[51] Paul Kocher. Timing attacks on implementations of Diffie-Hellman, RSA, DSS, and other systems. In Neal Koblitz, editor, *Advances in Cryptology—CRYPTO96*, volume 1109 of *Lecture Notes in Computer Science*, pages 104–113. Springer, 1996.

[52] Sandeep Kumar, Christof Paar, Jan Pelzl, Gerd Pfeiffer, and Manfred Schimmler. COPACOBANA A cost-optimized special-purpose hardware for code-breaking. In *FCCM*, pages 311–312. IEEE Computer Society, 2006.

[53] Serge Lang. *Algebra*. Springer, 3rd edition, 2002.

[54] Gregor Leander, Christof Paar, Axel Poschmann, and Kai Schramm. New lightweight DES variants. In Alex Biryukov, editor, *Fast Software Encryption, 14th International Workshop, FSE 2007, Luxembourg, Luxembourg, March 26-28, 2007, Revised Selected Papers*, volume 4593 of *Lecture Notes in Computer Science*, pages 196–210. Springer, 2007.

[55] Arjen K. Lenstra, Hendrik W. Lenstra, and László Lovász. Factoring polynomials with rational coefficients. *Mathematische Annalen*, 261(4):515–534, 1982.

[56] Yehuda Lindell and Benny Pinkas. Secure multiparty computation for privacy-preserving data mining. *The Journal of Privacy and Confidentiality*, 1(1):59–98, 2009.

[57] Alasdair McAndrew. Teaching cryptography with open-source software. In J. D. Dougherty, Susan H. Rodger, Sue Fitzgerald, and Mark Guzdial, editors, *Proceedings of the 39th SIGCSE Technical Symposium on Computer Science Education*, pages 325–329. ACM, 2008.

[58] Alasdair McAndrew. Using the Hill cipher to teach cryptographic principles. *International Journal of Mathematical Education in Science and Technology*, 39(7):967–979, 2008.

[59] Robert McEvoy, James Curran, Paul Cotter, and Colin Murphy. Fortuna: Cryptographically secure pseudo-random number generation in software and hardware. In Eugene Coyle, editor, *Irish Signals and Systems Conference*, pages 457–462, 2006.

[60] Willi Meier and Othmar Staffelbach. The self-shrinking generator. In Alfredo de Santis, editor, *Advances in Cryptology EUROCRYPT'94*, volume 950 of *Lecture Notes in Computer Science*, pages 205–214, 1994.

[61] Alfred J. Menezes, Paul C. van Oorschot, and Scott A. Vanstone, editors. *Handbook of Applied Cryptography*. CRC Press, 1996. Available at `http://www.cacr.math.uwaterloo.ca/hac/`.

[62] Cleve B. Moler. *Numerical Computing with MATLAB.* Society for Industrial and Applied Mathematics, 2004. Available at http://www.mathworks.com/moler/index_ncm.html.

[63] Yi Mu, Vijay Varadharajan, and Khanh Quoc Nguyen. Digital cash. In Weidong Kou, editor, *Payment Technologies for E-Commerce*, pages 171–194. Springer, 2003.

[64] Mohammad Musa, Edward F. Schaefer, and Stephen Wedig. A simplified AES algorithm and its linear and differential cryptanalysis. *CRYPTOLOGIA: Cryptologia*, 27, 2003.

[65] Moni Naor, Yael Naor, and Omer Reingold. Applied kid cryptography or how to convince your children you are not cheating, 1999. Available at http://www.wisdom.weizmann.ac.il/~naor/PAPERS/waldo.ps.

[66] National Institute of Standards and Technology. *FIPS PUB 186-3: Digital Signature Standard (DSS).* National Institute for Standards and Technology, 2009. Available at http://csrc.nist.gov/publications/fips/fips186-3/fips_186-3.pdf.

[67] National Institute of Standards and Technology. *FIPS PUB 197: Announcing the Advanced Encryption Standard (AES).* National Institute for Standards and Technology, 2001. Available at http://csrc.nist.gov/publications/fips/fips197/fips-197.pdf.

[68] Pascal Paillier. Public-key cryptosystems based on composite degree residuosity classes. In Jacques Stern, editor, *Advances in Cryptology: Proceedings of EUROCRYPT '99*, volume 1592 of *Lecture Notes in Computer Science*, pages 223–238. Springer-Verlag, 1999.

[69] Raphael Chung-Wei Phan. Mini Advanced Encryption Standard (Mini-AES): A testbed for cryptanalysis students. *Cryptologia*, 26(4):283–306, 2002.

[70] Josef Pieprzyk, Thomas Hardjono, and Jennifer Seberry. *Fundamentals of Computer Security.* Springer-Verlag, 2003.

[71] Stephen Pohlig and Martin Hellman. An improved algorithm for computing logarithms over GF(p) and its cryptographic significance. *IEEE Trans. Information Theory*, 24(1):106–110, 1978.

[72] Bart Preneel. Analysis and design of cryptographic hash functions, 1993. PhD Thesis, Katholieke Universiteit Leuven. Available at http://www.esat.kuleuven.ac.be/~preneel/phd_preneel_feb1993.pdf.

[73] Jean-Jacques Quisquater, Myriam Quisquater, Muriel Quisquater, Michael Quisquater, Louis Guillou, Marie Annick Guillou, Gaid Guillou, Anna Guillou, Gwenole Guillou, and Soazig Guillou. How to explain

zero-knowledge protocols to your children. In Gilles Brassard, editor, *Advances in Cryptology (CRYPTO '89)*, volume 435 of *Lecture Notes in Computer Science*, pages 628–631. Springer, 1990.

[74] Michael O. Rabin. How to exchange secrets by oblivious transfer, 1981. Available at http://eprint.iacr.org/2005/187.pdf.

[75] Peter Y. A. Ryan, David Bismark, James Heather, Steve Schneider, and Zhe Xia. Prêt à voter: a voter-verifiable voting system. *IEEE Transactions on Information Forensics and Security*, 4(4):662–673, 2009.

[76] Edward Schaefer. A simplified Data Encryption Standard algorithm. *Cryptologia*, 20(1):77–84, 1996.

[77] Bruce Schneier. *Applied Cryptography: Protocols, Algorithms, and Source Code in C*. John Wiley & Sons, 2nd edition, 1996.

[78] Jennifer Seberry, Xian-Mo Zhang, and Yuliang Zheng. Systematic generation of cryptographically robust S-boxes. In *ACM Conference on Computer and Communications Security*, pages 171–182, 1993. Available at http://www.uow.edu.au/~jennie/WEB/WEB69-93/max/194_1993.pdf.

[79] Claude Shannon. A mathematical theory of communication. *The Bell System Technical Journal*, 27(4):379–423 and 623–656, 1948. Available at http://cm.bell-labs.com/cm/ms/what/shannonday/shannon1948.pdf.

[80] Claude Shannon. Communication theory of secrecy systems. *The Bell System Technical Journal*, 28(4):656–715, 1949. Available at http://netlab.cs.ucla.edu/wiki/files/shannon1949.pdf.

[81] Claude E. Shannon. Prediction and entropy of printed English. *Bell System Technical Journal*, January 1951. Available at http://languagelog.ldc.upenn.edu/myl/Shannon1950.pdf.

[82] Abraham Sinkov. *Elementary Cryptanalysis: A Mathematical Approach*. New Mathematical Library. Random House, 1968.

[83] William Stallings. *Cryptography and Network Security: Principles and Practice*. Prentice Hall, 5th edition, 2007.

[84] William Stein. *Elementary Number Theory: Primes, Congruences, and Secrets*. Springer, 2009. Available for download at http://modular.math.washington.edu/ent/ent.pdf.

[85] William Stein et al. *Sage Mathematics Software, Version 4.3.1*. The Sage Development Team, 2010. http://www.sagemath.org.

[86] Douglas R. Stinson. *Cryptography: Theory and Practice.* Chapman & Hall/CRC Press, 3rd edition, 2006.

[87] Jean-Pierre Tillich and Gilles Zémor. Hashing with SL_2. In Yvo G. Desmedt, editor, *Advances in Cryptology—CRYPTO '94*, volume 839 of *Lecture Notes in Computer Science*, pages 40–49. Springer-Verlag, 21–25 1994.

[88] Wade Trappe and Lawrence C. Washington. *Introduction to Cryptography with Coding Theory.* Prentice Hall, 2nd edition, 2006.

[89] John von Neumann. Various techniques for use in connection with random digits. In *von Neumann's Collected Works*, volume 5, pages 768–770. Pergamon, 1963.

[90] Samuel S. Wagstaff. *Cryptanalysis of Number Theoretic Ciphers.* Chapman & Hall, CRC Press, 2002.

[91] Xiaoyun Wang, Yiqun Lisa Yin, and Hongbo Yu. Finding collisions in the full SHA-1. In Victor Shoup, editor, *Advances in Cryptology–CRYPTO 2005*, volume 3621 of *Lecture Notes in Computer Science*, pages 17–36. Springer, 2005.

[92] Lawrence C. Washington. *Elliptic Curves: Number Theory and Cryptography.* Chapman & Hall/CRC Press, 2003.

[93] Dominic Welsh. *Codes and Cryptography.* Oxford Science Publications. Oxford University Press, 1988.

[94] David J. Wheeler and Roger M. Needham. TEA, a tiny encryption algorithm. In Bart Preneel, editor, *IWFSE: International Workshop on Fast Software Encryption*, volume 1008 of *Lecture Notes in Computer Science*, pages 363–366, 1994.

[95] Hans Wussing and Abe Shenitzer. *The Genesis of the Abstract Group Concept.* MIT Press, 1984.

[96] Zhe Xia, Steve A. Schneider, James Heather, and Jacques Traoré. Analysis, improvement, and simplification of Prêt à Voter with Paillier encryption. In David L. Dill and Tadayoshi Kohno, editors, *USENIX/ACCURATE Electronic Voting Workshop*, 2008.

[97] Song Y. Yan. *Primality Testing and Integer Factorization in Public-Key Cryptography.* Kluwer, 2004.

[98] Andrew Chi-Chih Yao. Protocols for secure computations (extended abstract). In *23rd Annual Symposium on Foundations of Computer Science*, pages 160–164, 1982.

[99] Elias Yarrkov. Cryptanalysis of XXTEA. Available at `http://eprint.iacr.org/2010/254.pdf`, May 2010.

[100] Gideon Yuval. Reinventing the travois: Encryption/MAC in 30 ROM bytes. In Eli Biham, editor, *IWFSE: International Workshop on Fast Software Encryption*, volume 1267 of *Lecture Notes in Computer Science*, pages 205–209, 1997.

Index